W9-DJM-587

The Surprising Mathematics of Longest Increasing Subsequences

In a surprising sequence of developments, the longest increasing subsequence problem, originally mentioned as merely a curious example in a 1961 paper, has proven to have deep connections to many seemingly unrelated topics in mathematics, such as random matrices and interacting particle systems. The detailed and playful study of these connections makes this book suitable as a starting point for a wider exploration of elegant mathematical ideas that are of interest to every mathematician and to many computer scientists, physicists, and statisticians.

Among the specific topics covered are the Vershik–Kerov–Logan–Shepp limit shape theorem, the Baik–Deift–Johansson theorem, the Tracy–Widom distribution, and the corner growth process. This exciting body of work, encompassing important advances in probability and combinatorics over the last 40 years, is made accessible to a general graduate-level audience for the first time in a highly polished presentation.

DAN ROMIK is Professor of Mathematics at the University of California, Davis.

INSTITUTE OF MATHEMATICAL STATISTICS
TEXTBOOKS

IMS Textbooks give introductory accounts of topics of current concern suitable for advanced courses at master's level, for doctoral students, and for individual study. They are typically shorter than a fully developed textbook, often arising from material created for a topical course. Lengths of 100–290 pages are envisaged. The books typically contain exercises.

Other Books in the Series

1. *Probability on Graphs*, by Geoffrey Grimmett
2. *Bayesian Filtering and Smoothing*, by Simo Särkkä
3. *Stochastic Networks*, by Frank Kelly and Elena Yudovina

The Surprising Mathematics of Longest Increasing Subsequences

DAN ROMIK

CAMBRIDGE
UNIVERSITY PRESS

32 Avenue of the Americas, New York NY 10013-2473, USA

Cambridge University Press is part of the University of Cambridge.

It furthers the University's mission by disseminating knowledge in the pursuit of education, learning and research at the highest international levels of excellence.

www.cambridge.org
Information on this title: www.cambridge.org/9781107075832

First published 2015

A catalogue record for this publication is available from the British Library

Library of Congress Cataloguing in Publication data
Romik, Dan, 1976– author.
The surprising mathematics of longest increasing subsequences / Dan Romik.
pages cm
Includes bibliographical references and index.
ISBN 978-1-107-07583-2 (hardback) – ISBN 978-1-107-42882-9 (paperback)
1. Combinatorial analysis. 2. Probabilities. I. Title.
QA164.R66 2014
511′.6–dc23 2014023514

ISBN 978-1-107-07583-2 Hardback
ISBN 978-1-107-42882-9 Paperback

Contents

Preface

> "Good mathematics has an air of economy and an element of surprise."
>
> – Ian Stewart, *From Here to Infinity*

As many students of mathematics know, mathematical problems that are simple to state fall into several classes: there are those whose solutions are equally simple; those that seem practically impossible to solve despite their apparent simplicity; those that are solvable but whose solutions nonetheless end up being too complicated to provide much real insight; and finally, there are those rare and magical problems that turn out to have rich solutions that reveal a fascinating and unexpected structure, with surprising connections to other areas that lie well beyond the scope of the original problem. Such problems are hard, but in the most interesting and rewarding kind of way.

The problems that grew out of the study of longest increasing subsequences, which are the subject of this book, belong decidedly in the latter class. As readers will see, starting from an innocent-sounding question about random permutations we will be led on a journey touching on many areas of mathematics: combinatorics, probability, analysis, linear algebra and operator theory, differential equations, special functions, representation theory, and more. Techniques of random matrix theory, a sub-branch of probability theory whose development was originally motivated by problems in nuclear physics, will play a key role. In later chapters, connections to interacting particle systems, which are random processes used to model complicated systems with many interacting elements, will also surface. Thus, in this journey we will have the pleasure of tapping into a rich vein of mathematical knowledge, giving novices and experts alike fruitful avenues for exploration. And although the developments presented in this book are fairly modern, dating from the last 40 years, some of the tools we will need are based on classical 19th century mathematics. The fact that such old mathematics can be repurposed for use in new ways that could never have been imagined by its original discoverers is a delightful demonstration of what the physicist Eugene P. Wigner [146] (and later Hamming [55] and others) once famously described as the "unreasonable effectiveness" of mathematics.

Because the subject matter of this book involves such a diverse range of areas, rather than stick to a traditional textbook format I chose a style of presentation a bit similar in spirit to that of a travel guide. Each chapter is meant to take readers on an exploration of ideas covering a certain mathematical landscape, with the main goal being to prove some deep and difficult result that is the main "tourist attraction" of the subject being covered. Along the way, tools are developed, and sights and points of interest of less immediate importance are pointed out to give context and to inform readers where they might go exploring on their next visit.

Again because of the large number of topics touched upon, I have also made an effort to assume the minimum amount of background, giving quick overviews of relevant concepts, with pointers to more comprehensive literature when the need arises. The book should be accessible to any graduate student whose background includes graduate courses in probability theory and analysis and a modest amount of previous exposure to basic concepts from combinatorics and linear algebra. In a few isolated instances, a bit of patience and willingness to consult outside sources may be required by most readers to understand the finer points of the discussion. The dependencies between chapters are shown in the following diagram:

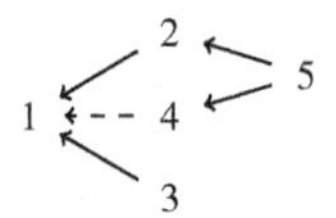

(Chapter 4 is only minimally dependent on Chapter 1 for some notation and definitions.)

The book is suitable for self-study or can be covered in a class setting in roughly two semester-long courses. Exercises at many levels of difficulty, including research problems of the "do not try this at home" kind, are included at the end of each chapter.

The subjects covered in the different chapters are as follows. Chapter 1 presents the **Ulam–Hammersley problem** of understanding the asymptotic behavior of the maximal length of an increasing subsequence in a uniformly random permutation as the permutation order grows. After developing the necessary tools the chapter culminates in the first solution of the problem by Vershik–Kerov and Logan–Shepp. Chapter 2 covers the beautiful **Baik–Deift–Johansson theorem** and its extension due to Borodin–Okounkov–Olshanski and Johansson – a major refinement of the picture revealed by Vershik–Kerov and Logan–Shepp that ties the problem of longest increasing subsequences to the Tracy–Widom distribution from random matrix theory and to other important concepts like determinantal point processes. Chapter 3 discusses **Erdős–Szekeres permutations**, a class of permutations possessing extremal behavior with respect to their maximal monotone subsequence lengths, which are analyzed by applying and extending the techniques developed in Chapter 1.

Chapters 4 and 5 are devoted to the study of the **corner growth process**, a random walk on Young diagrams that bears an important conceptual resemblance to another process introduced in Chapter 1. In Chapter 4 we prove the well-known limiting shape result of Rost and its extension to the case of corner growth in discrete time. Chapter 5 then develops a new approach to the problem, due to Johansson, that enables proving a much more precise fluctuation result, again involving the Tracy–Widom distribution.

I am grateful to the people and organizations who helped make this book possible. My work was supported by the National Science Foundation under grant DMS-0955584; by grant 228524 from the Simons Foundation; and of course by my excellent employer of the last 5 years, the University of California, Davis. I also received advice, suggestions, error reports, and encouragement from Arvind Ayyer, Eric Brattain-Morrin, Peter Chang, Alexander Coward, Ira Gessel, Geoffrey Grimmett, Indrajit Jana, Donald Knuth, Christian Krattenthaler, Greg Kuperberg, Isaac Lambert, Liron Mor Yosef, Vladimir Pchelin, Yuval Peres, Amir Sarid, Sasha Soshnikov, Perla Sousi, Mike Steele, and Peter Winkler. Ron Peled outdid everyone by sending me so many insightful suggestions for improvement that I had to beg him to stop, and deserves special thanks.

D. Romik
Davis
May 2014

0

A few things you need to know

0.1 Probability notation and prerequisites

The book assumes knowledge of the basic concepts of probability theory at the level of a first graduate course. For readers' convenience, we recall here a few standard definitions and notational conventions: first, throughout the book we use the following notation and abbreviations.

$\mathbb{P}(\cdot)$	Probability of an event
$\mathbb{E}(\cdot)$	Expectation of a random variable
$\mathbf{1}_{\{\cdot\}}$	The indicator (a.k.a. characteristic function) of an event/set
r.v.	random variable
i.i.d.	independent and identically distributed
a.s.	almost surely
$\stackrel{d}{=}$	equality in distribution
$\sim$	[a random variable] is distributed as [a distribution] (see below for examples)

Second, we make occasional use of the standard terminology regarding modes of convergence for sequences of random variables and probability distributions, which are defined as follows.

Almost sure convergence. We say that a sequence $(X_n)_{n=1}^{\infty}$ of random variables converges almost surely to a limiting random variable X, and denote $X_n \xrightarrow[n\to\infty]{\text{a.s.}} X$, if $\mathbb{P}(X_n \to X \text{ as } n \to \infty) = 1$.

Convergence in probability. We say that X_n converges in probability to X, and denote $X_n \xrightarrow[n\to\infty]{P} X$, if for any $\epsilon > 0$, $\mathbb{P}(|X_n - X| > \epsilon) \to 0$ as $n \to \infty$.

In a few places, the term "convergence in probability" is used in a broader sense that applies to convergence of random objects taking value in a more general space than the real line. In such cases, the meaning of the convergence statement is spelled out explicitly.

Convergence in distribution. We say that a sequence of distribution functions F_n converges in distribution to a limiting distribution function F, and denote $F_n \xrightarrow[n\to\infty]{d} F$, if $F_n(x) \to F(x)$ for any $x \in \mathbb{R}$ that is a continuity point of F; the same definition applies in the case when F_n and F are d-dimensional joint distribution functions. Similarly, we say that a sequence $(X_n)_{n=1}^{\infty}$ of r.v.s (or, more generally, d-dimensional random vectors) converges in distribution to F (a one-dimensional, or more generally d-dimensional, distribution function), and denote $X_n \xrightarrow[n\to\infty]{d} F$, if F_{X_n} converges in distribution to F, where for each n, F_{X_n} denotes the distribution function of X_n.

We will repeatedly encounter a few of the special distributions of probability theory, namely the **geometric**, **exponential** and **Poisson** distributions. The ubiquitous **Gaussian** (a.k.a. **normal**) distribution will also make a couple of brief appearances. For easy reference, here are their definitions.

The geometric distribution. If $0 < p < 1$, we say that an r.v. X has the geometric distribution with parameter p, and denote $X \sim \mathrm{Geom}(p)$, if

$$\mathbb{P}(X = k) = p(1-p)^{k-1}, \qquad (k = 1, 2, \ldots).$$

The exponential distribution. If $\alpha > 0$, we say that an r.v. X has the exponential distribution with parameter α, and denote $X \sim \mathrm{Exp}(\alpha)$, if

$$\mathbb{P}(X \geq t) = e^{-\alpha t}, \qquad (t \geq 0).$$

The Poisson distribution. If $\lambda > 0$, we say that an r.v. X has the Poisson distribution with parameter α, and denote $X \sim \mathrm{Poi}(\lambda)$, if

$$\mathbb{P}(X = k) = e^{-\lambda} \frac{\lambda^k}{k!}, \qquad (k = 0, 1, 2, \ldots).$$

The Gaussian distribution. If $\mu \in \mathbb{R}$ and $\sigma > 0$, we say that an r.v. X has the Gaussian distribution with mean μ and variance σ^2, and denote $X \sim N(\mu,\sigma^2)$, if

$$\mathbb{P}(a \leq X \leq b) = \frac{1}{\sqrt{2\pi}\sigma} \int_a^b e^{-(x-\mu)^2/2\sigma}\,dx, \qquad (a < b).$$

0.2 Little-*o* and big-*O* notation

Throughout the book, we are frequently concerned with asymptotic estimates for various quantities as a parameter (usually, but not always, a discrete parameter n) converges to a limit (usually ∞). We use the standard $o(\cdot)$ ("**little-*o***") and $O(\cdot)$ ("**big-*O***") notation conventions. In the typical case of a discrete parameter n converging to ∞ these are defined as follows. If a_n and b_n are functions of n, the statement

$$a_n = o(b_n) \quad \text{as } n \to \infty$$

means that $\lim_{n\to\infty} a_n/b_n = 0$. The statement

$$a_n = O(b_n) \quad \text{as } n \to \infty$$

means that there exists a constant $M > 0$ such that $|a_n/b_n| \leq M$ for all large enough values of n. Similarly, one can define statements such as "$f(x) = O(g(x))$ as $x \to L$" and "$f(x) = o(g(x))$ as $x \to L$"; we leave this variation to the reader to define precisely. Big-O and little-o notation can also be used more liberally in equations such as

$$a_n = \sqrt{n} + O(1) + O(\log n) + o(c_n) \quad \text{as } n \to \infty,$$

whose precise meaning is "$a_n - \sqrt{n}$ can be represented as a sum of three quantities x_n, y_n and z_n such that $x_n = O(1)$, $y_n = O(\log n)$ and $z_n = o(c_n)$." Usually such statements are derived from an earlier explicit description of the x_n, y_n, and z_n involved in such a representation. Frequently several big-O and little-o expressions can be combined into one, as in the equation

$$O(1) + O(\log n) + o(1/n) = O(\log n) \quad \text{as } n \to \infty.$$

As illustrated previously, asymptotic statements are usually accompanied by a qualifier like "as $n \to \infty$" indicating the parameter and limiting value with respect to which they apply. However, in cases when this specification is clear from the context it may on occasion be omitted.

More information regarding asymptotic estimation methods, along with many examples of the use of little-o and big-O notation, can be found in [49], [93].

0.3 Stirling's approximation

The canonical example of an interesting asymptotic relation is Stirling's approximation for $n!$. In the above notation it is written as

$$n! = (1 + o(1))\sqrt{2\pi n}(n/e)^n \quad \text{as } n \to \infty. \tag{0.1}$$

We make use of (0.1) on a few occasions. In some cases it is sufficient to use the more elementary (nonasymptotic) lower bound

$$n! \geq (n/e)^n \qquad (n \geq 1), \tag{0.2}$$

which is proved by substituting $x = n$ in the trivial inequality $e^x \geq x^n/n!$ valid for all $x \geq 0$. The relation (0.1) is harder (but not especially hard) to prove. A few different proofs can be found in [35, Section 6.3], [40], Sections II.9 and VII.3 of [41], [49, Section 9.6], [106], and p. 312 of this book.

1

Longest increasing subsequences in random permutations

Chapter summary. If σ is a permutation of n numbers, we consider the maximal length $L(\sigma)$ of an increasing subsequence of σ. For a permutation chosen *uniformly at random* from among all permutations of order n, how large can we expect $L(\sigma)$ to be? The goal of this chapter is to answer this question. The solution turns out to be rather complicated and will take us on a journey through a fascinating mathematical landscape of concepts such as **integer partitions**, **Young tableaux**, **hook walks**, **Plancherel measures**, **large deviation principles**, **Hilbert transforms**, and more.

1.1 The Ulam–Hammersley problem

We begin with a question about the asymptotic behavior of a sequence of real numbers. Let S_n denote the group of permutations of order n. If $\sigma \in S_n$ is a permutation, a **subsequence** of σ is a sequence $(\sigma(i_1), \sigma(i_2), \ldots, \sigma(i_k))$, where $1 \le i_1 < i_2 < \ldots < i_k \le n$. The subsequence is called an **increasing subsequence** if $\sigma(i_1) < \sigma(i_2) < \ldots < \sigma(i_k)$, a **decreasing subsequence** if $\sigma(i_1) > \sigma(i_2) > \ldots > \sigma(i_k)$, and a **monotone subsequence** if it is either increasing or decreasing. Define $L(\sigma)$ to be the maximal length of an increasing subsequence of σ. That is,

$$L(\sigma) = \max\left\{1 \le k \le n : \sigma \text{ has an increasing subsequence of length } k\right\}.$$

Similarly, define $D(\sigma)$ to be the maximal length of a *decreasing* subsequence of σ, i.e.,

$$D(\sigma) = \max\left\{1 \le k \le n : \sigma \text{ has a decreasing subsequence of length } k\right\}.$$

For example, if $\sigma = (3, 1, 6, 7, 2, 5, 4)$, then $L(\sigma) = 3$, since it has (several) increasing subsequences of length 3, but no increasing subsequence of length 4. Similarly, one can verify easily that $D(\sigma) = 3$.

Now define the sequence of numbers

$$\ell_n = \frac{1}{n!} \sum_{\sigma \in S_n} L(\sigma), \qquad (n = 1, 2, \ldots).$$

That is, ℓ_n is the average of $L(\sigma)$ over all permutations of order n. For example, the first few values in the sequence are $\ell_1 = 1$, $\ell_2 = 3/2$, $\ell_3 = 2$, $\ell_4 = 29/12$, $\ell_5 = 67/24$. We are interested in the problem of determining the asymptotic behavior of ℓ_n as n grows large. A version of the problem was first mentioned in a 1961 paper by Stanisław Ulam [138], a Polish-American mathematician better known for his work on the hydrogen bomb. In his paper, which concerned the Monte Carlo method for numerical computation (which Ulam pioneered), he discussed briefly the idea of studying the statistical distribution of the maximal monotone subsequence length in a random permutation; this was brought up as an example of the kinds of problems that can be attacked using Monte Carlo calculations. Subsequently, the question came to be referred to as "Ulam's problem" by some authors—starting with John M. Hammersley, who undertook (with some success) the first serious study of the problem, which he presented in a 1970 lecture and accompanying article [54].[1] To honor Hammersley's contribution to analyzing and popularizing Ulam's question, we refer to the problem here as the **Ulam–Hammersley problem**.

In this chapter and the next one we describe the developments leading up to a rather complete solution of Ulam and Hammersley's problem. The techniques developed along the way to finding this solution did much more than solve the original problem; in fact, they paved the way to many other interesting developments, some of which are described later in the book.

To avoid unnecessary suspense, one form of the "final answer," obtained in 1998 by Jinho Baik, Percy A. Deift, and Kurt Johansson [11], is as follows: as $n \to \infty$, we have

$$\ell_n = 2\sqrt{n} + cn^{1/6} + o(n^{1/6}), \tag{1.1}$$

where $c = -1.77108\ldots$ is a constant having a complicated definition in terms of the solution to a certain differential equation, the Painlevé equation of type II. We shall have to wait until Chapter 2 to see where this more exotic part of the asymptotics comes from. In this chapter our goal is to prove a first major result in this direction, which identifies only the leading asymptotic term $2\sqrt{n}$. The result, proved by Anatoly Vershik and Sergei Kerov [142], [143] and independently by Benjamin F. Logan and Lawrence A. Shepp [79] in 1977,[2] is the following.

Theorem 1.1 (The asymptotics of ℓ_n) *We have the limit*

$$\frac{\ell_n}{\sqrt{n}} \to 2$$

as $n \to \infty$. Furthermore, the limit is the same for the "typical" permutation of order n. That is, if for each n, σ_n denotes a uniformly random permutation in S_n, then $L(\sigma_n)/\sqrt{n} \to 2$ in probability as $n \to \infty$.

1.2 The Erdős–Szekeres theorem

To gain an initial understanding of the problem, let us turn to a classical result in combinatorics dating from 1935, the Erdős–Szekeres theorem.[3] Paul Erdős and George Szekeres observed that if a permutation has no long increasing subsequence, its elements must in some sense be arranged in a somewhat decreasing fashion, so it must have a commensurately long *decreasing* subsequence. The precise result is as follows.

Theorem 1.2 (Erdős–Szekeres theorem) *If $\sigma \in S_n$ and $n > rs$ for some integers $r, s \in \mathbb{N}$, then either $L(\sigma) > r$ or $D(\sigma) > s$.*

Proof We introduce the following variation on the permutation statistics $L(\cdot)$ and $D(\cdot)$: for each $1 \le k \le n$, let $L_k(\sigma)$ denote the maximal length of an increasing subsequence of σ that ends with $\sigma(k)$, and similarly let $D_k(\sigma)$ denote the maximal length of a decreasing subsequence of σ that ends with $\sigma(k)$.

Now consider the n pairs $(D_k(\sigma), L_k(\sigma))$, $1 \le k \le n$. The key observation is that they are all distinct. Indeed, for any $1 \le j < k \le n$, if $\sigma(j) < \sigma(k)$ then $L_j(\sigma) < L_k(\sigma)$, since we can take an increasing subsequence of σ that ends with $\sigma(j)$ and has length $L_j(\sigma)$, and append $\sigma(k)$ to it. If, on the other hand, $\sigma(j) > \sigma(k)$, then similarly we get that $D_j(\sigma) < D_k(\sigma)$, since any decreasing subsequence that ends with $\sigma(j)$ can be made longer by appending $\sigma(j)$ to it.

The conclusion from this observation is that for some $1 \leq k \leq n$, either $L_k(\sigma) > r$ or $D_k(\sigma) > s$, since otherwise the n distinct pairs $(D_k(\sigma), L_k(\sigma))$ would all be in the set $\{1, 2, \ldots, r\} \times \{1, 2, \ldots, s\}$, in contradiction to the assumption that $n > rs$. This proves the theorem. □

It is also interesting to note that the condition $n > rs$ in the theorem cannot be weakened. Indeed, it is easy to construct a permutation σ of order exactly rs for which $L(\sigma) = r$ and $D(\sigma) = s$; for example, define $\sigma(si + j) = si - j + s + 1$ for $0 \leq i < r$, $1 \leq j \leq s$ (this permutation has r "blocks," each comprising a decreasing s-tuple of numbers, with the ranges of successive blocks being increasing). In fact, it turns out that the set of permutations that demonstrate the sharpness of the condition has a very interesting structure; this topic is explored further in Chapter 3.

1.3 First bounds

From here on and throughout this chapter, σ_n denotes a uniformly random permutation of order n, so that in probabilistic notation we can write $\ell_n = \mathbb{E}L(\sigma_n)$. We can now use Theorem 1.2 to obtain a lower bound for ℓ_n.

Lemma 1.3 *For all $n \geq 1$ we have*

$$\ell_n \geq \sqrt{n}. \tag{1.2}$$

Proof Rephrasing Theorem 1.2 slightly, we can say that for each permutation $\sigma \in S_n$ we have $L(\sigma)D(\sigma) \geq n$. Now, ℓ_n is defined as the average value of $L(\sigma)$ over all $\sigma \in S_n$. However, by symmetry, clearly it is also the average value of $D(\sigma)$. By linearity of expectations of random variables, this is also equal to

$$\ell_n = \frac{1}{n!} \sum_{\sigma \in S_n} \frac{L(\sigma) + D(\sigma)}{2} = \mathbb{E}\left(\frac{L(\sigma_n) + D(\sigma_n)}{2} \right).$$

By the inequality of the arithmetic and geometric means, we get that

$$\ell_n \geq \mathbb{E}\left(\sqrt{L(\sigma_n)D(\sigma_n)} \right) \geq \sqrt{n}.$$ □

Comparing (1.2) with (1.1), we see that the bound gives the correct order of magnitude, namely $\sqrt{n}$, for ℓ_n, but with a wrong constant. What about an upper bound? As the following lemma shows, we can also fairly easily get an upper bound of a constant times $\sqrt{n}$, and thus establish that $\sqrt{n}$ is the correct order of magnitude for ℓ_n. This will give us a coarse, but still interesting, understanding of ℓ_n.

Lemma 1.4 *As $n \to \infty$ we have*

$$\limsup_{n\to\infty} \frac{\ell_n}{\sqrt{n}} \le e. \tag{1.3}$$

Proof For each $1 \le k \le n$, let $X_{n,k}$ denote the number of increasing subsequences of the random permutation σ_n that have length k. Now compute the expected value of $X_{n,k}$, noting that this is equal to the sum, over all $\binom{n}{k}$ subsequences of length k, of the probability for that subsequence to be increasing, which is $1/k!$. This gives

$$\mathbb{E}(X_{n,k}) = \frac{1}{k!}\binom{n}{k}.$$

This can be used to bound the probability that $L(\sigma_n)$ is at least k, by noting (using (0.2)) that

$$\begin{aligned}\mathbb{P}(L(\sigma_n) \ge k) &= \mathbb{P}(X_{n,k} \ge 1) \le \mathbb{E}(X_{n,k}) = \frac{1}{k!}\binom{n}{k} \\ &= \frac{n(n-1)\dots(n-k+1)}{(k!)^2} \le \frac{n^k}{(k/e)^{2k}}. \end{aligned} \tag{1.4}$$

Fixing some $\delta > 0$ and taking $k = \lceil (1+\delta)e\sqrt{n}\rceil$, we therefore get that

$$\mathbb{P}(L(\sigma_n) \ge k) \le \frac{n^k}{(k/e)^{2k}} \le \left(\frac{1}{1+\delta}\right)^{2k} \le \left(\frac{1}{1+\delta}\right)^{2(1+\delta)e\sqrt{n}},$$

a bound that converges to 0 at a rate exponential in $\sqrt{n}$ as $n \to \infty$. It follows (noting the fact that $L(\sigma) \le n$ for all $\sigma \in S_n$) that

$$\begin{aligned}\ell_n = \mathbb{E}(L(\sigma_n)) &\le \mathbb{P}(L(\sigma_n) < k)(1+\delta)e\sqrt{n} + \mathbb{P}(L(\sigma_n) \ge k)n \\ &\le (1+\delta)e\sqrt{n} + O(e^{-c\sqrt{n}}),\end{aligned}$$

where c is some positive constant that depends on δ. This proves the claim, since δ was an arbitrary positive number. □

Note that the proof of Lemma 1.4 actually gave slightly more information than what was claimed, establishing the quantity $(1+\delta)e\sqrt{n}$ as a bound not just for the *average* value of $L(\sigma_n)$, but also for the *typical* value, namely the value that is attained with a probability close to 1 for large n. Furthermore, the bounds we derived also yielded the fact (which will be useful later on) that the probability of large fluctuations of $L(\sigma_n)$ from its typical value decays like an exponential function of $\sqrt{n}$. We record these observations in the following lemma.

Lemma 1.5 *For any $\alpha > e$ we have for all n that*

$$\mathbb{P}(L(\sigma_n) > \alpha\sqrt{n}) \le Ce^{-c\sqrt{n}}$$

for some constants $C, c > 0$ that depend on α but not on n.

It is interesting to compare this with the argument that was used to prove Lemma 1.3, which really only bounds the average value of $L(\sigma_n)$ and not the typical value, since it does not rule out a situation in which (for example) approximately half of all permutations might have a value of $L(\sigma)$ close to 0 and the other half have a value close to $2\sqrt{n}$. However, as we shall see in the next section, in fact the behavior of $L(\sigma_n)$ for a typical permutation σ_n is asymptotically the same as that of its average value.

1.4 Hammersley's theorem

Our goal in this section is to prove the following result, originally due to Hammersley [54].

Theorem 1.6 (Hammersley's convergence theorem for the maximal increasing subsequence length) *The limit $\Lambda = \lim_{n\to\infty} \frac{\ell_n}{\sqrt{n}}$ exists. Furthermore, we have the convergence $L(\sigma_n)/\sqrt{n} \to \Lambda$ in probability as $n \to \infty$.*

Hammersley's idea was to reformulate the problem of studying longest increasing subsequences in permutations in a more geometric way. Denote by $\preceq$ a partial order on $\mathbb{R}^2$ where the relation $(x_1, y_1) \preceq (x_2, y_2)$ holds precisely if $x_1 \le x_2$ and $y_1 \le y_2$. For a set $A = ((x_k, y_k))_{k=1}^n$ of n points in the plane, an **increasing subset** of A is a subset any two of whose elements are comparable in the order $\preceq$ (in the context of partially ordered sets such a subset of A would be called a **chain**). See Fig. 1.1. Denote by $L(A)$ the maximal length of an increasing subset of A. Note that this generalizes the definition of $L(\sigma)$ for a permutation $\sigma \in S_n$, since in that case $L(\sigma) = L(G_\sigma)$, where $G_\sigma = \{(i, \sigma(i)) : 1 \le i \le n\}$ is the graph of σ.

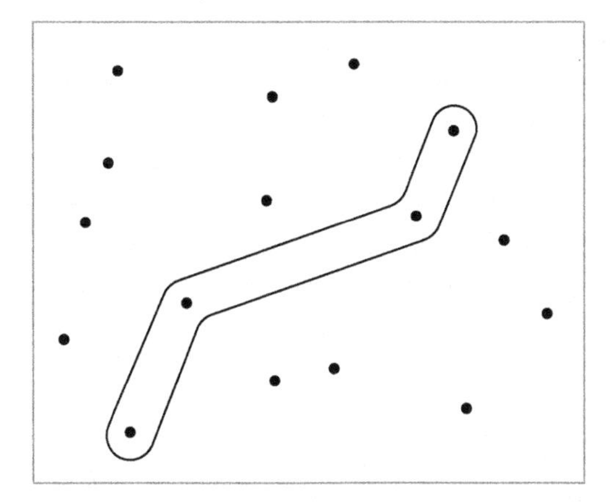

Figure 1.1 Points in the plane and an increasing subset (a.k.a. chain).

With this geometric outlook, a second natural step is to replace the random permutation by random points chosen independently in some square, or more generally rectangular, region. To understand why this works, first recall that a convenient way to sample a uniformly random permutation σ_n in S_n is to start with a sequence $X_1, \ldots, X_n$ of independent and identically distributed random variables distributed according to the uniform distribution $U[a, b]$ on some interval $[a, b]$, and to let σ_n encode the relative rankings of $X_1, \ldots, X_n$; that is, defining $\sigma_n(j)$ for each $1 \leq j \leq n$ to be the number k such that X_j is the kth smallest among $X_1, \ldots, X_n$ (this is well defined as long as the X_j take distinct values, which happens with probability 1). The permutation σ_n defined in this way is sometimes referred to as the **order structure** associated with $X_1, \ldots, X_n$. It is an easy exercise to check that σ_n defined in this way is indeed uniformly distributed in S_n.

Now, let $A_n(s, t)$ denote a random set of n points chosen independently and uniformly at random from a rectangle $[0, s] \times [0, t]$, where $s, t > 0$. We claim that the maximal increasing subset size $L(A_n(s, t))$ is a random variable with the same distribution as $L(\sigma_n)$. Indeed, we can represent such a set as $A_n(s, t) = \{(X_k, Y_k)\}_{k=1}^n$, where $X_1, \ldots, X_n, Y_1, \ldots, Y_n$ are independent random variables, with the X_j being uniformly distributed in $[0, s]$ and the Y_j being uniform in $[0, t]$. Denoting by π_n and η_n the permutations representing the respective order structures associated with the sequences $X_1, \ldots, X_n$ and $Y_1, \ldots, Y_n$, and denoting $\sigma_n = \pi_n \circ \eta_n^{-1}$ (clearly also a uniformly random permutation of order n), we leave to readers to verify that $L(A_n(s,t)) = L(\sigma_n)$.

The next step is to recall that probabilists have a clever way of constructing a single probability space that contains many random sets of the form $A_n(s,t)$ as defined previously, coupled together for all possible values of $s, t > 0$. This is called the **Poisson point process** (see box on the opposite page). If we take a Poisson point process of unit intensity in $\mathbb{R}_+^2 = [0,\infty) \times [0,\infty)$, which can be represented as a random discrete countable subset $\mathbf{\Pi}$ of $\mathbb{R}_+^2$, one of its defining properties is that for any $s, t > 0$, the random variable $N(s,t) = |\mathbf{\Pi} \cap ([0,s] \times [0,t])|$ (where $|\cdot|$ denotes the cardinality of a set) has the Poisson distribution with mean $s \cdot t$. A well-known, and easy to prove, property of the Poisson point process (see [29], Exercise 2.1.6, p. 24) is that, conditioned on the event $N(s,t) = n$, the distribution of the set $\mathbf{\Pi} \cap ([0,s] \times [0,t])$ is exactly that of the random set $A_n(s,t)$ discussed previously.

Now, for any $t > s > 0$, consider the random variable

$$Y_{s,t} = L(\mathbf{\Pi} \cap ([s,t) \times [s,t))). \tag{1.5}$$

It is easy to see directly from the definitions that we have

$$Y_{0,m} + Y_{m,n} \leq Y_{0,n}, \qquad 0 < m < n,$$

since the left-hand side represents the maximal length of a possible increasing subset in $\mathbf{\Pi} \cap ([0,n] \times [0,n])$ that is formed by combining an increasing subset in $\mathbf{\Pi} \cap ([0,m] \times [0,m])$ of length $Y_{0,m}$ with an increasing subset in $\mathbf{\Pi} \cap ([m,n] \times [m,n])$ of length $Y_{m,n}$. This *superadditivity* property brings to mind a well-known result in probability theory, **Kingman's subadditive ergodic theorem**, discussed in the Appendix. In particular, condition 1 of Theorem A.3 from the Appendix is satisfied for the r.v.'s $(-Y_{m,n})_{m,n}$. It is straightforward to verify (Exercise 1.2) that the other conditions of that theorem are satisfied as well. Therefore we get that, almost surely,

$$\frac{Y_{0,n}}{n} \to \Lambda = \sup_{m \geq 1} \frac{\mathbb{E}(Y_{0,m})}{m} \quad \text{as } n \to \infty. \tag{1.6}$$

Note that we only deduced convergence along integer values of n, but

The Poisson point process

Physicists once hypothesized that the universe is infinite, with stars distributed more or less at random with a uniform large-scale density throughout all of space. Aside from the fact that this turns out not to be a realistic model of the universe, it is interesting to ask how one might construct a rigorous mathematical model that would describe such a hypothetical universe. The Poisson point process is just such a concept: it models how points can be spread at random with uniform density throughout d-dimensional space (or some given subset of it) in the "most random" or "most independent" way. This is suitable for describing all kinds of phenomena, such as: the distribution of impurities in a material; times of decay events in a chunk of radioactive material; times of service requests to a customer call center; and many others.

Formally, the d-dimensional Poisson point process can be defined as the unique distribution of points in $\mathbb{R}^d$ with constant mean density $\lambda > 0$ per unit of volume, such that for any disjoint measurable sets $A_1, \ldots, A_k \subset \mathbb{R}^d$ with finite volumes, the numbers of random points falling in each of the sets A_i are independent random variables, each having the Poisson distribution whose mean is λ times the volume of the corresponding A_i. For more details, see [29], [33].

since $Y_{0,t}$ is monotone nondecreasing in t, we also have this convergence if n is considered as a real-valued parameter.

Next, we relate this limit back to the behavior of $L(\sigma_n)$ using two key observations. First, for each $n \geq 1$, define a random variable T_n by

$$T_n = \inf\{t > 0 \,:\, |\mathbf{\Pi} \cap ([0,t] \times [0,t])| = n\}, \tag{1.7}$$

and consider the scaled set $\frac{1}{T_{n+1}} (\mathbf{\Pi} \cap [0, T_{n+1}) \times [0, T_{n+1}))$. This is a random set of n points in the unit square $[0,1] \times [0,1]$, and it is easy to check (using the conditioning property of Poisson processes mentioned previously) that its joint distribution is exactly that of n independent uniformly random points in $[0,1]^2$. In particular, the random variable

$$Y_{0,T_{n+1}} = L(\mathbf{\Pi} \cap ([0, T_{n+1}) \times [0, T_{n+1})))$$

is equal in distribution to $L(\sigma_n)$.

The second observation concerns the asymptotic behavior of T_n. Denote

$S_0 = 0$ and $S_n = T_n^2$ for $n \geq 1$, and rewrite (1.7) as

$$S_n = \inf\{s > 0 : |\mathbf{\Pi} \cap ([0, \sqrt{s}] \times [0, \sqrt{s}])| = n\}$$
$$= \inf\{s > 0 : M(s) = n\},$$

where $M(s) = |\mathbf{\Pi} \cap ([0, \sqrt{s}] \times [0, \sqrt{s}])|$. It is immediate from the definition of the Poisson point process that $(M(s))_{s\geq 0}$ is a one-dimensional Poisson process of unit intensity on $[0, \infty)$. (For one-dimensional Poisson processes, instead of denoting the process as a random set it is customary to denote it by a family of random variables $(M(s))_{s\geq 0}$ such that $M(s)$ denotes the number of points from the process in the interval $[0, s]$.) Therefore, by a standard fact from probability theory (see [33], pp. 133–134), the increments $W_k = S_n - S_{n-1}$ of the sequence S_n are i.i.d. random variables with the exponential distribution Exp(1). It follows using the strong law of large numbers that almost surely

$$\frac{1}{n} S_n = \frac{1}{n} \sum_{k=1}^{n} W_k \to 1 \text{ as } n \to \infty,$$

or equivalently that $T_n / \sqrt{n} \to 1$ almost surely as $n \to \infty$. Combining this with (1.6), we get that

$$\frac{Y_{0,T_{n+1}}}{\sqrt{n}} = \frac{T_{n+1}}{\sqrt{n}} \cdot \frac{Y_{0,T_{n+1}}}{T_{n+1}} \to \Lambda \quad \text{almost surely as } n \to \infty.$$

Since $Y_{0,T_{n+1}}$ is equal in distribution to $L(\sigma_n)$, and almost sure convergence implies convergence in probability, it follows that $L(\sigma_n)/\sqrt{n}$ converges in probability to Λ.

We have proved convergence in probability of $L(\sigma_n)/\sqrt{n}$ to a constant limit Λ, which was the second claim of Theorem 1.6. We can now deduce the first claim regarding the convergence to Λ of the expected values $\ell_n/\sqrt{n}$. Intuitively, convergence in probability of a sequence of random variables to a limiting constant implies convergence of the expected values to the same limit, as long as one can bound the "tail behavior" of the random variables to rule out a situation in which large fluctuations from the typical value, which occur with low probability, can still have a non-negligible effect on the expected values. In our case, the knowledge that $L(\sigma_n)$ is always at most n (which is only a polynomial factor larger than the typical scale of $\sqrt{n}$), together with the exponential bound of Lemma 1.5, will be enough to prove the claim.

To make the idea more precise, fix some $\delta > 0$. We have

$$\begin{aligned}\left|\frac{\ell_n}{\sqrt{n}} - \Lambda\right| &= \left|n^{-1/2}\mathbb{E}L(\sigma_n) - \Lambda\right| \leq \mathbb{E}\left|n^{-1/2}L(\sigma_n) - \Lambda\right| \\ &= \mathbb{E}\left(\left|n^{-1/2}L(\sigma_n) - \Lambda\right| \mathbf{1}_{\{|n^{-1/2}L(\sigma_n)-\Lambda|\leq\delta\}}\right) \\ &\quad + \mathbb{E}\left(\left|n^{-1/2}L(\sigma_n) - \Lambda\right| \mathbf{1}_{\{|n^{-1/2}L(\sigma_n)-\Lambda|>\delta, L(\sigma_n)\leq 3\sqrt{n}\}}\right) \\ &\quad + \mathbb{E}\left(\left|n^{-1/2}L(\sigma_n) - \Lambda\right| \mathbf{1}_{\{|n^{-1/2}L(\sigma_n)-\Lambda|>\delta, L(\sigma_n)>3\sqrt{n}\}}\right),\end{aligned}$$

where the notation $\mathbf{1}_A$ is used to denote the indicator random variable of an event A. The first summand in the last expression is at most δ. The second summand satisfies

$$\begin{aligned}\mathbb{E}\left(\left|n^{-1/2}L(\sigma_n) - \Lambda\right|\mathbf{1}_{\{|n^{-1/2}L(\sigma_n)-\Lambda|>\delta, L(\sigma_n)\leq 3\sqrt{n}\}}\right) \\ \leq (3+\Lambda)\mathbb{P}(|n^{-1/2}L(\sigma_n) - \Lambda| > \delta),\end{aligned}$$

which converges to 0 as $n \to \infty$ by the result we proved on convergence in probability. The third summand satisfies

$$\mathbb{E}\left(\left|n^{-1/2}L(\sigma_n) - \Lambda\right|\mathbf{1}_{\{|n^{-1/2}L(\sigma_n)-\Lambda|>\delta, L(\sigma_n)>3\sqrt{n}\}}\right) \leq C(\sqrt{n} + \Lambda)e^{-c\sqrt{n}}$$

by Lemma 1.5. Combining the three bounds we see that

$$\limsup_{n\to\infty}\left|\frac{\ell_n}{\sqrt{n}} - \Lambda\right| \leq \delta.$$

Since δ was an arbitrary positive number, the claim follows. □

An alternative approach to proving Theorem 1.6, which does not rely on knowledge of Kingman's subadditive ergodic theorem, is outlined in Exercises 1.3 and 1.4. This approach is based on deriving an upper bound on the variance of $L(\sigma_n)$, which is also of independent interest, since it provides an explicit measure of the concentration of $L(\sigma_n)$ around its mean.

1.5 Patience sorting

We continue our pursuit of an understanding of the numbers ℓ_n and the permutation statistic $L(\sigma)$. Looking back at the results in the previous sections, we see that the proofs, both of the concrete bounds (1.2) and (1.3) and of the convergence result, Theorem 1.6, used only relatively general and not very detailed information about the behavior of $L(\sigma)$ as input for the proofs. It is therefore not so surprising that the knowledge we gained was in turn relatively meager.

We now turn to the task of developing a more detailed understanding of the statistic $L(\cdot)$ at the combinatorial level. We start in this section by answering perhaps the most fundamental question of all: given a permutation σ, how may one compute $L(\sigma)$? It turns out that this can be done using a simple algorithm called **patience sorting**, invented by A.S.C. Ross around 1960 and named by Colin L. Mallows [84], who further analyzed its behavior in [85] (see also [3]). The name of the algorithm alludes to its application to sorting a deck of cards. It consists of scanning sequentially through the permutation values in order, and piling them in a linear array of "stacks," according to the following rules:

1. Each new value x will be placed at the top of an existing stack, or will form a new stack by itself positioned to the right of all existing stacks.
2. The stack on which each new value x is placed is the leftmost stack from among those whose current top number is bigger than x. If there are no such stacks, x forms a new stack.

As an example, consider the permutation $(4, 1, 2, 7, 6, 5, 8, 9, 3)$; running the algorithm with this permutation as input leads to the sequence of stack arrays shown in Fig. 1.2. Note that the different stacks are aligned at their tops – in other words, imagine each stack as being "pushed down" whenever a new number is added to it.

Lemma 1.7 *When patience sorting is applied to a permutation $\sigma \in S_n$, the number of stacks at the end is equal to $L(\sigma)$.*

Proof Denote by s the number of stacks at the end of the run, and let $(\sigma(i_1), \sigma(i_2), \dots, \sigma(i_k))$ be an increasing subsequence of σ of maximal length $k = L(\sigma)$. It is a simple observation that the values $\sigma(i_1), \dots, \sigma(i_k)$ must all be in different stacks, since each successive value $\sigma(i_{j+1})$ must be placed in a stack to the right of the stacks containing the existing values $\sigma(i_1), \dots, \sigma(i_j)$. This shows that $s \geq L(\sigma)$. Conversely, one can form an increasing subsequence of length s by judiciously choosing one number

$$
4 \;\to\; \begin{array}{l}1\\4\end{array} \;\to\; \begin{array}{ll}1&2\\4&\end{array} \;\to\; \begin{array}{lll}1&2&7\\4&&\end{array} \;\to\; \begin{array}{lll}1&2&6\\4&&7\end{array}
$$

$$
\to\; \begin{array}{lll}1&2&5\\4&&6\\&&7\end{array} \;\to\; \begin{array}{llll}1&2&5&8\\4&&6&\\&&7&\end{array}
$$

$$
\to\; \begin{array}{lllll}1&2&5&8&9\\4&&6&&\\&&7&&\end{array} \;\to\; \begin{array}{lllll}1&2&3&8&9\\4&&5&&\\&&6&&\\&&7&&\end{array}
$$

Figure 1.2 The patience sorting algorithm applied to the permutation $(4, 1, 2, 7, 6, 5, 8, 9, 3)$.

from each stack. This is done as follows: start with the top number $x = x_s$ in the rightmost stack, and now repeatedly go to the value in the stack to its left that was at the top of that stack at the time that x was added, to obtain x_{s-1}, x_{s-2} and so on down to x_1. It is easy to see that $x_1, \ldots, x_s$ forms an increasing subsequence. □

1.6 The Robinson–Schensted algorithm

Although the patience sorting algorithm of the previous section gives us a practical way of computing $L(\sigma)$, it does not seem to offer any immediate help in analyzing the probabilistic behavior of the permutation statistic $L(\sigma_n)$, where σ_n is chosen uniformly at random in S_n. We have simply traded one question, about the behavior of $L(\sigma_n)$, for another, about the behavior of the number of stacks in patience sorting applied to a random permutation. But how is the latter problem approached?

In fact, the combinatorial power of the algorithm is truly unleashed when it is applied recursively, leading to a remarkable procedure known as the **Robinson–Schensted algorithm**.

To understand this idea, first note that the array of stacks formed during the application of patience sorting actually contains much more information than we need to simply compute $L(\sigma)$, the number of stacks. To that end, it is only the top numbers of the stacks (the first row of the array of stacks in our visualization scheme) that are important at any point in the execution, since only they enter into the decision of where to place a new number. What we shall do is take the additional "useless" information in the rows below the first row, and package it in an aesthetically more pleasing (and mathematically more revealing) form, by taking the numbers pushed, or "bumped," down to the second row and using them as input for an additional patience sorting algorithm, and so on recursively for the third row, fourth row, and so on.

To make this more precise, consider what happens whenever a new number is inserted. The number settles down into the first row at the top of one of the stacks (now more appropriately considered simply as abstract columns of numbers). As it does so, it either starts its own new column to the right of the existing ones, in which case the process ends there and we move on to the next permutation value; or, if it settles down at the top of an existing column, instead of pushing the entire column down as we did before, we simply "bump" the previous top entry from that column down to the second row, where it now itself searches for a column to settle down in, following similar rules to the standard patience sorting algorithm: namely, either it settles in an empty space to the right of all other second-row numbers if it is bigger than all the numbers currently in the second row, or otherwise it settles down in the leftmost column having a second-row entry bigger than itself.

In the event that the number settling down in the second row replaces an existing number, that number now gets bumped down to the third row, and the same patience sorting-like process gets repeated in that row. This continues until the bumping stops when one of the bumped numbers finally settles down in a previously unoccupied space. When this happens, we move on to the next value of the permutation and repeat the procedure. We call each such cycle consisting of an insertion of a new value and the resulting cascade of bumped values an **insertion step**.

To illustrate this, Fig. 1.3 shows the sequence of arrays formed by running this procedure on our previous sample permutation. For example, during the sixth insertion step the number 5 was inserted into the first row. This bumped the number 6 down to the second row, where it took the place of the number 7, which in turn was bumped down to the third row.

$$
4 \;\to\; \begin{array}{l} 1 \\ 4 \end{array} \;\to\; \begin{array}{ll} 1 & 2 \\ 4 & \end{array} \;\to\; \begin{array}{lll} 1 & 2 & 7 \\ 4 & & \end{array} \;\to\; \begin{array}{lll} 1 & 2 & 6 \\ 4 & 7 & \end{array}
$$

$$
\to\; \begin{array}{lll} 1 & 2 & 5 \\ 4 & 6 & \\ 7 & & \end{array} \;\to\; \begin{array}{llll} 1 & 2 & 5 & 8 \\ 4 & 6 & & \\ 7 & & & \end{array}
$$

$$
\to\; \begin{array}{lllll} 1 & 2 & 5 & 8 & 9 \\ 4 & 6 & & & \\ 7 & & & & \end{array} \;\to\; \begin{array}{lllll} 1 & 2 & 3 & 8 & 9 \\ 4 & 5 & & & \\ 6 & & & & \\ 7 & & & & \end{array}
$$

Figure 1.3 The "recursive patience sorting" algorithm (essentially the Robinson–Schensted algorithm) applied to the permutation $(4, 1, 2, 7, 6, 5, 8, 9, 3)$.

For a reason that will be made clear in the next section, it is useful to note that such an insertion step can be reversed: by starting from the last value in the bumping sequence (in the preceding example, the number 7 now sitting in the third row), we can undo the sequence of bumping events to recover the array of numbers as it existed before the insertion, and separately the number that was inserted that is now "bumped out" of the array. It is easy to see that the position of each "reverse bumping" operation is uniquely determined. Such an inverse operation is called a **deletion step**.

To summarize, we have defined a computational procedure that refines the patience sorting algorithm by applying patience sorting recursively to the rows below the first row. Its input is a permutation and its output is an interesting-looking two-dimensional array of numbers, where in particular the length of the first row in the array is exactly $L(\sigma)$. The array also has some useful monotonicity properties that are not present in the output of simple patience sorting. This is almost the Robinson–Schensted algorithm in its final form. One more refinement is needed, to make the algorithm *reversible*. We describe it in the next section after introducing some useful new terminology, and then study some of its remarkable properties.

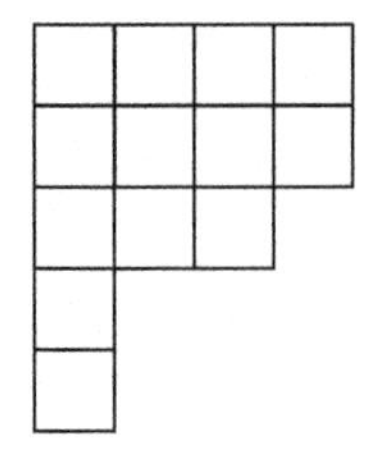

Figure 1.4 The Young diagram $(4, 4, 3, 1, 1)$.

1.7 Young diagrams and Young tableaux

At this point we introduce some important terminology. Let $n \in \mathbb{N}$. A **partition** (also called **integer partition**, or **unordered partition**) of n is, loosely speaking, a way to represent n as the sum of positive integers, without consideration for the order of the summands. More precisely, we say that λ is a partition of n, and denote this $\lambda \vdash n$, if λ is a vector of the form $\lambda = (\lambda_1, \lambda_2, \ldots, \lambda_k)$, where $\lambda_1, \ldots, \lambda_k$ are positive integers, $\lambda_1 \geq \lambda_2 \geq \ldots \geq \lambda_k$, and $n = \lambda_1 + \ldots + \lambda_k$. The integers $\lambda_1, \ldots, \lambda_k$ are called the **parts** of the partition. We also call n the **size** of λ, and denote $n = |\lambda|$. We denote the set of partitions of n by $\mathcal{P}(n)$.

The relevance of this to our discussion is that the row lengths in the two-dimensional array of numbers produced by the algorithm we defined form the parts of a partition of n (with n being the order of the permutation). Next, for a partition $\lambda \vdash n$, define the **Young diagram** of λ (named after Alfred Young, an early 20th-century English mathematician) to be a graphical diagram representing the partition λ as a two-dimensional array of boxes, or **cells**, where the jth row (counting from the top down) has λ_j cells, and the rows are left-justified. The cells of the diagram will serve as placeholders for the array of numbers produced by the algorithm. For example, Fig. 1.4 shows the Young diagram corresponding to the partition $(4, 4, 3, 1, 1)$.

Since Young diagrams are simply schematic devices representing integer partitions, we shall often identify a Young diagram with its associated partition and refer to a given λ interchangeably as either a partition or a Young diagram.

If $\lambda \vdash n$, the **conjugate partition** of λ, denoted λ', is the partition obtained by reading the lengths of the *columns* of the Young diagram of λ, instead of the rows. In other words, the Young diagram of λ' is obtained by reflecting the Young diagram of λ along its principal diagonal. For example, the partition conjugate to $\lambda = (4, 4, 3, 1, 1)$ is $\lambda' = (5, 3, 3, 2)$.

Finally, a **Young tableau**[4] (often referred to as a **standard Young tableau**, or **standard tableau**) consists of a Young diagram $\lambda \vdash n$ together with a filling of the cells of λ with the numbers $1, 2, \ldots, n$, such that the numbers in each row and each column are arranged in increasing order. Note that this is a property possessed by the array of numbers that is output by the algorithm we defined. We call the diagram λ the **shape** of the tableau. If P denotes a Young tableau of shape λ, its **transpose** is the Young tableau $P^\top$ of shape λ' obtained by reflecting P along the principal diagonal.

We are now ready to finish the definition of the Robinson–Schensted algorithm. The algorithm we defined so far can be thought of as taking a permutation $\sigma \in S_n$, and computing a partition $\lambda \vdash n$ together with a Young tableau P of shape λ. We call P the **insertion tableau**. Now note that there is a nice way to obtain from the algorithm another Young tableau Q of shape λ, by recording in each cell of the Young diagram λ the number k if that cell first became occupied in P during the kth insertion step during the execution of the algorithm (that is, after inserting the number $\sigma(k)$). In other words, Q records the order in which different cells were added to the shape λ as it "grew" from an empty diagram into its final form during the execution of the algorithm. It is trivial to see that Q is also a Young tableau. We call it the **recording tableau**. The Robinson–Schensted algorithm is defined as the mapping taking a permutation σ to the triple (λ, P, Q). See Fig. 1.5 for an illustration.

Why is the recording tableau interesting? A crucial fact is that knowing it allows us to reverse the action of the algorithm to reproduce from the insertion tableau P the original permutation σ. To see this, consider how we can recover just the *last* value $\sigma(n)$. When we inserted $\sigma(n)$ into the tableau-in-progress P, this resulted in a sequence of bumping operations ($\sigma(n)$ settling down in the first row and possibly bumping another number to the second row, that number settling down in the second row and bumping another number to the third row, and so on). If we knew the last place where a number being bumped down a row ended up settling, we would

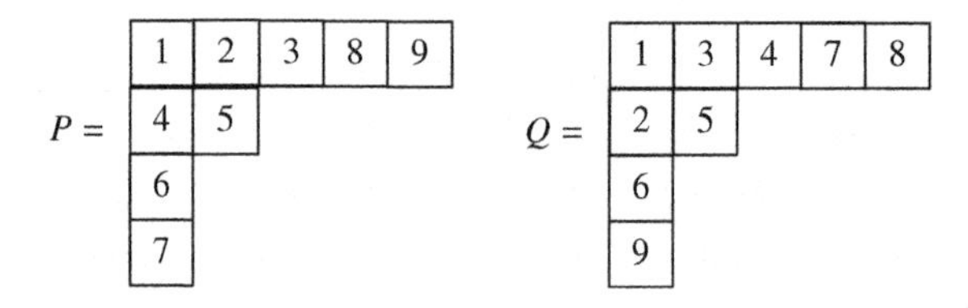

Figure 1.5 The insertion and recording tableaux generated by the Robinson–Schensted algorithm for the permutation $(4, 1, 2, 7, 6, 5, 8, 9, 3)$.

be able to perform a deletion step (as described in the previous section), and recover the state of the tableau P before this last insertion. But this is exactly the information encoded in the recording tableau Q – specifically, this position is exactly the cell where Q has its maximal entry. So, to recover $\sigma(n)$ we look for that cell, perform a deletion step starting at that cell, and recover the previous state of P, together with the value $\sigma(n)$ which is the entry that was "bumped out" of the tableau P at the end of the deletion step.

We can now proceed by induction, by deleting the maximal entry from Q, and repeating the same procedure to recover $\sigma(n-1)$, followed by $\sigma(n-2)$, and so on all the way down to $\sigma(1)$. This is sometimes referred to as applying the **inverse Robinson–Schensted algorithm**.

We summarize this last discussion in a formal statement that will be extremely important in what follows.

Theorem 1.8 *The Robinson–Schensted algorithm (also called the* ***Robinson–Schensted correspondence****) is a mapping that takes a permutation $\sigma \in S_n$ and returns a triple (λ, P, Q), where: $\lambda \vdash n$, and P, Q are two Young tableaux of shape λ. This mapping is a bijection between S_n and the set of such triples. Under this bijection, we have the identity*

$$L(\sigma) = \lambda_1 \quad \textit{(the length of the first row of } \lambda\textit{)}. \tag{1.8}$$

It is recommended to apply the algorithm by hand in a couple of examples to get a sense of how it works; see Exercises 1.7 and 1.8.

If $\lambda \vdash n$, denote by d_λ the number of Young tableaux of shape λ. Because of a connection to representation theory that is not important for our

Robinson, Schensted, Knuth, and RSK

The Robinson–Schensted algorithm is named after Gilbert de Beauregard Robinson and Craige E. Schensted. Robinson described an equivalent but less explicit version of the algorithm in a 1938 paper [104] in connection with a problem in representation theory. Schensted [113] rediscovered it in 1961, described it in a purely combinatorial way, and – importantly – pointed out the connection to longest increasing subsequences.

In 1970, Donald E. Knuth discovered [70] a significant additional generalization of the algorithm, which we will discuss in Chapter 5. This has become known as the **Robinson–Schensted–Knuth algorithm**, or **RSK** for short. In the literature the RSK acronym is often used even in reference to the more restricted Robinson–Schensted algorithm.

purposes but is explained later (see p. 63), the number d_λ is sometimes referred to as the **dimension** of λ, and denoted $\dim \lambda$ or f^λ. Pay attention to this definition: the mapping $\lambda \mapsto d_\lambda$ is extremely important in the combinatorics of Young diagrams and longest increasing subsequences, and plays a central role in many of the developments in the book.

As a corollary to Theorem 1.8, we get a curious fact involving the enumeration of Young tableaux of various shapes, by simply comparing the cardinalities of the two sets that the bijection maps between.

Corollary 1.9 *For each $n \geq 1$ we have*

$$\sum_{\lambda \vdash n} d_\lambda^2 = n!, \tag{1.9}$$

where the sum is over all partitions of n.

Note that our definition of the Robinson–Schensted algorithm treats rows and columns differently. Another apparent asymmetry of the algorithm is that the tableaux P and Q play very different roles. It is therefore surprising that the algorithm does have a symmetry property corresponding to the interchange of rows and columns and another symmetry related to interchanging P and Q. This is explained in the following theorem,[5] whose proof is outlined in Exercises 1.9 and 1.10.

Theorem 1.10 *Let* $\sigma = (\sigma(1), \dots, \sigma(n)) \in S_n$ *be a permutation, and let* (λ, P, Q) *be the triple associated to* σ *by the Robinson–Schensted algorithm. Then:*

(a) *If* (μ, P', Q') *is the triple associated by the Robinson–Schensted algorithm with the permutation* $(\sigma(n), \dots, \sigma(1))$ *("the reverse of* σ*") then* $\mu = \lambda'$ *and* $P' = P^\top$ *(the transpose of* P*).*[6]

(b) *The triple associated by the Robinson–Schensted algorithm with* σ^{-1} *is* (λ, Q, P)*.*

As a corollary, we get that the shape λ also encodes information about the maximal *decreasing* subsequence length of σ.

Corollary 1.11 *If* (λ, P, Q) *is the triple associated to a permutation* $\sigma \in S_n$ *by the Robinson–Schensted algorithm, then in addition to the relation* (1.8) *we have the symmetric relation*

$$D(\sigma) = \lambda'_1 \tag{1.10}$$

between the maximal decreasing subsequence length of σ *and the length of the first column of* λ*.*

Proof If we denote $\sigma' = (\sigma(n), \dots, \sigma(1))$, then clearly $D(\sigma) = L(\sigma')$, so the claim follows from part (a) of Theorem 1.10. □

Another corollary of Theorem 1.10 is an elegant formula for the total number of Young tableaux of order n; see Exercise 1.11.

1.8 Plancherel measure

The Robinson–Schensted algorithm finally allows us to reformulate our original question on the asymptotic behavior of the permutation statistic $L(\sigma)$ in a way that is truly useful. It does so by giving us an alternative way of looking at permutations (replacing them with the triples (λ, P, Q) computed by the algorithm), where the statistic $L(\sigma)$ appears as a natural quantity: the length of the first row of the Robinson–Schensted shape λ. In fact, note that for the purposes of computing λ_1, the tableaux P and Q don't actually play any role. It therefore seems natural to simply forget about them, and to focus on the random Young diagram $\lambda^{(n)}$ obtained when applying the Robinson–Schensted algorithm to the random permutation σ_n.

How does this random $\lambda^{(n)}$ behave, probabilistically speaking? Clearly, for any $\lambda \vdash n$, the probability that $\lambda^{(n)} = \lambda$ is $1/n!$ times the number of permutations $\sigma \in S_n$ whose Robinson–Schensted shape is λ. By the properties of the algorithm, this can be written as

$$\mathbb{P}(\lambda^{(n)} = \lambda) = \frac{d_\lambda^2}{n!}. \tag{1.11}$$

Note that the fact that these probabilities all sum to 1 is simply a restatement of equation (1.9). The probability measure on the set of integer partitions of n that assigns measure $d_\lambda^2/n!$ to any partition λ is known as **Plancherel measure** (of order n). Historically, it has its origins in the work of the Swiss mathematician Michel Plancherel in representation theory, dating to the early 20th century, but here we see that it appears naturally in connection with the purely combinatorial problem of understanding $L(\sigma)$: to understand how $L(\sigma_n)$ behaves for a uniformly random permutation in S_n, we can look instead at the random variable $\lambda_1^{(n)}$ (the length of the first row) for a random Young diagram $\lambda^{(n)}$ chosen according to Plancherel measure. Equation (1.11), together with Theorem 1.8, implies that these two random variables are equal in distribution.

1.9 The hook-length formula

All of the foregoing analysis would be in vain if we did not have a good way of analyzing Plancherel measure. Fortunately, we do. A key element is the existence of a remarkable formula, known as the **hook-length formula**, for computing d_λ, the number of Young tableaux of shape $\lambda \vdash n$. It was proved in 1954 by Frame, Thrall, and Robinson [43], who deduced it from an earlier and less convenient formula due to Frobenius.

If (i, j) is a cell in the Young diagram of $\lambda \vdash n$ (where i is the row index and j is the column index, as in matrix notation in linear algebra), define the **arm** of (i, j) to be the collection of cells in λ of the form (i, x) where $j \le x \le \lambda_i$ – that is, the cell (i, j) together with the cells in the same row that lie to its right. Define the **leg** of (i, j) to be the collection of cells in λ of the form (y, j) for $i \le y \le \lambda'_j$ ((i, j) and the cells in its column that lie below it). Define the **hook** of (i, j) to be the union of its arm and its leg, and denote this set of cells by $H_\lambda(i, j)$.

Define the **hook-length** of (i, j) as the number of cells in the hook, and

8	5	4	2
7	4	3	1
5	2	1	
2			
1			

Figure 1.6 A hook-length tableau.

denote it by $h_\lambda(i, j)$. Formally, it is easy to see that we can write

$$h_\lambda(i, j) = |H_\lambda(i, j)| = \lambda_i - j + \lambda'_j - i + 1.$$

Theorem 1.12 (The hook-length formula) *If $\lambda \vdash n$, then we have*

$$d_\lambda = \frac{n!}{\prod_{(i,j)} h_\lambda(i, j)}, \tag{1.12}$$

where the product is over all cells (i, j) in λ.

As an illustration, given $\lambda = (4, 4, 3, 1, 1)$, a Young diagram of order 13, to compute d_λ we start by tabulating the hook-lengths in the diagram shown in Fig. 1.6. (Call the resulting array of numbers a **hook-length tableau.**) The number of Young tableaux d_λ can then be computed as

$$d_\lambda = \frac{13!}{1 \cdot 1 \cdot 1 \cdot 2 \cdot 2 \cdot 2 \cdot 3 \cdot 4 \cdot 4 \cdot 5 \cdot 5 \cdot 7 \cdot 8} = 11583.$$

Theorem 1.12 has many proofs, including some that are "bijective," that is, are based on establishing a combinatorial bijection between two sets that immediately implies (1.12); see [44], [71, Section 5.1.4], [92], [96]. Here, we give a beautiful probabilistic proof due to Greene, Nijenhuis, and Wilf [50]. This proof has the added advantage of producing an elegant and efficient algorithm for sampling a uniformly random Young tableau of shape λ. This sampling algorithm is described in the next section.

Proof of Theorem 1.12 To start the proof, denote the right-hand side of (1.12) by e_λ. Our goal is to show that $d_\lambda = e_\lambda$ for all integer partitions λ. Note that d_λ satisfies a simple recurrence relation, namely

$$d_\lambda = \sum_{\mu \nearrow \lambda} d_\mu,$$

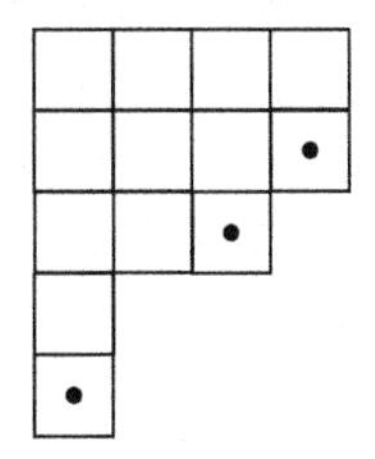

Figure 1.7 The corners of the Young diagram $(4, 4, 3, 1, 1)$.

where the notation $\mu \nearrow \lambda$ indicates that the Young diagram of λ can be obtained from the Young diagram of μ by the addition of a single cell (and the summation is over all μ such that $\mu \nearrow \lambda$). This is explained by the fact that, starting from any Young tableau of shape λ, by deleting the cell containing the maximal entry we are left with a Young tableau of shape μ for some $\mu \nearrow \lambda$. This recurrence, together with the "initial condition" $d_\emptyset = 1$ (for the "empty" diagram of size 0), determines all values of d_λ uniquely. Since e_μ trivially satisfies $e_\emptyset = 1$, it will be enough to show that it also satisfies the same recurrence, that is, that

$$e_\lambda = \sum_{\mu \nearrow \lambda} e_\mu, \qquad (\lambda \vdash n, \;\; n = 1, 2, \ldots).$$

To recast this in a more probabilistic form, we rewrite it in the form

$$\sum_{\mu \nearrow \lambda} \frac{e_\mu}{e_\lambda} = 1.$$

This seems to hint that for each Young diagram λ, there ought to exist a "natural" probability measure on the diagrams μ that satisfy $\mu \nearrow \lambda$, assigning probability exactly e_μ / e_λ to each such μ. The proof consists in constructively defining a random process leading to just such a measure.

First, let us examine the set of μ such that $\mu \nearrow \lambda$. These are clearly in bijection with cells (i, j) that can be removed from λ to obtain a new valid Young diagram μ. We refer to such cells as the **corners** of λ; they are cells that are simultaneously the last in their row and in their column, as illustrated in Fig. 1.7.

Fix the diagram λ. We now define a probabilistic process – a kind of random walk, dubbed the **hook walk** by the authors of [50] – that leads to a random choice of corner. First, we choose an initial cell for the walk to be a uniformly random cell in λ, out of the $|\lambda|$ possibilities. Next, we perform a random walk, successively replacing the current cell with a new one until we reach a corner, according to the following rule: given the current cell (i, j), its successor is chosen uniformly at random from the hook $H_\lambda(i, j) \setminus \{(i, j)\}$ (the current cell is excluded to avoid staying in the same place, although allowing it as a possible choice will only make the walk last longer, but will not change the probability distribution of its terminal point). Of course, the corners are exactly the cells that are the only elements of their hook, so once we reach a corner there is nowhere else to move to, and the process terminates.

Denote by $\mathbf{c}$ the terminating corner of the hook walk. From the foregoing discussion, we see that the proof of the hook-length formula has been reduced to proving the following claim.

Proposition 1.13 *For any corner cell (a, b) of λ, we have*

$$\mathbb{P}\Big(\mathbf{c} = (a, b)\Big) = \frac{e_\mu}{e_\lambda}, \tag{1.13}$$

where $\mu = \lambda \setminus \{(a, b)\}$.

Letting $n = |\lambda|$, we can compute the right-hand side of (1.13) as

$$\frac{e_\mu}{e_\lambda} = \frac{(n-1)!/\prod_{(i,j)} h_\mu(i, j)}{n!/\prod_{(i,j)} h_\lambda(i, j)} = \frac{1}{n} \prod_{(i,j)} \frac{h_\lambda(i, j)}{h_\mu(i, j)}.$$

In this product of hook-length ratios, all the terms cancel out except those in the same row or column as the corner cell (a, b), so this can be rewritten as

$$\begin{aligned}\frac{e_\mu}{e_\lambda} &= \frac{1}{n} \prod_{i=1}^{a-1} \frac{h_\lambda(i, b)}{h_\lambda(i, b) - 1} \prod_{j=1}^{b-1} \frac{h_\lambda(a, j)}{h_\lambda(a, j) - 1} \\ &= \frac{1}{n} \prod_{i=1}^{a-1} \left(1 + \frac{1}{h_\lambda(i, b) - 1}\right) \prod_{j=1}^{b-1} \left(1 + \frac{1}{h_\lambda(a, j) - 1}\right).\end{aligned}$$

This last expression can be further expanded in the form

$$\frac{1}{n}\sum_{A\subseteq\{1,\ldots,a-1\}}\sum_{B\subseteq\{1,\ldots,b-1\}}\left(\prod_{i\in A}\frac{1}{h_\lambda(i,b)-1}\prod_{j\in B}\frac{1}{h_\lambda(a,j)-1}\right)$$
$$=\frac{1}{n}\sum_{A\subseteq\{1,\ldots,a-1\}}\sum_{B\subseteq\{1,\ldots,b-1\}}Q(A,B), \qquad (1.14)$$

where for sets $A \subseteq \{1,\ldots,a-1\}$ and $B \subseteq \{1,\ldots,b-1\}$ we denote $Q(A,B)=\prod_{i\in A}\frac{1}{h_\lambda(i,b)-1}\prod_{j\in B}\frac{1}{h_\lambda(a,j)-1}$. We now show that $Q(A,B)$ itself can be given a probabilistic interpretation. For a possible trajectory $(x_1,y_1)\to\ldots\to(x_k,y_k)$ of cells visited by a hook walk, define its horizontal and vertical projections to be the sets $\{y_1,\ldots,y_k\}$, $\{x_1,\ldots,x_k\}$, respectively. Now for the sets A and B as earlier define a quantity $P(A,B)$ to be the conditional probability for a hook walk, conditioned to start from the initial cell $(\min(A\cup\{a\}),\min(B\cup\{b\}))$, to eventually reach the corner (a,b), while passing through a trajectory $(x_1,y_1)\to\ldots\to(x_k,y_k)=(a,b)$ whose vertical projection is $A\cup\{a\}$ and whose horizontal projection is $B\cup\{b\}=\{y_1,y_2,\ldots,y_k\}$. We claim that $P(A,B)=Q(A,B)$. If we prove this, the theorem will follow, since the probability $\mathbb{P}\big(\mathbf{c}=(a,b)\big)$ on the left-hand side of (1.13) can be expanded using the total probability formula as a sum of conditional probabilities given the initial cell (x_1,y_1) and the vertical and horizontal projection sets V and H of the walk, times the probability $1/n$ for each initial position. This gives

$$\mathbb{P}\big(\mathbf{c}=(a,b)\big)=\frac{1}{n}\sum_{x_1=1}^{a}\sum_{y_1=1}^{b}\sum_{\substack{\emptyset\subsetneq H\subseteq\{x_1,\ldots,a\}\\ \min(H)=x_1,\max(H)=a}}\sum_{\substack{V\subseteq\{y_1,\ldots,b\},\\ \min(V)=y_1,\max(V)=b}}P(H\setminus\{a\},V\setminus\{b\}),$$
$$=\frac{1}{n}\sum_{x_1=1}^{a}\sum_{y_1=1}^{b}\sum_{\substack{\emptyset\subsetneq H\subseteq\{x_1,\ldots,a\}\\ \min(H)=x_1,\max(H)=a}}\sum_{\substack{V\subseteq\{y_1,\ldots,b\},\\ \min(V)=y_1,\max(V)=b}}Q(H\setminus\{a\},V\setminus\{b\}),$$

which can then be easily seen to equal the right-hand side of (1.14).

To prove that $P(A,B)=Q(A,B)$, use induction on k. The induction base $k=1$ (corresponding to $A=B=\emptyset$) is trivial. For the inductive step, assume that both A and B are nonempty (the case when only one of them is empty is similar and is left to the reader) and denote $i_1=\min(A)$ and

$j_1 = \min(B)$. By the definition of the hook walk, we have that

$$P(A,B) = \frac{1}{h_\lambda(i_1,j_1)-1}\left(P(A\setminus\{i_1\},B) + P(A,B\setminus\{j_1\})\right).$$

By induction, this is equal to

$$\begin{aligned}
&\frac{1}{h_\lambda(i_1,j_1)-1}\left(Q(A\setminus\{i_1\},B) + Q(A,B\setminus\{j_1\})\right)\\
&\quad = \frac{1}{h_\lambda(i_1,j_1)-1}\left(\prod_{i\in A\setminus\{i_1\}}\frac{1}{h_\lambda(i,b)-1}\prod_{j\in B}\frac{1}{h_\lambda(a,j)-1}\right.\\
&\qquad\qquad\qquad\left. + \prod_{i\in A}\frac{1}{h_\lambda(i,b)-1}\prod_{j\in B\setminus\{j_1\}}\frac{1}{h_\lambda(a,j)-1}\right)\\
&\quad = \frac{(h_\lambda(i_1,b)-1)+(h_\lambda(a,j_1)-1)}{h_\lambda(i_1,j_1)-1}\left(\prod_{i\in A}\frac{1}{h_\lambda(i,b)-1}\prod_{j\in B}\frac{1}{h_\lambda(a,j)-1}\right).
\end{aligned}$$

But, fortuitously, we have that $(h_\lambda(i_1,b)-1)+(h_\lambda(a,j_1)-1) = h_\lambda(i_1,j_1)-1$, so the last expression in the above chain of equalities is equal to $Q(A,B)$. This completes the proof of Proposition 1.13, and hence of the hook-length formula. □

1.10 An algorithm for sampling random Young tableaux

The proof of Theorem 1.12 gives a probabilistic interpretation to the ratio e_μ/e_λ as the probability distribution of the terminal cell of a hook walk on the Young diagram of λ. However, the ratio d_μ/d_λ (which we now know is equal to e_μ/e_λ) also has a fairly obvious, but different, probabilistic interpretation: it is the probability for a Young tableau chosen uniformly at random from the set of Young tableaux of shape λ (out of the d_λ possible tableaux), to have its maximal entry (namely, n, the size of the diagram λ) located in the corner cell which is the difference between λ and μ. This is so because Young tableaux of shape λ having their maximal entry in this corner cell are in obvious correspondence with Young tableaux of shape μ.

This observation gives rise to a simple algorithm to sample a uniformly random Young tableau of shape λ: First, choose the cell that will contain the maximal entry n, by performing a hook walk (which by the preceding observation leads to the correct probability distribution for the location of the maximal entry corner). Write n in this cell, and then delete it to obtain a smaller Young diagram μ; then repeat the process with μ by performing another hook walk to choose where to write $n-1$, then repeat to choose where to write $n-2$, and so on, until all the cells of the original diagram are filled with the numbers $1, \ldots, n$. By induction it follows immediately that the Young tableau thus generated is uniformly random in the set of tableaux of shape λ.

As an exercise, I suggest that readers write a computer program to perform this algorithm, and run entertaining simulations, perhaps discovering new facts about random Young tableaux.

1.11 Plancherel measure and limit shapes

With the combinatorial theory in the preceding sections, we have paved the way for an analysis of the asymptotic behavior of random partitions of n chosen according to Plancherel measure, as $n \to \infty$. This analysis will use tools of a more, well, *analytical*, nature, since we are now interested in asymptotic results. In particular, as we shall see, the problem of understanding the asymptotic behavior of Plancherel-random partitions will lead us to a certain minimization problem in the calculus of variations, whose solution will ultimately require the use of some techniques of harmonic analysis.

Before starting the rigorous analysis, it is worthwhile to look at a computer simulation to develop some intuition. And why not, really? It is easy to generate a Plancherel-random partition $\lambda^{(n)}$ of size n, by simply applying the Robinson–Schensted algorithm to a uniformly random permutation $\sigma_n \in S_n$ (another nice algorithm based on a variant of the hook walk is outlined in Exercise 1.23). The results for two different partition sizes are shown in Fig. 1.8.

As we can see, for a large value of n the random partition, a discrete object, takes on a nearly continuous (maybe even smooth) shape. One can run the simulation many times and still get a shape that looks approximately the same as in the figure. It turns out that this is not a coincidence, and in fact this is our first instance of a **limit shape** phenomenon – namely, the phenomenon of some geometric object defined on a class of random discrete objects converging to a limiting continuous shape as a size parameter

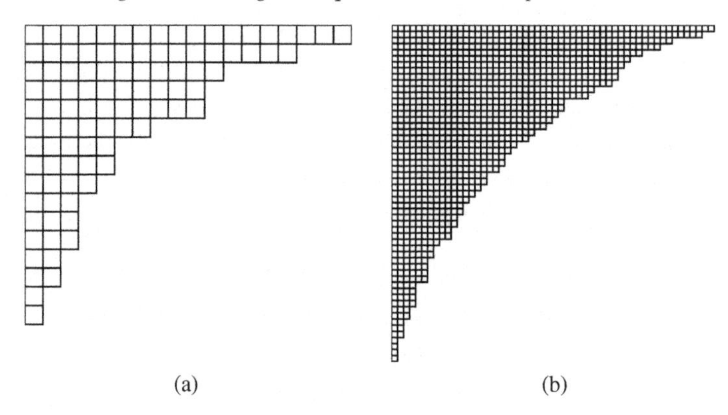

(a) (b)

Figure 1.8 Random Young diagrams sampled from Plancherel measures of orders (a) $n = 100$ and (b) $n = 1000$.

tends to infinity. We will see more examples of limit shapes later in the book.

It therefore seems that our goal should be to rigorously prove the existence of this limit shape, in an appropriate limiting sense. With a bit of luck, we might even find a precise formula for it. Note that such a result will give us much more information than what we originally bargained for, which concerned only the length of the first row of the random partition $\lambda^{(n)}$. (This illustrates the principle that in mathematics, when one is trying to solve a difficult problem, the right thing to do is often to try to solve a *much harder* problem.)

To begin, our problem involves a probability measure on the set $\mathcal{P}(n)$ of partitions of n, so it will be useful to know roughly how big this set is. Define

$$p(n) = |\mathcal{P}(n)|.$$

The function $p(n)$ (called the **partition function**) is an important special function of number theory. Its first few values are $1, 2, 3, 5, 7, 11, 15, \ldots$. It has many remarkable arithmetic properties and deep connections to many branches of mathematics, but for our purposes we are interested only in its rate of growth. It is easy to derive some rough estimates for this, namely that there are constants $C, c > 0$ such that the bounds

$$ce^{c\sqrt{n}} \leq p(n) \leq Ce^{C\sqrt{n}} \tag{1.15}$$

hold for all $n \geq 1$; see Exercises 1.13 and 1.15. In qualitative terms, this means that $p(n)$ is roughly of exponential order in the square root of n. This will be sufficient for our needs, but for the curious-minded, a more precise answer was derived by G. H. Hardy and S. Ramanujan, who showed in 1918 [56] (see also [91]) that[7]

$$p(n) = (1 + o(1))\frac{1}{4\sqrt{3}n}e^{\pi\sqrt{2n/3}} \quad \text{as } n \to \infty. \tag{1.16}$$

(Exercise 1.16 describes a simple method for proving an upper bound for $p(n)$ that gives some of indication of where the constant $\pi\sqrt{2/3}$ comes from. A proof of (1.16) and additional related results is sketched in Exercises 4.9 and 4.10 in Chapter 4.)

Next, since we will be talking about the convergence of a sequence of Young diagrams to a limiting shape, we need to define the space of shapes on which our analysis will take place, and the notion of convergence (i.e., a topology or metric structure) on the space. Since integer partitions are described by a nonincreasing sequence of nonnegative integers that sum to a given integer n, we can describe the analogous limiting objects as nonnegative nonincreasing functions on the positive real line that integrate to a given real number. Define $\mathcal{F}$ to be the space of functions $f\colon [0,\infty) \to [0,\infty)$ such that:

1. f is nonincreasing;
2. $\int_0^\infty f(x)\,dx = 1$;
3. f has compact support, that is, $\sup\{x \geq 0 : f(x) > 0\} < \infty$.

The requirement for f to integrate to 1 is arbitrary and is made for obvious reasons of convenience. Condition 3 is also added for convenience, and although it is not a priori obvious that it should be included, we will see later that adding it simplifies the analysis and does not restrict the generality of the results. We refer to elements of $\mathcal{F}$ as **continual Young diagrams**.

If $\lambda \in \mathcal{P}(n)$, we can embed λ in $\mathcal{F}$ by associating with it a function $\phi_\lambda \in \mathcal{F}$, defined by

$$\phi_\lambda(x) = n^{-1/2}\,\lambda'_{\lfloor\sqrt{n}x\rfloor+1}, \qquad (x \geq 0). \tag{1.17}$$

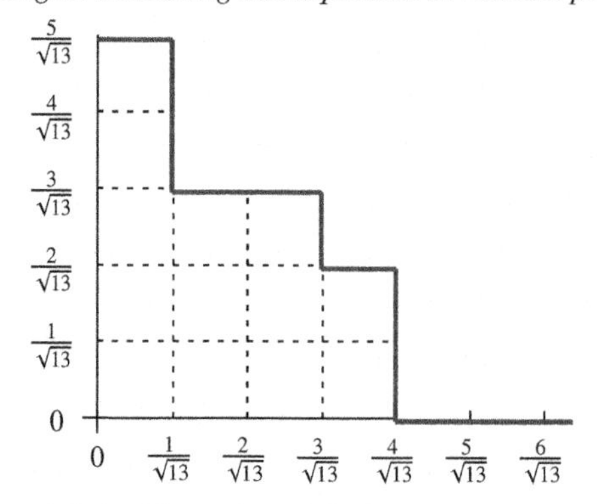

Figure 1.9 An illustration of the function ϕ_λ encoding a Young diagram $(4, 4, 3, 1, 1)$.

In other words, we represent the graph of the Young diagram as a nonincreasing function, and scale both the horizontal and positive axes by a factor of $\sqrt{n}$ each, to scale the area of the Young diagram from n down to 1. The use of the conjugate partition λ', rather than λ, is not essential, but is designed to reconcile the two somewhat incompatible coordinate systems used in the combinatorics of Young diagrams and in calculus (see box on the next page). Fig. 1.9 illustrates the function ϕ_λ.

We defer for the moment the decision about the topology on $\mathcal{F}$; the correct notion of convergence will become apparent in the course of the analysis. The important thing to remember is that as $n \to \infty$, we expect the scaled graph of the typical Plancherel-random partition $\lambda^{(n)}$ of order n to become, with asymptotically high probability, close (in some limiting sense that is yet to be determined) to a certain fixed and nonrandom function in $\mathcal{F}$.

In the next section we take a closer look at the term d_λ^2 that appears in the formula (1.11) for Plancherel measure, and show that it can be approximated by a certain analytic expression involving the function ϕ_λ.

1.12 An asymptotic hook-length formula

The product of hook-lengths of the cells of λ that is incorporated within the term d_λ^2 in (1.11) is clearly the most interesting part of the formula for Plancherel measure, so getting a better understanding of this product is

English, French, and Russian coordinates

The literature on Young diagrams and Young tableaux uses several different coordinate systems to represent Young diagrams and Young tableaux. The standard way of depicting them, sometimes referred to as "English" notation, is the one we have used earlier, where the diagram rows (in order of decreasing size) are drawn from top to bottom. The reference to England may allude to the legacy of the English mathematicians Alfred Young and Norman Macleod Ferrers (after whom are named the so-called **Ferrers diagrams**, an earlier variant of Young diagrams using disks instead of squares) who pioneered the use of these diagrams. In the work of some French authors a different convention has taken root in which the diagrams are drawn from the bottom up. This "French" coordinate system leads to drawings that are more compatible with the standard Cartesian coordinate system used in calculus and much of the rest of mathematics. (Well, Descartes was French after all...)

A third coordinate system, which we refer to as "Russian coordinates" (a.k.a. "rotated coordinates"), has emerged in which diagrams are presented rotated counterclockwise by 45 degrees with respect to the French convention. This turns out to be a useful trick that simplifies their analysis for many purposes (including ours). The use of this coordinate system was especially popularized in the 1980s and 1990s through the works of the Russian mathematician Sergei Kerov and other authors building on his work.

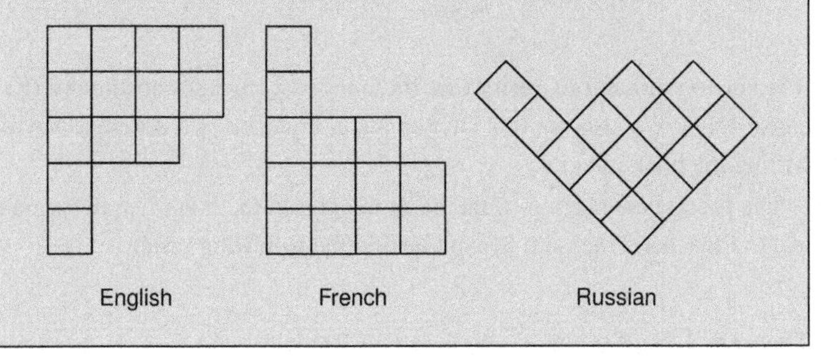

English French Russian

an obvious goal on the way to proving a limit shape result. Because of the simple form of the product, it turns out that we can, without much difficulty, define an analogous analytical expression that appears in connection with functions $f \in \mathcal{F}$, and that can be used to get a very good approximation of the hook-length product for large Young diagrams.

Let $f \in \mathcal{F}$, and let $(x, y) \in [0, \infty) \times [0, \infty)$ be some point that is bounded under the graph of f; that is, we have $y \le f(x)$ (if f is thought of as a limiting notion of a Young diagram, then (x, y) corresponds to the limiting location of a cell). We define the **arm** of (x, y) to be the set $\{(x', y) : x \le x' \le f^{-1}(y)\}$. We define the **leg** of (x, y) to be the set $\{(x, y') : y \le y' \le f(x)\}$. Let the **hook** of (x, y), denoted $H_f(x, y)$, be the union of the arm and the leg of (x, y). Define the **hook-length** of (x, y), denoted $h_f(x, y)$, to be the length of $H_f(x, y)$, namely

$$h_f(x, y) = f(x) - y + f^{-1}(y) - x,$$

where $f^{-1}(y)$ is defined by

$$f^{-1}(y) = \inf\{x \ge 0 \,:\, f(x) \le y\}.$$

Define the **hook integral** of f by the expression

$$I_{\text{hook}}(f) = \int_0^\infty \int_0^{f(x)} \log h_f(x, y)\, dy\, dx. \tag{1.18}$$

It is not too difficult to verify that the integral converges absolutely (Exercise 1.17). We also see this later as a consequence of a different way of writing the hook integral.

The precise connection of the hook integral to the hook-length formula and to Plancherel measure is explained in the following result.

Theorem 1.14 (Asymptotic hook-length formula) *As $n \to \infty$, we have, uniformly over all partitions $\lambda \in \mathcal{P}(n)$, that*

$$\frac{d_\lambda^2}{n!} = \exp\left[-n\left(1 + 2I_{hook}(\phi_\lambda) + O\left(\frac{\log n}{\sqrt{n}}\right)\right)\right]. \tag{1.19}$$

Proof First, by elementary manipulations and Stirling's approximation (0.1), we can bring the formula for Plancherel measure to a form more suitable for doing asymptotics, as follows:

$$\begin{aligned}
\frac{d_\lambda^2}{n!} &= \frac{1}{n!} \cdot \frac{(n!)^2}{\left(\prod_{(i,j)} h_\lambda(i,j)\right)^2} = n! \prod_{(i,j)} h_\lambda(i,j)^{-2} \\
&= \exp\left((n+\tfrac{1}{2})\log n - n + \tfrac{1}{2}\log(2\pi) - 2\sum_{(i,j)} \log h_\lambda(i,j) + o(1)\right) \\
&= \exp\left(-2\sum_{(i,j)} \log\left(\frac{h_\lambda(i,j)}{\sqrt{n}}\right) - n + \tfrac{1}{2}\log(2\pi n) + o(1)\right) \\
&= \exp\left[-n\left(1 + 2\cdot\frac{1}{n}\sum_{(i,j)} \log\left(\frac{h_\lambda(i,j)}{\sqrt{n}}\right) - \frac{1}{2n}\log(2\pi n) + o(1/n)\right)\right].
\end{aligned}$$

Here, the products and sums are over all cells (i,j) of λ. As a consequence of this, we see that to prove (1.19), it will be enough to show that

$$\frac{1}{n}\sum_{(i,j)} \log\left(\frac{h_\lambda(i,j)}{\sqrt{n}}\right) = I_{\text{hook}}(\phi_\lambda) + O\left(\frac{\log n}{\sqrt{n}}\right). \tag{1.20}$$

This already starts to look reasonable, since we can decompose the hook integral too as a sum over the cells (i,j) of λ, namely

$$I_{\text{hook}}(\phi_\lambda) = \sum_{(i,j)} \int_{(j-1)/\sqrt{n}}^{j/\sqrt{n}} \int_{(i-1)/\sqrt{n}}^{i/\sqrt{n}} \log(\phi_\lambda(x) + \phi_\lambda^{-1}(y) - x - y)\, dy\, dx.$$

Fix a cell (i,j), and denote $h = h_\lambda(i,j)$. We estimate the error in the approximation arising from the cell (i,j), namely

$$E_{i,j} := \int_{(j-1)/\sqrt{n}}^{j/\sqrt{n}} \int_{(i-1)/\sqrt{n}}^{i/\sqrt{n}} \log(\phi_\lambda(x) + \phi_\lambda^{-1}(y) - x - y)\, dy\, dx - \frac{1}{n}\log\left(\frac{h}{\sqrt{n}}\right).$$

Note that from the definition of ϕ_λ it follows that when x ranges over the interval $[(j-1)/\sqrt{n}, j/\sqrt{n}]$, the value of $\phi_\lambda(x)$ is constant and equal to $\frac{1}{\sqrt{n}}\lambda'_j$. Similarly, when y ranges over $[(i-1)/\sqrt{n}, i/\sqrt{n}]$, the value of $\phi_\lambda^{-1}(y)$ is constant and equal to $\frac{1}{\sqrt{n}}\lambda_i$. Therefore $E_{i,j}$ can be rewritten as

$$E_{i,j} = \int_{(j-1)/\sqrt{n}}^{j/\sqrt{n}} \int_{(i-1)/\sqrt{n}}^{i/\sqrt{n}} \log\left(\frac{\lambda_i + \lambda'_j}{\sqrt{n}} - x - y\right) dy\, dx - \frac{1}{n}\log\left(\frac{h}{\sqrt{n}}\right).$$

Noting that $h = \lambda_i + \lambda'_j - i - j + 1$, by a simple change of variable we see that this can be rewritten in the form

$$E_{i,j} = \frac{1}{n}\int_{-1/2}^{1/2}\int_{-1/2}^{1/2} (\log(h-x-y) - \log h)\,dy\,dx$$
$$= \frac{1}{n}\int_{-1/2}^{1/2}\int_{-1/2}^{1/2} \log\left(1 - \frac{x+y}{h}\right)dy\,dx.$$

When $h \geq 2$, this can be easily bounded: since $|\log(1-t)| \leq 2|t|$ for $t \in [-1/2, 1/2]$, we have

$$|E_{i,j}| \leq \frac{2}{n \cdot h}. \tag{1.21}$$

When $h = 1$, on the other hand, $E_{i,j}$ is exactly equal to

$$\frac{1}{n}\int_{-1/2}^{1/2}\int_{-1/2}^{1/2} \log(1-x-y)\,dy\,dx = \frac{1}{n}(2\log 2 - 3/2) \approx -\frac{0.11}{n},$$

and in particular the bound (1.21) also holds in this case.

Finally, the total approximation error that we were trying to bound is

$$\left| I_{\text{hook}}(\phi_\lambda) - \frac{1}{n}\sum_{(i,j)} \log\left(\frac{h_\lambda(i,j)}{\sqrt{n}}\right)\right| = \left|\sum_{(i,j)} E_{i,j}\right| \leq \sum_{(i,j)} |E_{i,j}| \leq \frac{2}{n}\sum_{(i,j)} \frac{1}{h_\lambda(i,j)},$$

so (1.20) reduces to showing that for any partition $\lambda \in \mathcal{P}(n)$ we have

$$\sum_{(i,j)} \frac{1}{h_\lambda(i,j)} = O(\sqrt{n}\log n), \tag{1.22}$$

with a constant concealed by the big-O symbol that is uniform over all partitions of n. To prove this claim, fix $1 \leq m \leq n$, and note that the cells (i, j) of λ with a given hook-length $h_\lambda(i, j) = m$ all lie on different rows and different columns. This implies that there are at most $\sqrt{2n}$ of them, since otherwise the sum of their column indices (which are all distinct) would be greater than n – in contradiction to the fact that, because their row indices are all distinct, this sum represents the number of cells that are directly to the left of such an "m-hook" cell, and is therefore $\leq n$.

Since this is true for each m, it follows that the sum of the hook-length reciprocals is at most what it would be if the hook-lengths were crowded down to the smallest possible values under this constraint, namely

$$\overbrace{\tfrac{1}{1} + \ldots + \tfrac{1}{1}}^{\lceil\sqrt{2n}\rceil} + \overbrace{\tfrac{1}{2} + \ldots + \tfrac{1}{2}}^{\lceil\sqrt{2n}\rceil} + \overbrace{\tfrac{1}{3} + \ldots + \tfrac{1}{3}}^{\lceil\sqrt{2n}\rceil} + \ldots,$$

where we terminate the sum as soon as n summands have been included. This clearly results in a bound of order $\sqrt{n}\log n$, proving (1.22). □

1.13 A variational problem for hook integrals

Theorem 1.14 can be interpreted qualitatively as giving us a means of computing, for each given shape $f \in \mathcal{F}$, how unlikely this shape (or one approximating it) is to be observed in a Plancherel-random partition of high order. The measure of unlikelihood for this event is given by the expression $1 + 2I_{\text{hook}}(f)$ involving the hook integral, which is the main term in the exponent in (1.19). More precisely, this quantity is an *exponential* measure of unlikelihood, since the integral appears in the exponent; that is, each change of f that leads to an increase of $I_{\text{hook}}(f)$ by an additive constant translates to a *decrease* in probability by a *multiplicative* constant. The factor of n appearing in front of this expression in the exponent magnifies the effect so that it becomes more and more pronounced as n grows larger.

In probability theory one often encounters results of similar nature concerning an exponential measure of unlikelihood of an event. They are called **large deviation principles** (since they measure probabilities of extremely unlikely events, i.e., large deviations from the typical behavior). A general **theory of large deviations** has been developed that facilitates proving and using such theorems (see box on the next page). This theory becomes useful in relatively abstract situations. Here, since the setting involves a fairly concrete combinatorial structure, we can accomplish our goals with more transparent ad hoc arguments, so there will be no need to develop the general theory.

An important idea in the theory of large deviations is that in many cases when we have a large deviation principle, the behavior of the random model that is observed typically (i.e., with asymptotically high probability) is similar to the behavior that is the *most likely* – that is, the behavior where the "exponential measure of unlikelihood" (usually referred to as the **rate function**, or **rate functional**) is at its minimum. This favorable state of affairs does not occur in all situations,[9] but for our problem we shall see later that it does. Thus, the key to the limit shape result will be to identify the element $f \in \mathcal{F}$ for which the hook integral $I_{\text{hook}}(f)$ takes its minimal value. For an integral functional such as the hook integral defined on some

Large deviations

The theory of large deviations studies probabilities of extremely rare events. For example, what can be said about the probability that a sequence of 10,000 tosses of a fair coin will yield the result "heads" 6000 times? To be sure, "almost 0" would be a pretty good answer, but often it is desirable to have more accurate answers, since such "exponentially unlikely" events, if they have an exponentially large influence on the outcome of an experiment, can have a significant effect on the average result. For this reason, the Swedish actuary Frederick Esscher was among the first people to study such problems in the 1930s, motivated by questions of risk analysis for the insurance industry [139].

The theory of large deviations was later extensively developed in the second half of the 20th century with the works of notable mathematicians such as Cramér, Donsker, Sanov, and especially Srinivasa S. R. Varadhan, who formalized and unified the theory[8]; see [31], [139]. The theory has found many applications. One particularly useful idea (which we use here) is that the analysis of the large deviation properties of a system can often give insight into its *typical* behavior. This is somewhat counterintuitive, since understanding the typical behavior requires one to identify events that are very likely to happen, whereas the large deviation analysis is concerned with measuring the probabilities of *extremely unlikely* events. But, to paraphrase Sherlock Holmes, once you have eliminated the exponentially improbable, whatever remains must be the truth!

(generally infinite-dimensional) space of functions, the problem of finding its minimum (or more generally its stationary points) is known as a **variational problem**. The area of mathematics that concerns the systematic study of such problems is called the **calculus of variations** (see the box opposite).

The variational problem for hook integrals. *Find the function $f \in \mathcal{F}$ for which $I_{hook}(f)$ is minimized, and compute the value $I_{hook}(f)$ at the minimizing function.*

Our goal for the next few sections will be to solve this variational problem. This will proceed through several stages of simplification and analysis, culminating in an explicit formula for the minimizer. Finally, once the problem is solved, we will deduce from its solution our limit shape theorem for Plancherel-random partitions. From there, the way will be paved

The calculus of variations

The calculus of variations is concerned with problems involving the minimization or maximization of real-valued "functions of functions" (which are often called **functionals**, to avoid an obvious source of confusion). Such numerical quantities depending on an entire function (e.g., the length of a curve or energy of a vibrating string) usually involve integration of some local parameter along the graph of the function.

Historically, the first examples motivating the development of the general theory came from physics; some famous ones are the **tautochrone problem** of finding a curve along which a ball rolling under the influence of gravity will oscillate around the minimum point of the curve with a period independent of its starting position, and the related **brachistochrone problem** of finding a curve connecting two given points such that a ball rolling from the top point to the bottom will make the descent in the fastest time. The former question was solved in 1657 by Christiaan Huygens (who then unsuccessfully attempted to use the solution to construct a more precise version of his famous pendulum clock). The latter was solved by Johann Bernoulli in 1696. The solutions to both problems are curves known as cycloids. Later, in 1755, Leonhard Euler and Joseph-Louis Lagrange generalized the techniques originating in the tautochrone problem to derive a much more general principle for finding stationary points of integral functionals, which became known as the **Euler–Lagrange equation**. This laid the foundation for the emerging field of the calculus of variations.

To this day, the calculus of variations is of fundamental importance in physics. One reason for this is the fact that many physical laws, though usually derived as differential equations, have equivalent formulations as **variational principles** – namely, statements to the effect that the physical system "chooses" its behavior so as to minimize the integral of a certain quantity over the entire history of the system.

for solving our original problem of determining the first-order asymptotics of ℓ_n, the average length of a longest increasing subsequence of a uniformly random permutation.

1.14 Transformation to hook coordinates

The hook integral is rather messy as it involves the expression $h_f(x, y)$ for the hook-lengths (which, annoyingly, involves both the function f and its inverse), and is therefore difficult to analyze. It turns out that it can be simplified and brought to a much cleaner form that will give us consider-

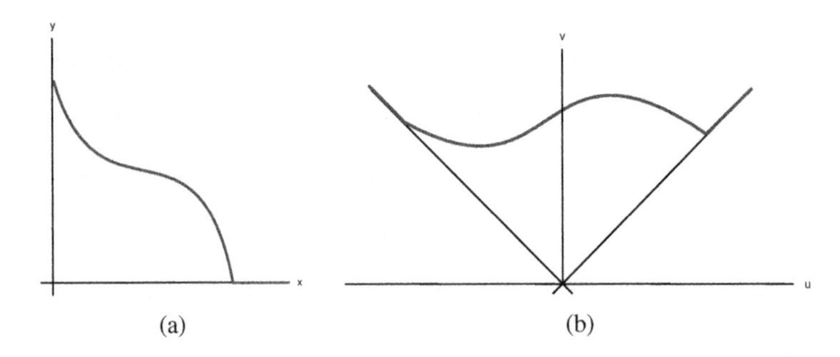

Figure 1.10 A continual Young diagram in standard and rotated coordinates.

able insight into its behavior, using several clever transformations. First, we change our coordinate system from the "standard" or "English" coordinate system to the "Russian" coordinates

$$u = \frac{x-y}{\sqrt{2}}, \qquad v = \frac{x+y}{\sqrt{2}}. \tag{1.23}$$

The u–v coordinates are simply an orthogonal coordinate system rotated by 45 degrees relative to the standard x–y coordinates. Note that some authors divide by 2 instead of $\sqrt{2}$, which results in the change of variables being a nonorthogonal transformation. This results in some cleaner formulas but seems a bit less intuitive, so we prefer the convention given here.

Having changed the coordinate system, we can define a new space of functions $\mathcal{G}$ that describe the limiting shapes of Young diagrams in this new coordinate system. More precisely, to each continual Young diagram $f \in \mathcal{F}$ there corresponds a unique function $g \in \mathcal{G}$ whose graph is the graph of f drawn in the new coordinate system. We still refer to g as a continual Young diagram, leaving the coordinate system to be understood from the context. For example, Fig. 1.10 shows a continual Young diagram $f \in \mathcal{F}$ and its associated rotated version $g \in \mathcal{G}$. Formally, f and g are related to each other by

$$v = g(u) \quad \Longleftrightarrow \quad \frac{v-u}{\sqrt{2}} = f\left(\frac{v+u}{\sqrt{2}}\right). \tag{1.24}$$

What exactly is the new space of functions $\mathcal{G}$ obtained by transforming

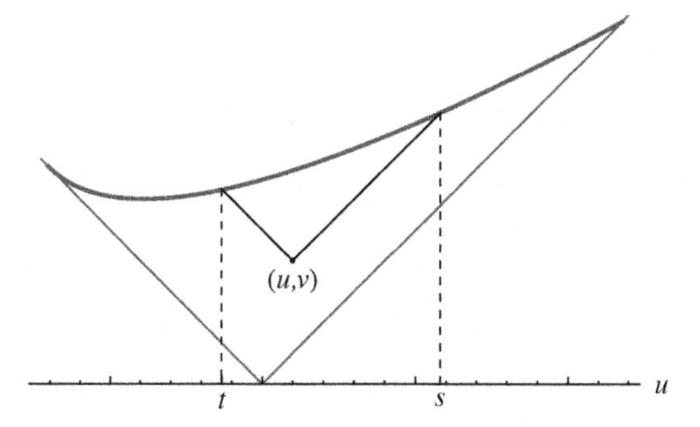

Figure 1.11 Hook coordinates.

functions in f in this way? It is not hard to see that $\mathcal{G}$ is exactly the space of functions $g: \mathbb{R} \to [0, \infty)$ satisfying the following conditions:

1. g is 1-Lipschitz, that is, $|g(s) - g(t)| \leq |s - t|$ for all $s, t \in \mathbb{R}$;
2. $g(u) \geq |u|$ for all $u \in \mathbb{R}$;
3. $\int_{-\infty}^{\infty} (g(u) - |u|)\, du = 1$;
4. $g(u) - |u|$ is supported on a compact interval $[-I, I]$.

Note that in particular, in this new coordinate system the functions being considered as candidates to be the limit shape are now continuous, and in fact also (being Lipschitz functions) almost everywhere differentiable. This is already somewhat of an advantage over the previous coordinate system.

Fix $f \in \mathcal{F}$, and let $g \in \mathcal{G}$ be the function corresponding to f in the new coordinate system. The hook integral is defined as a double integral over the region bounded under the graph of f – or, in terms of u–v coordinates, over the region bounded between the graph of g and the graph of the absolute value function $v = |u|$. We now define a second transformation that parametrizes this region in terms of yet another set of coordinates, which we call the **hook coordinates**. These are defined as follows: if (u, v) is a point such that $|u| \leq v \leq g(u)$, consider the arm and the leg which together form the hook of the point (u, v). In the Russian coordinate system, the arm and the leg extend from (u, v) in the northeast and northwest directions, respectively, until they intersect the graph of g. Denote the points of intersection with the graph of g by $(s, g(s))$ and $(t, g(t))$. The hook coordinates are simply the pair (t, s); see Fig. 1.11

Formally, t and s can be defined using the equations

$$g(s) - s = v - u, \qquad g(t) + t = v + u. \tag{1.25}$$

The hook coordinates can be thought of as a system of nonlinear coordinates, mapping the region bounded between the graphs of g and $|u|$ to some region Δ in the t–s plane. This allows us to rewrite the hook integral as a double integral over the region Δ. The integrand $\log h_f(x, y)$ simply transforms to $\log\left(\sqrt{2}(s-t)\right)$, so the integral will be

$$I_{\text{hook}}(f) = \iint_\Delta \log\left(\sqrt{2}(s-t)\right) \left|\frac{D(u, v)}{D(t, s)}\right| \, ds\, dt,$$

where $D(u, v)/D(t, s)$ is the Jacobian of the transformation $(t, s) \to (u, v)$. Let us compute this Jacobian, or rather (since this is a bit more convenient) its reciprocal, $D(t, s)/D(u, v)$, which is the Jacobian of the inverse transformation $(u, v) \to (t, s)$. To do this, take partial derivatives with respect to both the variables u and v of both equations in (1.25), and solve for the four partial derivatives $\partial s/\partial u, \partial s/\partial v, \partial t/\partial u, \partial t/\partial v$, to get easily that

$$\begin{aligned} \frac{\partial t}{\partial u} &= \frac{\partial t}{\partial v} = \frac{1}{1 + g'(t)}, \\ \frac{\partial s}{\partial u} &= -\frac{\partial s}{\partial v} = \frac{1}{1 - g'(s)}. \end{aligned}$$

Therefore we get that

$$\frac{D(t, s)}{D(u, v)} = \det\begin{pmatrix} \frac{\partial t}{\partial u} & \frac{\partial t}{\partial v} \\ \frac{\partial s}{\partial u} & \frac{\partial s}{\partial v} \end{pmatrix} = \frac{2}{(g'(s) - 1)(g'(t) + 1)}. \tag{1.26}$$

So, we have shown so far that the hook integral from (1.18) can be expressed as

$$\begin{aligned} I_{\text{hook}}(f) &= \tfrac{1}{2} \iint_\Delta \log\left(\sqrt{2}(s-t)\right) (1 + g'(t))(1 - g'(s))\, dt\, ds \\ &= \tfrac{1}{2} \iint_{-\infty<t<s<\infty} \log\left(\sqrt{2}(s-t)\right) (1 + g'(t))(1 - g'(s))\, dt\, ds, \end{aligned} \tag{1.27}$$

where the last equality comes from the observation that outside of Δ we always have that either $1 + g'(t) = 0$ (when t becomes sufficiently negative so that $g(t) = -t$) or $1 - g'(s) = 0$ (when s is sufficiently positive so that $g(s) = t$), so the integrand vanishes.

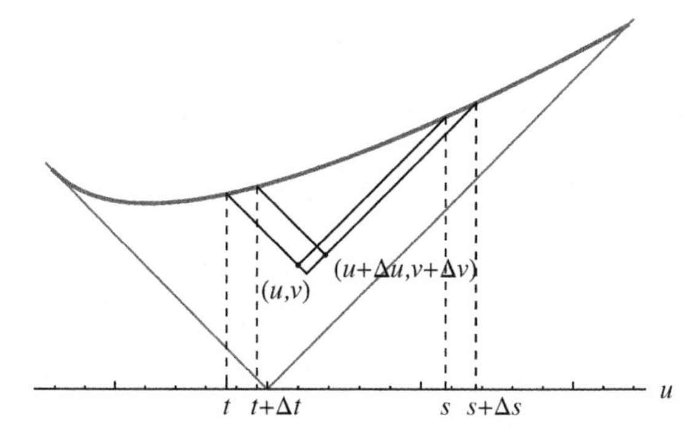

Figure 1.12 Perturbing the hook coordinates as a geometric way to compute the Jacobian of the transformation to hook coordinates.

As a side remark, note that there is another, more geometric, way to understand the Jacobian factor of the transformation $(u, v) \to (s, t)$. This is illustrated in Fig. 1.12, which shows the effect of perturbing the t and s coordinates slightly. When Δt and Δs become small, the Jacobian (ignoring its sign) emerges as the approximate ratio of the area of the rotated rectangle with opposite corners (u, v) and $(u + \Delta u, v + \Delta v)$ shown in the figure, to the differential $\Delta t\, \Delta s$. It is not hard to see by elementary geometric considerations that this is approximately equal to $\frac{1}{2}(1 - g'(s))(1 + g'(t))$.

To simplify (1.27) some more, first note that the $\sqrt{2}$ term in the logarithm can be handled separately, and adds a constant contribution of $\frac{1}{2}\log(2)$ (since the constant is integrated against a weight function that integrates to 1); second, define

$$h(u) = g(u) - |u|, \tag{1.28}$$

so that we have $h'(u) = g'(u) - \operatorname{sgn}(u)$ (where $\operatorname{sgn}(u)$ is the signum function), and observe that it is convenient to represent the integral

$$\iint_{t<s} \log(s - t)(1 + g'(t))(1 - g'(s))\, dt\, ds \tag{1.29}$$

as a sum of two terms, the first one of which being

$$\iint_{t<s} \log(s-t)(g'(t)-\operatorname{sgn}(t))(\operatorname{sgn}(s)-g'(s))\,dt\,ds$$
$$= -\iint_{t<s} \log(s-t)h'(t)h'(s)\,dt\,ds, \tag{1.30}$$

and the second being the difference of (1.29) and this last integral. This difference can be computed by splitting the domain of integration into several parts according to the signs of t and s. The part where $t < 0 < s$ disappears because of cancellation, and we are left with

$$\iint_{-\infty<t<s<0} \log(s-t)\,h'(t)(1-(-1))\,dt\,ds$$
$$+ \iint_{0<t<s<\infty} \log(s-t)(1-(-1))(-h'(s))\,dt\,ds$$
$$= 2\int_{-\infty}^{0} h'(t)\left(\int_{t}^{0} \log(s-t)\,ds\right)dt - 2\int_{0}^{\infty} h'(s)\left(\int_{0}^{s} \log(s-t)\,dt\right)ds$$
$$= -2\int_{-\infty}^{\infty} h'(u)\,(u\log|u| - u)\;du.$$

We can now write our integral functional $I_{\text{hook}}(f)$ in the final form that we will need for our analysis. Define

$$B(h_1,h_2) = -\tfrac{1}{2}\int_{-\infty}^{\infty}\int_{-\infty}^{\infty} \log|s-t|\,h_1'(t)h_2'(s)\,dt\,ds,$$
$$Q(h) = B(h,h) = -\tfrac{1}{2}\int_{-\infty}^{\infty}\int_{-\infty}^{\infty} \log|s-t|\,h'(t)h'(s)\,dt\,ds,$$
$$L(h) = -2\int_{-\infty}^{\infty} h'(u)\,(u\log|u| - u)\;du.$$

Then combining (1.27) with the computations above, and noting that $Q(h)$ is a symmetrized form of (1.30), we have shown that

$$2I_{\text{hook}}(f) = \log 2 + Q(h) + L(h) =: J(h). \tag{1.31}$$

Note that $B(h_1,h_2)$ is a bilinear form in h_1' and h_2'; $Q(h)$ is a quadratic form in h'; and $L(h)$ is a linear functional of h'. The representation (1.31) provides an alternative method of showing that the hook integral converges absolutely; see Exercise 1.17.

The function h, by its definition in (1.28) as a transformed version of g, now ranges over a new space of functions, which we denote by $\mathcal{H}$. By the connection between the Lipschitz property and the notion of absolute continuity, this can be described as the space of functions $h\colon \mathbb{R} \to [0,\infty)$ satisfying:

1. h is absolutely continuous (that is, almost everywhere differentiable and satisfies the Newton–Leibniz formula $h(x) = h(0) + \int_0^x h'(u)\,du$ for all x);
2. $\int_{-\infty}^{\infty} h(u)\,du = 1$;
3. $\operatorname{sgn}(u) \cdot h'(u) \in [-2, 0]$ wherever $h'(u)$ is defined;
4. h is supported on a compact interval $[-I, I]$.

The problem of minimizing $I_{\text{hook}}(f)$ on the function space $\mathcal{F}$ has thus been reformulated to that of minimizing $J(h)$ on the new space $\mathcal{H}$. This is addressed in the next section.

1.15 Analysis of the functional

Our new representation of the hook integral in the form $J(h) = Q(h) + L(h) + \log(2)$, consisting of a constant, a linear term, and a quadratic term, should give us hope that the minimization problem that we are trying to solve should actually not be too difficult. (To be sure, many clever steps were already required just to get to this point in the analysis!) In fact, it is interesting to note that our minimization problem is similar to variational problems that appear in several other branches of mathematics – notably in the analysis of stationary distributions of electrostatic charges in certain physical systems, in potential theory, and in random matrix theory; see [58], Section 5.3. In other words, we are in fairly well-charted territory and can use more or less standard techniques, although this will require using some nontrivial facts from analysis, and in particular some of the properties of the Hilbert transform (see box on the next page).

We start by examining the quadratic part $Q(h)$ of the hook integral functional. A key property is that it is positive-definite.

Proposition 1.15 *For any Lipschitz function $h\colon \mathbb{R} \to \mathbb{R}$ with compact support, we have that $Q(h) \geq 0$, with equality if and only if $h \equiv 0$.*

Proof Let h be a function satisfying the assumptions of the proposition.

The Hilbert transform

The Hilbert transform, named after David Hilbert, is an important analytic transform that maps a function $f\colon \mathbb{R} \to \mathbb{R}$ to the function $\tilde{f}$ defined by

$$\tilde{f}(x) = \frac{1}{\pi} \int_{-\infty}^{\infty} \frac{f(y)}{x-y}\, dy,$$

where the integral is required to converge almost everywhere in the sense of the **Cauchy principal value** (i.e., the limit as $\epsilon \downarrow 0$ of the integrals where an ϵ-neighborhood centered around the singularity point $y = x$ is removed from the domain of integration). When the Hilbert transform exists (e.g., if $f \in L^2(\mathbb{R})$), we have the inversion relation

$$f(x) = -\frac{1}{\pi} \int_{-\infty}^{\infty} \frac{\tilde{f}(y)}{x-y}\, dy$$

(i.e., $f(x) = -\tilde{\tilde{f}}(x)$), and we say that f and $\tilde{f}$ are a **Hilbert transform pair**. This concept is closely related to the notion of **conjugate harmonic functions** in complex analysis: it can be shown that if $F(z) = u(z) + iv(z)$ is a complex analytic function in the upper half-plane $\mathbb{H} = \{z \in \mathbb{C} : \operatorname{Im} z > 0\}$ (so that u and v are a pair of conjugate harmonic functions on $\mathbb{H}$) such that $\lim_{z \in \mathbb{H}, |z| \to \infty} F(z) = 0$, then, under some additional integrability conditions, the limits

$$f(x) = \lim_{y \downarrow 0} u(x + iy), \qquad g(x) = \lim_{y \downarrow 0} v(x + iy)$$

exist for almost every $x \in \mathbb{R}$, and $g(x) = \tilde{f}(x)$.

Another interesting way of looking at the Hilbert transform is in terms of how it acts in the frequency domain. It is known that $\tilde{f}$ can be obtained from f by shifting the phase of all positive frequencies by $-\pi/2$ and shifting the phase of all negative frequencies by $\pi/2$. More precisely, we have the equation

$$\hat{\tilde{f}}(s) = -i \operatorname{sgn}(s) \hat{f}(s),$$

where $\hat{g}$ denotes the Fourier transform of g. This fact is roughly equivalent to the identities

$$\widetilde{\sin}(x) = -\cos(x), \qquad \widetilde{\cos}(x) = \sin(x),$$

which can be checked without too much difficulty, and also offers an easy way to see that the Hilbert transform is an isometry of $L^2(\mathbb{R})$ onto itself. For more details, refer to [68].

Recall that the **Fourier transform** of h is defined by

$$\hat{h}(x) = \int_{-\infty}^{\infty} h(u) e^{-ixu}\, du$$

(where $i = \sqrt{-1}$). The **Hilbert transform** of h is defined by

$$\tilde{h}(x) = \frac{1}{\pi} \int_{-\infty}^{\infty} \frac{h(u)}{x-u} \, du,$$

where the integral is interpreted in the sense of the Cauchy principal value (see box). Given our assumptions on h it is easy to see that this integral converges for every $x \in \mathbb{R}$. By integration by parts (which is also easy to justify under our assumptions), $\tilde{h}(x)$ can also be expressed as

$$\tilde{h}(x) = \frac{1}{\pi} \int_{-\infty}^{\infty} h'(u) \log |x-u| \, du.$$

It follows that $Q(h)$ can in turn be written as

$$Q(h) = -\frac{\pi}{2} \int_{-\infty}^{\infty} h'(u) \, \tilde{h}(u) \, du = -\frac{\pi}{2} \int_{-\infty}^{\infty} h'(u) \, \overline{\tilde{h}(u)} \, du,$$

where an overline is used to denote the complex conjugate (note that since h is real-valued, so is its Hilbert transform). The integral is the inner product in $L^2(\mathbb{R})$ of the derivative of h and its Hilbert transform, so, by Parseval's theorem in Fourier analysis, we have that

$$Q(h) = -\tfrac{1}{4} \int_{-\infty}^{\infty} \hat{h'}(x) \overline{\hat{\tilde{h}}(x)} \, dx. \tag{1.32}$$

Here, the Fourier transforms of h' and $\tilde{h}$ are given by[10]

$$\hat{h'}(x) = (i\,x)\hat{h}(x), \tag{1.33}$$

$$\hat{\tilde{h}}(x) = (-i \ \mathrm{sgn}(x))\hat{h}(x). \tag{1.34}$$

These relations correspond to the well-known facts that in the frequency domain the derivative operation and Hilbert transform both appear as simple operations of multiplication by the functions $i\,x$ and $-i \ \mathrm{sgn}(x)$, respectively. So we get from (1.32) that

$$Q(h) = \tfrac{1}{4} \int_{-\infty}^{\infty} |x| \cdot \left|\hat{h}(x)\right|^2 dx. \tag{1.35}$$

This immediately implies our claim. □

Next, we prove an explicit computational criterion that will allow us to deduce that a given element of $\mathcal{H}$ is a minimizer for $J(\cdot)$.

Theorem 1.16 *Let $h_0 \in \mathcal{H}$. Assume that for some constant $\lambda \in \mathbb{R}$, the function $p\colon \mathbb{R} \to \mathbb{R}$ defined by*

$$p(u) = -\int_{-\infty}^{\infty} h_0'(s) \log|s-u|\,ds - 2\,(u \log|u| - u) + \lambda u \tag{1.36}$$

has the property that, for almost every $u \in \mathbb{R}$,

$$p(u) \text{ is } \begin{cases} = 0, & \text{if } \operatorname{sgn}(u) \cdot h_0'(u) \in (-2, 0), \\ \leq 0, & \text{if } h_0'(u) + \operatorname{sgn}(u) = 1, \\ \geq 0, & \text{if } h_0'(u) + \operatorname{sgn}(u) = -1. \end{cases} \tag{1.37}$$

Then for any $h \in \mathcal{H}$ we have that

$$J(h) \geq J(h_0) + Q(h - h_0). \tag{1.38}$$

In particular (since $Q(h-h_0) \geq 0$), it follows that h_0 is the unique minimizer for $J(\cdot)$ on $\mathcal{H}$.

Proof Let $h \in \mathcal{H}$. Because we know that h satisfies $|h'(u) + \operatorname{sgn}(u)| \leq 1$, the condition (1.37) implies that

$$(h'(t) - h_0'(t))p(t) \geq 0$$

holds almost everywhere. It follows by integration that

$$\int_{-\infty}^{\infty} h'(t)p(t)\,dt \geq \int_{-\infty}^{\infty} h_0'(t)p(t)\,dt.$$

By the definition of p, this translates to

$$2B(h, h_0) + L(h) + \lambda \int_{-\infty}^{\infty} uh'(u)\,du \geq 2B(h_0, h_0) + L(h_0) + \lambda \int_{-\infty}^{\infty} uh_0'(u)\,du.$$

Equivalently, because we know that $\int_{-\infty}^{\infty} uh'(u)\,du = -\int_{-\infty}^{\infty} h(u)\,du = -1$ and similarly for h_0, we can write this as $2B(h - h_0, h_0) \geq L(h_0 - h)$. It follows that

$$\begin{aligned} J(h) &= Q(h) + L(h) + \log 2 = Q(h_0 + (h - h_0)) + L(h) + \log 2 \\ &= Q(h_0) + Q(h - h_0) + 2B(h_0, h - h_0) + L(h) + \log 2 \\ &\geq Q(h_0) + Q(h - h_0) + L(h_0) + \log 2 = J(h_0) + Q(h - h_0), \end{aligned}$$

as claimed. □

It is worth mentioning that there is some methodology involved in formulating the condition (1.37). It is a version of the so-called **complementary slackness** condition from the theory of convex optimization – essentially a fancy application of the standard technique of Lagrange multipliers from calculus. The constant λ plays the role of the Lagrange multiplier.

It now remains to exhibit a function $h_0 \in \mathcal{H}$ for which the assumptions of Theorem 1.16 are satisfied. We follow a slightly simplified version of the argument from Logan and Shepp's paper [79].

First, we make an informed guess about the form of the solution h_0. Remember that the variational problem originally arose out of the problem of finding the asymptotic shape of a typical Young diagram chosen according to Plancherel measure of high order. Specifically, $h_0(u)$ is of the form $g_0(u) - |u|$, where g_0 represents the limit shape in the Russian coordinate system. Looking back at the simulations from p. 32, we see that we should expect h_0 to have compact support, and to be symmetric about the origin (i.e., an even function). In particular, we should have that

$$h_0'(u) = 0 \qquad \text{if and only if } |u| \geq \beta,$$

for some value $\beta > 0$. In fact, the value $\beta = \sqrt{2}$ is a strong candidate to be the correct value, since, after taking the scaling and rotation into account, that would mean (heuristically, at least) that the graph of the Plancherel-random Young diagram of order n would have a first row of length approximately $2\sqrt{n}$ for n large. This corresponds, via the Robinson–Schensted algorithm, to the empirical observation that uniformly random permutations of order n have a longest increasing subsequence of length about $2\sqrt{n}$, which is the result that we are ultimately aiming to prove. So we assume that $\beta = \sqrt{2}$.

Having made these assumptions, note that the first condition in (1.37) translates to the equation

$$-\int_{-\sqrt{2}}^{\sqrt{2}} h_0'(s) \log|s-u|\, ds = 2(u\log|u| - u) - \lambda u, \qquad (|u| \leq \sqrt{2});$$

or, differentiating with respect to u, we get the condition

$$-\pi(\tilde{h_0'})(u) = -\int_{-\sqrt{2}}^{\sqrt{2}} \frac{h_0'(s)}{u-s}\, ds = 2\log|u| - \lambda \tag{1.39}$$

that should hold for $|u| \leq \sqrt{2}$. In other words, we need at the very least to find a function that is supported on $[-\sqrt{2}, \sqrt{2}]$ and whose Hilbert transform coincides up to an additive constant with $(-2/\pi)\log|u|$ on the interval $[-\sqrt{2}, \sqrt{2}]$. This is accomplished using the following lemma.

Lemma 1.17 *Define functions $f, g\colon \mathbb{R} \to \mathbb{R}$ by*

$$f(x) = \begin{cases} \sin^{-1}(x) - \frac{\pi}{2}\operatorname{sgn}(x) & \text{if } |x| \le 1, \\ 0 & \text{if } |x| > 1, \end{cases}$$

$$g(x) = \begin{cases} -\log(2|x|) & \text{if } |x| \le 1, \\ -\log(2|x|) - \log\left(|x| - \sqrt{x^2-1}\right) & \text{if } |x| > 1. \end{cases}$$

Then g is the Hilbert transform of f.

Proof We use the notion of a **Hilbert transform pair** (see box on p. 48). The claim will follow by showing that the pointwise limit

$$f(x) + ig(x) = \lim_{y\downarrow 0} F(x+iy) \tag{1.40}$$

holds for almost all $x \in \mathbb{R}$, where $F(z)$ is a suitable analytic function defined on the upper half-plane $\mathbb{H} = \{z \in \mathbb{C} : \operatorname{Im} z > 0\}$.

Consider the following simple computations. Let $G(z) = \operatorname{Log}(-iz)$, where $\operatorname{Log}(w)$ denotes the principal value of the complex logarithm function (whose imaginary value is defined to be in $(-\pi,\pi]$). Then $G(z)$ is analytic on $\mathbb{H}$, and for all $x \in \mathbb{R}\setminus\{0\}$ we have

$$\lim_{y\downarrow 0} G(x+iy) = \log|x| - \frac{\pi i}{2}\operatorname{sgn}(x). \tag{1.41}$$

Next, let $H(z) = \arcsin(z)$, a branch of the arcsine function considered as an analytic function on $\mathbb{H}$. Formally, $H(z) = -i\operatorname{Log}(iz + \sqrt{1-z^2})$, where $\sqrt{w}$ is the branch of the square root function defined as $\sqrt{w} = \exp(\frac{1}{2}\operatorname{Log} w)$. It is easy to verify that

$$\lim_{y\downarrow 0} \sqrt{1-(x+iy)^2} = \begin{cases} \sqrt{1-x^2} & \text{if } |x| \le 1, \\ -i\operatorname{sgn}(x)\sqrt{x^2-1} & \text{if } |x| > 1, \end{cases}$$

from which it follows that

$$\lim_{y\downarrow 0} H(x+iy) = \begin{cases} \sin^{-1}(x) & \text{if } |x| \le 1, \\ \frac{\pi}{2}\operatorname{sgn}(x) - i\log\left(|x| - \sqrt{x^2-1}\right) & \text{if } |x| > 1. \end{cases} \tag{1.42}$$

Finally, define

$$F(z) = H(z) - iG(z) - i\log(2) = \arcsin(z) - i\,\mathrm{Log}(-iz) - i\log(2). \quad (1.43)$$

By combining (1.41) and (1.42) we get exactly (1.40).

Finally, according to the theory of the Hilbert transform, to conclude from (1.40) that f and g are a Hilbert transform pair, it is still necessary to check that $F(z)$ is a "good" function; for example, knowing that it is in the Hardy space $H^2(\mathbb{H})$ of the upper half-plane would suffice. This requires checking that

$$\sup_{y\in(0,\infty)} \int_{-\infty}^{\infty} |F(x+iy)|^2\,dx < \infty,$$

whereupon Theorem 95 of [135] (a version of a result known as the Paley–Wiener theorem) implies our claim. An alternative and perhaps easier method is to show first that $\lim_{z\in\mathbb{H},|z|\to\infty} F(z) = 0$, and then follow the derivation in [68], Section 3.4, which produces the desired result as long as one separately verifies a slightly modified version of the convergence we claimed, namely the fact that

$$\int_a^b |F(x+iy) - (f(x) + ig(x))|\,dx \xrightarrow[y\searrow 0]{} 0$$

for any $-\infty < a < b < \infty$. Either of these claims are straightforward and left as exercises for readers with an affinity for complex analysis. □

Theorem 1.18 *The function*

$$h_0(u) = \begin{cases} \frac{2}{\pi}\left(u\sin^{-1}\left(\frac{u}{\sqrt{2}}\right) + \sqrt{2-u^2}\right) - |u| & \text{if } |u| \le \sqrt{2}, \\ 0 & \text{if } |u| > \sqrt{2}, \end{cases} \quad (1.44)$$

is the unique minimizer of the functional $J(\cdot)$ on $\mathcal{H}$.

Proof Note that h_0 is continuous and piecewise differentiable, with its derivative given by

$$h_0'(u) = \begin{cases} \frac{2}{\pi}\sin^{-1}\left(\frac{u}{\sqrt{2}}\right) - \mathrm{sgn}(u) & \text{if } |u| \le \sqrt{2}, \\ 0 & \text{if } |u| > \sqrt{2}. \end{cases}$$

First, we verify that $h_0 \in \mathcal{H}$. Trivially, $0 \le h_0'(u) \le 1$ for $u \le 0$ and

$-1 \le h_0'(u) \le 0$ for $u \ge 0$, so also $h_0(u) \ge 0$ for all u. Furthermore,

$$\begin{aligned}
\int_{-\infty}^{\infty} h_0(u)\,du &= \int_{-\sqrt{2}}^{\sqrt{2}} h_0(u)\,du = -\int_{-\sqrt{2}}^{\sqrt{2}} u h_0'(u)\,du \\
&= \int_{-\sqrt{2}}^{\sqrt{2}} |u|\,du - \frac{2}{\pi}\int_{-\sqrt{2}}^{\sqrt{2}} u \sin^{-1}(u/\sqrt{2})\,du \\
&= 2 - \frac{4}{\pi}\int_{-1}^{1} x\sin^{-1}(x)\,dx = 2 - \frac{4}{\pi}\cdot\frac{\pi}{4} = 1,
\end{aligned}$$

where the last integral is evaluated using the indefinite integral evaluation

$$\int x\sin^{-1}(x)\,dx = \left(\tfrac{1}{2}x^2 - \tfrac{1}{4}\right)\sin^{-1}(x) + \tfrac{1}{4}x\sqrt{1-x^2} + C.$$

So h_0 is indeed in $\mathcal{H}$.

Next, we verify that the conditions in (1.37) are satisfied for a suitable choice of constant λ. Differentiating (1.36), we get that

$$p'(u) = -\int_{-\sqrt{2}}^{\sqrt{2}} \frac{h_0'(s)}{u-s}\,ds - 2\log|u| + \lambda = -\pi(\tilde{h_0'})(u) - 2\log|u| + \lambda.$$

But the Hilbert transform of $h_0'(u)$ is given by Lemma 1.17, up to a small scaling operation (recall that the Hilbert transform is linear and commutes with dilations, or simply manipulate the integral to bring it to the form of the lemma). So we get after a short computation that

$$p'(u) = \begin{cases} \log(2) + \lambda & \text{if } |u| < \sqrt{2}, \\ \log(2) + \lambda + 2\log\left(\frac{|u|}{\sqrt{2}} - \sqrt{\frac{u^2}{2} - 1}\right) & \text{if } |u| > \sqrt{2}, \end{cases}$$

or, now choosing $\lambda = -\log(2)$,

$$p'(u) = \begin{cases} 0 & \text{if } |u| < \sqrt{2}, \\ 2\log\left(\frac{|u|}{\sqrt{2}} - \sqrt{\frac{u^2}{2} - 1}\right) & \text{if } |u| > \sqrt{2}. \end{cases}$$

In particular, $p(u)$ is constant on $[-\sqrt{2}, \sqrt{2}]$, and it is easy to check that $p'(u) \le 0$ for all u. Finally, $p(0) = 0$, since for $u = 0$ the integrand in (1.36) is an odd function. Combining these facts gives that (1.37) is indeed satisfied, which completes the proof. ◻

1.16 Computation of $J(h_0)$

We proved that the function h_0 given in (1.44) is the minimizer for the functional $J(\cdot)$, but we do not yet know the actual minimum value $J(h_0)$. This is not a very difficult computation, given the results we already derived.

Theorem 1.19 $J(h_0) = -1$.

Proof First, we compute $L(h_0)$, as follows. We already saw that h_0 satisfies $\int_{-\sqrt{2}}^{\sqrt{2}} h_0'(u)u\,du = -1$, so

$$\begin{aligned} L(h_0) &= -2\int_{-\sqrt{2}}^{\sqrt{2}} h_0'(u)(u\log|u| - u)\,du \\ &= -2 - 4\int_0^{\sqrt{2}} h_0'(u)u\log(u)\,du \\ &= -2 + 4\int_0^{\sqrt{2}} u\log(u)\,du - \frac{8}{\pi}\int_0^{\sqrt{2}} u\log(u)\sin^{-1}(u/\sqrt{2})\,du \\ &= -2 + (2\log(2) - 2) + (2 - 3\log(2)) = -2 - \log(2), \end{aligned}$$

where we make use of the definite integral evaluation

$$\int_0^1 x\log(x)\sin^{-1}(x)\,dx = \frac{\pi}{8}(\log(2) - 1), \tag{1.45}$$

(a proof of this slightly nontrivial evaluation is sketched in Exercise 1.18). Next, note that the fact that $p(u) = 0$ on $[-\sqrt{2}, \sqrt{2}]$ implies that

$$0 = \int_{-\sqrt{2}}^{\sqrt{2}} p(u)h_0'(u)\,du = 2Q(h_0) + L(h_0) - \lambda = 2Q(h_0) - 2,$$

so $Q(h_0) = 1$. From this we get finally that

$$J(h_0) = Q(h_0) + L(h_0) + \log(2) = -1. \qquad \square$$

1.17 The limit shape theorem

We now go back to the problem of the limit shape of Plancherel-random partitions. As before, for each $n \geq 1$ we let $\lambda^{(n)}$ denote a random partition of n chosen according to Plancherel measure of order n. Let $\phi_n(x)$ be the function in the function space $\mathcal{F}$ corresponding to $\lambda^{(n)}$ as in (1.17). Since it will be natural to formulate our results in terms of the rotated coordinate system, let ψ_n be the element of the function space $\mathcal{G}$ corresponding to ϕ_n in the rotated coordinate system, with ϕ_n and ψ_n being related as f and g in (1.24).

Our first formulation of the limit shape theorem will be in terms of a certain nonstandard topology on $\mathcal{G}$ related to the quadratic functional $Q(\cdot)$. For any compactly supported Lipschitz function $h\colon \mathbb{R} \to [0,\infty)$, denote

$$\|h\|_Q = Q(h)^{1/2}. \tag{1.46}$$

Because Q is positive-definite (Proposition 1.15), this norm-like function induces a metric d_Q on $\mathcal{G}$, defined by

$$d_Q(g_1, g_2) = \|g_1 - g_2\|_Q = Q(g_1 - g_2)^{1/2} \qquad (g_1, g_2 \in \mathcal{G}).$$

Theorem 1.20 *As $n \to \infty$, the random function ψ_n converges in probability in the metric d_Q to the limiting shape given by*

$$\Omega(u) = \begin{cases} \frac{2}{\pi}\left(u \sin^{-1}\left(\frac{u}{\sqrt{2}}\right) + \sqrt{2-u^2}\right) & \text{if } |u| \le \sqrt{2}, \\ |u| & \text{if } |u| > \sqrt{2}. \end{cases} \tag{1.47}$$

That is, for all $\epsilon > 0$ we have

$$\mathbb{P}\left(\|\psi_n - \Omega\|_Q > \epsilon\right) \xrightarrow[n\to\infty]{} 0.$$

Proof For each partition $\lambda \in \mathcal{P}(n)$, denote by g_λ the element of $\mathcal{G}$ corresponding to the function $\phi_\lambda \in \mathcal{F}$, and denote $h_\lambda(u) = g_\lambda(u) - |u|$. Denote by $\mathcal{M}_n$ the set of partitions $\lambda \in \mathcal{P}(n)$ for which $\|g_\lambda - \Omega\|_Q > \epsilon$. By (1.38), for each $\lambda \in \mathcal{M}_n$ we have

$$J(h_\lambda) \ge J(h_0) + Q(h_\lambda - h_0) = -1 + Q(h_\lambda - h_0) > -1 + \epsilon^2.$$

By Theorem 1.14 and (1.31), this implies that

$$\mathbb{P}(\lambda^{(n)} = \lambda) \le \exp\left(-\epsilon^2 n + O(\sqrt{n}\log n)\right),$$

where the O-term is uniform over all partitions of n.

Note that we also know that $|\mathcal{M}_n| \le |\mathcal{P}(n)| \le e^{C\sqrt{n}}$, by (1.15). So altogether we get that

$$\mathbb{P}(\lambda^{(n)} \in \mathcal{M}_n) = \sum_{\lambda\in\mathcal{M}_n} \mathbb{P}(\lambda^{(n)} = \lambda) \le C\exp(-\epsilon^2 n + C\sqrt{n} + O(\log n\sqrt{n})),$$

which indeed converges to 0 as $n \to \infty$. □

One problem with Theorem 1.20 is that it proves the convergence to the limit shape in an exotic topology whose connection to more standard notions of convergence in function spaces is unclear. In addition to being an aesthetic flaw, this makes it difficult to apply the theorem to our needs. However, with a bit more work this flaw can be corrected, and in fact we can get convergence in the more familiar uniform norm (a.k.a. supremum norm or L^∞ norm), defined by

$$\|f\|_\infty = \sup_{u\in\mathbb{R}} |f(u)|.$$

The necessary tool is the following lemma.[11]

Lemma 1.21 *Let $A, L > 0$, and let $\mathrm{Lip}(L, A)$ denote the space of all functions $f\colon \mathbb{R} \to \mathbb{R}$ which are Lipschitz with constant L and are supported on the interval $[-A, A]$. For any $f \in \mathrm{Lip}(L, A)$ we have*

$$\|f\|_\infty \le C\, Q(f)^{1/4} \tag{1.48}$$

where $C > 0$ is some constant that depends on A and L.

Proof We prove this using standard facts from the **fractional calculus**. For completeness, we include a self-contained argument, but for more background and motivation see the box on the next page.

First, to see why the fractional calculus may prove useful in this context, note that, by (1.35) and the Plancherel theorem from Fourier analysis, the quantity $Q(f)^{1/2}$ can be thought of, somewhat speculatively, as being proportional to the L^2-norm of a function $g\colon \mathbb{R} \to \mathbb{C}$ whose Fourier transform satisfies $|\hat{g}(s)| = |s|^{1/2}\big|\hat{f}(s)\big|$. Indeed, such a function exists and is exactly the **half-derivative**, or derivative of order $1/2$, a well-understood object from the fractional calculus. We can construct it explicitly as

$$g(x) = \frac{1}{2\sqrt{\pi}} \int_0^\infty \frac{f(x) - f(x-t)}{t^{3/2}}\, dt = \frac{1}{2\sqrt{\pi}} \int_{-\infty}^x \frac{f(x) - f(y)}{(x-y)^{3/2}}\, dy. \tag{1.49}$$

It is easy to see using the Lipschitz property and the fact that f is supported on $[-A, A]$ that the integral converges absolutely for any x, and furthermore that we can represent $g(x)$ more explicitly as

$$g(x) = \begin{cases} 0 & \text{if } x \le -A, \\ \frac{f(x)}{\sqrt{\pi(x+A)}} + \frac{1}{2\sqrt{\pi}} \int_{-A}^x \frac{f(x)-f(y)}{(x-y)^{3/2}}\, dy & \text{if } -A < x < A, \\ -\frac{1}{2\sqrt{\pi}} \int_{-A}^A \frac{f(y)}{(x-y)^{3/2}}\, dy & \text{if } x \ge A. \end{cases}$$

Fractional calculus

The **fractional calculus** studies the various senses in which differentiation and integration operations can be applied to an arbitrary fractional (real or complex) order α. For example, a **half-derivative** operator $D_{1/2}$ should satisfy the property that $D_{1/2}D_{1/2}f = f'$ in the appropriate function space. In contrast to the usual calculus operations of differentiation and integration applied an integer number of times, there is no unique way to define such an operation, and the meaning that it can be given, as well as the sense in which the fractional integral operator is inverse to the fractional derivative operator of the same order, will depend on the particular space of functions and type of fractional differential and integral operators used. One popular approach starts with the **Riemann–Liouville integral**, which is the fractional integral operator of order α taking a function f to

$$(I_\alpha f)(x) = \frac{1}{\Gamma(\alpha)} \int_0^x f(t)(x-t)^{-1+\alpha}\, dt,$$

where $\Gamma(\cdot)$ denotes the Euler gamma function. This operator is defined and well-behaved if $\mathrm{Re}(\alpha) > 0$. It can be shown that under certain conditions, the inverse operation is given by the **Marchaud fractional derivative operator** D_α, defined by

$$(D_\alpha g)(x) = \frac{\alpha}{\Gamma(1-\alpha)} \int_0^\infty \frac{f(x) - f(x-t)}{t^{1+\alpha}}\, dt.$$

For more details, refer to [89].

From this representation it follows easily that g is bounded and for large positive x satisfies a bound of the form $|g(x)| \le Cx^{-3/2}$ where C is some positive constant. In particular, g is in $L^1(\mathbb{R})$ (and also in $L^2(\mathbb{R})$) and therefore has a Fourier transform, which can be computed as follows:

$$\begin{aligned}
\hat{g}(s) &= \int_{-\infty}^\infty g(x)e^{-isx}\, dx = \frac{1}{2\sqrt{\pi}} \int_0^\infty \frac{1}{t^{3/2}} \int_{-\infty}^\infty (f(x) - f(x-t))e^{-isx}\, dx\, dt \\
&= \frac{1}{2\sqrt{\pi}} \int_0^\infty \frac{1}{t^{3/2}} \left(\hat{f}(s) - \hat{f}(s)e^{-ist}\right) dt \\
&= \frac{1}{2\sqrt{\pi}} |s|^{1/2} \hat{f}(s) \int_0^\infty \frac{1 - e^{-i\,\mathrm{sgn}(s) u}}{u^{3/2}}\, du = |s|^{1/2} e^{-\pi i\,\mathrm{sgn}(s)/4} \hat{f}(s),
\end{aligned}$$

where we use the integral evaluation

$$\int_0^\infty \frac{1 - e^{iu}}{u^{3/2}}\, du = 2e^{-\pi i/4}\sqrt{\pi} \tag{1.50}$$

(Exercise 1.19). This was exactly the property we wanted, since it implies that $\|g\|_2^2 = \|\hat{g}\|_2^2/2\pi = (2/\pi)Q(f)$.

Next, to compare $\|f\|_\infty$ with $Q(f) = (\pi/2)\|g\|_2^2$, we need to know how to recover f from g. This is done by performing a fractional *integration* of order 1/2, defined by

$$h(x) = \frac{1}{\sqrt{\pi}} \int_{-\infty}^{x} \frac{g(t)}{\sqrt{x-t}}\, dt = \begin{cases} 0 & \text{if } x < -A, \\ \frac{1}{\sqrt{\pi}} \int_{-A}^{x} \frac{g(t)}{\sqrt{x-t}}\, dt & \text{if } x \geq -A. \end{cases} \tag{1.51}$$

The properties of g ensure that the integral defining $h(x)$ converges absolutely for all x. The Fourier transform of h is

$$\begin{aligned} \hat{h}(s) &= \int_{-A}^{\infty} h(x) e^{-isx}\, dx = \frac{1}{\sqrt{\pi}} \int_{-A}^{\infty} e^{-isx} \int_{-A}^{x} \frac{g(t)}{\sqrt{x-t}}\, dt\, dx \\ &= \frac{1}{\sqrt{\pi}} \int_{-A}^{\infty} g(t) \int_{t}^{\infty} \frac{e^{-isx}}{\sqrt{x-t}}\, dx\, dt = \frac{1}{\sqrt{\pi}} \int_{-A}^{\infty} g(t) e^{-ist}\, dt \int_{0}^{\infty} \frac{e^{-isu}}{\sqrt{u}}\, du \\ &= \frac{1}{\sqrt{\pi}} |s|^{-1/2} \hat{g}(s) \int_{0}^{\infty} \frac{e^{-i\,\mathrm{sgn}(s)x}}{\sqrt{x}}\, dx = |s|^{-1/2} e^{\pi i\,\mathrm{sgn}(s)/4} \hat{g}(s) = \hat{f}(s), \end{aligned}$$

for all $s \neq 0$, where we use another integral evaluation (actually equivalent to (1.50) – see Exercise 1.19)

$$\int_{0}^{\infty} \frac{e^{iu}}{u^{1/2}}\, du = e^{\pi i/4} \sqrt{\pi}. \tag{1.52}$$

This result ought to imply that $h = f$ almost everywhere, that is, that the half-integral operation (1.51) is indeed inverse to the half-differentiation in (1.49). Note, however, that this is merely a formal computation, since a priori we do not know that $h \in L^1(\mathbb{R})$ and therefore the integral defining its Fourier transform may not be defined, and the change in the order of integration we performed is also unjustified. Nonetheless, the conclusion that $h = f$ a.e. is correct and is not hard to justify (Exercise 1.20).

Equipped with these latest observations, we can now bound the L^1-norm of f in terms of $||g||_2 = \sqrt{2/\pi}Q(f)^{1/2}$, as follows:

$$\begin{aligned}||f||_1 &= \int_{-A}^{A} |f(x)|\,dx \le \frac{1}{\sqrt{\pi}} \int_{-A}^{A}\int_{-A}^{x} \frac{|g(t)|}{\sqrt{x-t}}\,dt\,dx \\ &= \frac{1}{\sqrt{\pi}} \int_{-A}^{A} |g(t)| \left(\int_{t}^{A} \frac{1}{\sqrt{x-t}}\,dx \right) dt = \frac{1}{\sqrt{\pi}} \int_{-A}^{A} |g(t)| 2\sqrt{A-t}\,dt \\ &\le \frac{2\sqrt{2A}}{\sqrt{\pi}} \int_{-A}^{A} |g(t)|\,dt \le \frac{4A}{\sqrt{\pi}} \left(\int_{-A}^{A} g(t)^2\,dt \right)^{1/2} \\ &\le \frac{4A}{\sqrt{\pi}} ||g||_2 = \frac{4\sqrt{2}}{\pi} A Q(f)^{1/2}. \end{aligned} \tag{1.53}$$

Finally, we convert this bound to a bound on the uniform norm of f. If $x_0 \in [-A, A]$ is such that $|f(x_0)| = ||f||_\infty$, then by the Lipschitz property we get that

$$\begin{aligned}||f||_1 &= \int_{-A}^{A} |f(x)|\,dx = \int_{-\infty}^{\infty} |f(x)|\,dx \\ &\ge \int_{-\infty}^{\infty} \max\left(0, |f(x_0)| - L|x - x_0|\right) dx = \frac{|f(x_0)|^2}{L} = \frac{||f||_\infty^2}{L}. \end{aligned} \tag{1.54}$$

Combining (1.53) and (1.54) we get the claim of the lemma. □

As a consequence, we can finally prove the celebrated 1977 theorem of Vershik, Kerov, Logan, and Shepp.

Theorem 1.22 (Limit shape theorem for Plancherel-random partitions) *As $n \to \infty$, the random function ψ_n converges in probability in the norm $||\cdot||_\infty$ to the limiting shape Ω defined in* (1.47). *That is, for all $\epsilon > 0$ we have*

$$\mathbb{P}\left(\sup_{u \in \mathbb{R}} |\psi_n(u) - \Omega(u)| > \epsilon \right) \xrightarrow[n\to\infty]{} 0.$$

Proof By Lemma 1.5, combined with the Robinson–Schensted algorithm, we see that, with probability that converges to 1 as $n \to \infty$, the partition $\lambda^{(n)}$ has a first row of length $< 3\sqrt{n}$, and by the symmetry between rows and columns this bound also applies with high probability to the length of its first column. These bounds imply that the corresponding function ψ_n is supported on the compact interval $[-3\sqrt{2}, 3\sqrt{2}]$. Therefore the convergence in the metric d_Q given in Theorem 1.20 immediately translates, via Lemma 1.21, to a convergence in the uniform norm. □

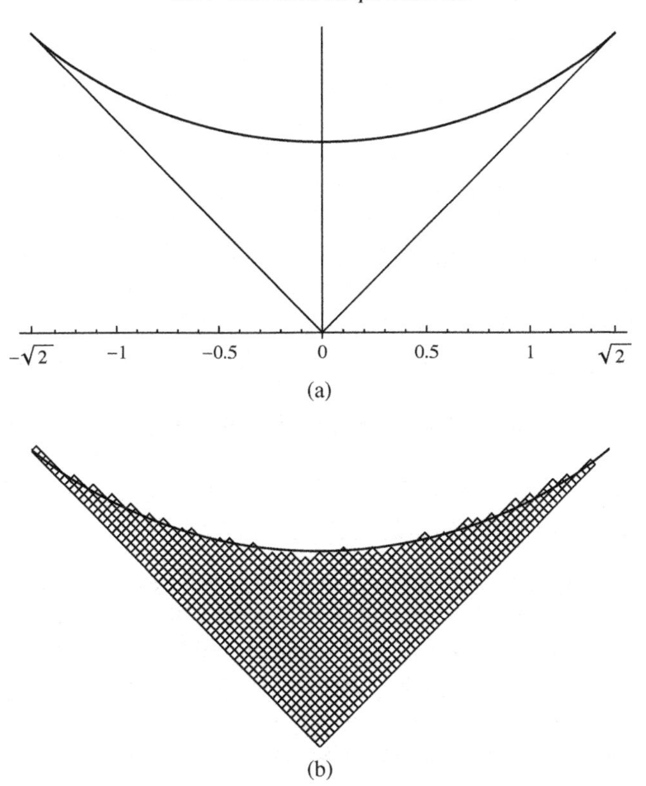

Figure 1.13 (a) The Logan–Shepp–Vershik–Kerov limit shape Ω. (b) The limit shape superposed for comparison (after correct scaling) on a simulated Plancherel-random Young diagram of order $n = 1000$.

Theorem 1.22 is a landmark achievement in the development of the mathematics of longest increasing subsequences; the result, and the ideas used in its proof, have had a considerable impact on research in the field beyond their immediate applicability to the Ulam–Hammersley problem. Take a moment to appreciate this beautiful result with a look at Fig. 1.13.

As a first application of Theorem 1.22, we can prove a first asymptotically sharp bound on ℓ_n, the average maximal length of a longest increasing subsequence of a random permutation of order n.

Theorem 1.23 *If* $\Lambda = \lim_{n\to\infty} \ell_n/\sqrt{n}$ *is the constant whose existence is proved in Theorem 1.6, then we have*

$$\Lambda \geq 2.$$

Proof As we already noted, the Robinson–Schensted algorithm implies that the random variable $L(\sigma_n)$, where σ_n is a uniformly random element of S_n, is equal in distribution to the length $\lambda_1^{(n)}$ of the first row of a Plancherel-random partition $\lambda^{(n)}$ of order n. But by Theorem 1.22, with asymptotically high probability $\lambda^{(n)}$ must have a first row of length at least $(2 - o(1))\sqrt{n}$: for, if this were not the case, then for infinitely many values of n the corresponding function ψ_n encoding its diagram in the Russian coordinate system would coincide with the absolute value function $|u|$ on some interval $[-\sqrt{2}, -\sqrt{2} + \epsilon]$, for some fixed $\epsilon > 0$ that does not depend on n. Clearly this would prevent ψ_n from converging uniformly to the limit shape Ω, in contradiction to Theorem 1.22. □

It is instructive to try to similarly prove the matching *upper* bound $\Lambda \leq 2$ using the limit shape theorem. Such an attempt will not succeed. Indeed, while the assumption that $\Lambda > 2$ would lead to the conclusion that ψ_n differs from the absolute value function (and hence from the limit shape $\Omega(u)$) on some interval $\left[-\sqrt{2} - \epsilon, -\sqrt{2}\right]$, this does *not* contradict the limit shape theorem, nor does it even imply that $J(\psi_n - |u|) > J(h_0) + \delta$ for some fixed $\delta > 0$ (which was the more fundamental fact that we used to prove the limit shape theorem). Thus, proving the upper bound $\Lambda \leq 2$ in order to conclude our proof of Theorem 1.1 requires a new idea. However, this part turns out to be quite easy and will be done in Section 1.19 after a brief digression into representation theory.

1.18 Irreducible representations of S_n with maximal dimension

The limit shape theorem for Plancherel measure is a powerful and elegant result in combinatorial probability. In the next section we shall use it to prove Theorem 1.1, which will finish the first main part of our analysis of the maximal length of increasing subsequences of random permutations. In this section, we digress briefly from this original motivation to note that Theorem 1.22, or more precisely the analysis leading to it, actually provides the answer to another question in pure mathematics concerning the

Representation theory and the representations of S_n

Representation theory aims to understand algebraic structures, for example (in the simplest case) finite groups, by studying their **representations**. The representations of a finite group G are homomorphisms $\varphi: G \to \mathrm{GL}(n, \mathbb{C})$, where $\mathrm{GL}(n, \mathbb{C})$ is the group of invertible matrices of order n over $\mathbb{C}$. The number n is called the **dimension** of the representation. A given finite group will have infinitely many representations, but only a finite number (in fact, equal to the number of conjugacy classes of G) of **irreducible representations**, which are special representations that cannot be decomposed (in some natural sense) into a sum of smaller representations. Each representation can itself be decomposed in a unique way into a sum of irreducible representations. The irreducible representations thus play a role in representation theory somewhat akin to that of the prime numbers in multiplicative arithmetic.

The representations of complicated groups can be complicated to understand. It is therefore no surprise that the representation theory of the symmetric group S_n, developed in the works of Frobenius, Young, Weyl, von Neumann, and others, is rich and nontrivial. A few fundamental facts, however, are simple to state: it is known that the irreducible representations of S_n are in canonical bijection with partitions of n, and that the dimension of the irreducible representation corresponding to a given partition $\lambda \in \mathcal{P}(n)$ is exactly d_λ, the number of standard Young tableaux of shape λ. In fact, this was the original motivation for studying Young tableaux, and many combinatorial facts about Young tableaux find an expression in terms of representations of S_n. For more details, see [24], [112].

representations of the symmetric group. Specifically, the question is to identify (approximately, or precisely if possible) the irreducible representation of the symmetric group of order n having the largest dimension. For readers unfamiliar with representation theory, see the box for a summary of the relevant background facts.

The question of finding the irreducible representation of maximal dimension of a given group is natural and simple to state,[12] but (it turns out) can be difficult to answer. In the case of the symmetric group S_n, since the dimension of the irreducible representation corresponding to a Young diagram $\lambda \in \mathcal{P}(n)$ is equal to d_λ, the number of standard Young tableaux of shape λ, the question is equivalent to finding the Young diagram $\lambda \in \mathcal{P}(n)$ for which d_λ is maximized. This is exactly where Theorem 1.14 comes in: what we interpreted probabilistically as a large deviation principle for the behavior of Plancherel measure can be thought of alternatively as a first-order asymptotic expansion for the mapping $\lambda \mapsto d_\lambda$. Using our identification of the minimizer for the hook functional I_{hook}, it is an easy matter to deduce that the asymptotic shape of the Young diagram corresponding to the irreducible representation of maximal dimension is the same limiting shape as the one we found for Plancherel-random partitions. The precise result is as follows.

Theorem 1.24 *For each $n \geq 1$, let $\mu^{(n)} \in \mathcal{P}(n)$ be the Young diagram corresponding to the maximal dimension irreducible representation of S_n (if there is more than one such, we take $\mu^{(n)}$ to denote an arbitrary choice of one among them). Let $\phi_n(x)$ be the function in the function space $\mathcal{F}$ corresponding to $\mu^{(n)}$. Let ψ_n be the element of the function space $\mathcal{G}$ corresponding to ϕ_n in the rotated coordinate system. Then we have*

$$\max_{u\in\mathbb{R}} |\psi_n(u) - \Omega(u)| \xrightarrow[n\to\infty]{} 0.$$

Note that the theorem gives only an approximate answer to the question that is valid in the limit of large n. More detailed results have been derived on the irreducible representation of maximal dimension and on its dimension, and on the related question of better understanding the asymptotics of the typical dimension of a Plancherel-random Young diagram (see [23], [140], [143]), but these results are still of an asymptotic nature. Some interesting open problems in this context would be to say anything precise (that is, nonasymptotic) about the maximal dimension shape for finite values of n (e.g., in the form of some kind of explicit description or even just an efficient algorithm for computing this shape); to prove that the maximal dimension shape is unique; and to improve the asymptotic bounds in the papers cited earlier.

1.19 The Plancherel growth process

We now want to prove that $\Lambda \leq 2$ and thereby finish the proof of Theorem 1.1. This will involve introducing a new and interesting way to think about the family of Plancherel measures. First, observe that it will be enough to prove that

$$\ell_n - \ell_{n-1} \leq \frac{1}{\sqrt{n}}, \qquad (n \geq 1), \tag{1.55}$$

which by induction will imply that the bound

$$\ell_n \le 2\sqrt{n}$$

holds for all $n \ge 1$. The left-hand side of (1.55) is the difference of the expectations of $\lambda_1^{(n)}$ and $\lambda_1^{(n-1)}$. Note that until now we considered each of the random partitions $\lambda^{(n)}$ as existing in its own probability space, but an elegant new idea that enters here, and has clear relevance to the discussion, is to consider the sequence of Plancherel-random partitions as a random *process* $(\lambda^{(n)})_{n=1}^{\infty}$. That is, if we had a natural coupling of all the $\lambda^{(n)}$ for all values of n simultaneously on a single probability space, we would have a potential way of approaching inequalities such as (1.55). Fortunately, there is a natural such coupling that emerges directly from the combinatorial interpretation of the $\lambda^{(n)}$ in terms of the Robinson–Schensted correspondence. Recall that we originally obtained Plancherel measure, the distribution of $\lambda^{(n)}$, as the distribution of the Young diagram that is output (along with two standard Young tableaux, which we discarded, whose shape is the Young diagram) by the Robinson–Schensted algorithm applied to a uniformly random permutation σ_n of order n. But it is well known that there is a simple way to couple all the uniform measures on the symmetric groups $(S_n)_{n=1}^{\infty}$. Perhaps the most convenient way to describe it is by taking a sequence $X_1, X_2, X_3, \ldots$ of independent and identically distributed random variables with the uniform distribution $U[0, 1]$, and defining σ_n for each n to be the order structure of the finite sequence $X_1, \ldots, X_n$. (See the related discussion on p. 11.) Alternatively, one can give an algorithm that, given the random permutation σ_n in S_n, modifies it (adding some new randomness) to produce the random permutation σ_{n+1} that is distributed uniformly in S_{n+1}. The algorithm chooses a random integer k uniformly from $\{1, \ldots, n+1\}$, then returns the new permutation

$$\sigma_{n+1}(j) = \begin{cases} \sigma_n(j) & \text{if } j \le n \text{ and } \sigma_n(j) < k, \\ \sigma_n(j) + 1 & \text{if } j \le n \text{ and } \sigma_n(j) \ge k, \\ k & \text{if } j = n+1. \end{cases}$$

It is a simple matter to check that this produces a uniformly random element of S_{n+1}, and that this construction is equivalent to the previous one, in the sense that the joint distribution of the sequence $(\sigma_n)_{n=1}^{\infty}$ generated using this construction is the same as for the first construction.

Having defined a process version $(\sigma_n)_{n=1}^\infty$ of the sequence of uniformly random permutations, by applying the Robinson–Schensted algorithm to each permutation in the sequence (and discarding the Young tableaux, keeping just their shape) we now have a process of random partitions $(\lambda^{(n)})_{n=1}^\infty$, where for each n, $\lambda^{(n)}$ is distributed according to Plancherel measure of order n. This process is called the **Plancherel growth process.**[13]

To understand this process, note that by the construction, the Robinson–Schensted shape $\lambda^{(n)}$ of the permutation σ_n is obtained from $\lambda^{(n-1)}$ by attaching an additional cell in one of the positions where a cell may be added to the Young diagram $\lambda^{(n-1)}$ to result in a new Young diagram (such positions may be called **external corners**, in analogy with the internal corners discussed in Section 1.9 in connection with the hook-length formula). In the notation of Section 1.9, we have that $\lambda^{(n-1)} \nearrow \lambda^{(n)}$. If we define the **Young graph** (also called the **Young lattice**) to be the directed graph whose vertex set is the set $\cup_{n=1}^\infty \mathcal{P}(n)$ of all nonempty integer partitions, and whose edges are the ordered pairs (μ,λ) with $\mu \nearrow \lambda$, then the Plancherel growth process defines a random infinite path on this graph, starting from the trivial partition $\lambda^{(1)} = (1) \in \mathcal{P}(1)$. The first few levels of the Young graph are shown in Fig. 1.14.

As it turns out, the Plancherel growth process has a simple description as a **Markov chain**, or in other words a kind of weighted random walk on the Young graph. This is explored in Exercise 1.22. For our purposes, we need only the following simple lemma on conditional probabilities.

Lemma 1.25 *For any $\mu \in \mathcal{P}(n-1)$ and $\lambda \in \mathcal{P}(n)$ satisfying $\mu \nearrow \lambda$, we have*

$$\mathbb{P}(\lambda^{(n)} = \lambda \,|\, \lambda^{(n-1)} = \mu) = \frac{d_\lambda}{n\, d_\mu}. \tag{1.56}$$

Proof By definition, we have

$$\mathbb{P}(\lambda^{(n)} = \lambda \,|\, \lambda^{(n-1)} = \mu) = \frac{\mathbb{P}(\{\lambda^{(n-1)} = \mu\} \cap \{\lambda^{(n)} = \lambda\})}{\mathbb{P}(\lambda^{(n-1)} = \mu)}$$
$$= \frac{\mathbb{P}(\{\lambda^{(n-1)} = \mu\} \cap \{\lambda^{(n)} = \lambda\})}{d_\mu^2/(n-1)!}.$$

In this expression, the numerator is given by $1/n!$ times the number of permutations $\sigma \in S_n$ such that after applying the first $n-1$ insertion steps in the Robinson–Schensted algorithm to the sequence $(\sigma(1), \ldots, \sigma(n-1))$

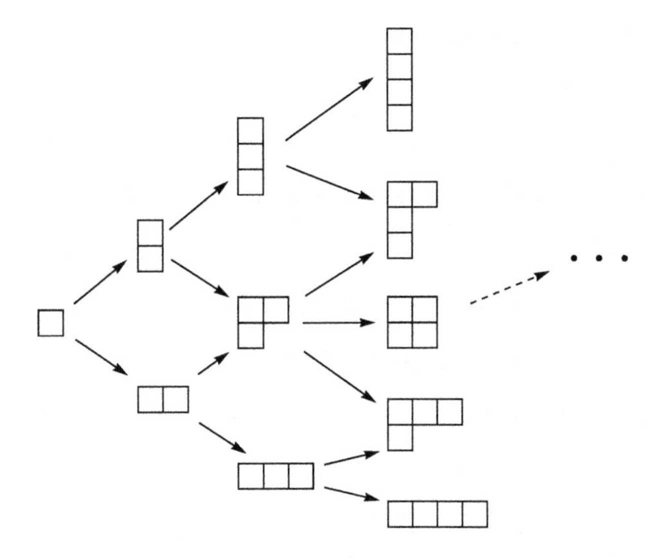

Figure 1.14 The Young graph.

the resulting pair of Young tableaux has shape μ, and after in addition performing the last insertion step where the value $\sigma(n)$ is inserted, the resulting shape is λ. Note that for such a permutation, if we denote the output of the Robinson–Schensted algorithm by (λ, P, Q), where P is the insertion tableau and Q is the recording tableau, then Q has its maximal entry exactly in the corner cell $\mathbf{c}$ of λ such that $\{\mathbf{c}\} = \lambda \setminus \mu$, since before inserting the last value $\sigma(n)$ the shape was μ. The number of standard Young tableaux of shape λ whose maximal entry is in position $\mathbf{c}$ is trivially equal to d_μ. Therefore, the number of permutations we are trying to count is exactly the product $d_\mu d_\lambda$, and we get that

$$\mathbb{P}(\lambda^{(n)} = \lambda \mid \lambda^{(n-1)} = \mu) = \frac{d_\mu d_\lambda / n!}{d_\mu^2/(n-1)!} = \frac{d_\lambda}{n\, d_\mu},$$

as claimed. □

Proof of (1.55) We identify the left-hand side of (1.55) with the expected value $\mathbb{E}(\lambda_1^{(n)} - \lambda_1^{(n-1)})$, where $\lambda^{(n-1)}$ and $\lambda^{(n)}$ are now coupled according to the Plancherel growth process. In this realization, the random variable

$\lambda_1^{(n)} - \lambda_1^{(n-1)}$ is the indicator function of the event

$$E_n = \left\{\lambda^{(n)} = \text{grow}_1\left(\lambda^{(n-1)}\right)\right\},$$

where for a diagram $\mu \in \mathcal{P}(n-1)$ we denote by $\text{grow}_1(\mu)$ the diagram obtained from μ by adding a square at the end of the first row of μ. To bound the probability of this event, we use Lemma 1.25 to write

$$\begin{aligned}\mathbb{P}(E_n) &= \sum_{\mu \vdash n-1} \frac{d_\mu^2}{(n-1)!} \mathbb{P}(E_n \mid \lambda^{(n-1)} = \mu) \\ &= \sum_{\mu \vdash n-1} \frac{d_\mu^2}{(n-1)!} \mathbb{P}(\lambda^{(n)} = \text{grow}_1(\mu) \mid \lambda^{(n-1)} = \mu) \\ &= \sum_{\mu \vdash n-1} \frac{d_\mu^2}{(n-1)!} \cdot \frac{d_{\text{grow}_1(\mu)}}{n\, d_\mu}.\end{aligned}$$

This is an average of the quantity $d_{\text{grow}_1(\mu)}/nd_\mu$ with respect to Plancherel measure, so we can apply the Cauchy–Schwarz inequality to this average, and deduce that

$$\mathbb{P}(E_n) \le \left[\sum_{\mu \vdash n-1} \frac{d_\mu^2}{(n-1)!} \left(\frac{d_{\text{grow}_1(\mu)}}{n\, d_\mu}\right)^2\right]^{1/2} = \frac{1}{\sqrt{n}} \left(\sum_{\mu \vdash n-1} \frac{d_{\text{grow}_1(\mu)}^2}{n!}\right)^{1/2}.$$

In this last expression, the sum inside the square root represents the probability that $\lambda^{(n)}$ is of the form $\text{grow}_1(\mu)$ for *some* $\mu \vdash n-1$, and hence is at most 1. So we have shown that $\mathbb{P}(E_n) = \mathbb{E}(\lambda_1^{(n)} - \lambda_1^{(n-1)}) \le 1/\sqrt{n}$, as claimed. □

1.20 Final version of the limit shape theorem

We have achieved our goal of proving Theorem 1.1, the main result of this chapter. We conclude the chapter by reformulating one of our other main results, the limit shape theorem for Plancherel-random Young diagrams (Theorem 1.22), in a way that incorporates the observations of the last section. This will result in a stronger version of the theorem. The new form is also more suitable for comparison with an important analogous result we will prove in Chapter 4.

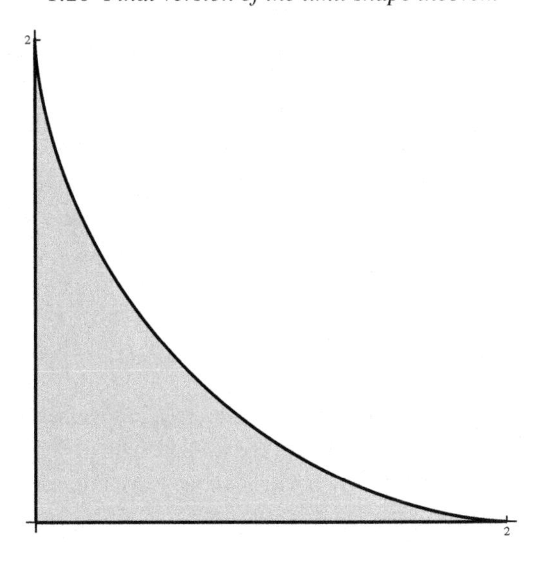

Figure 1.15 The limit shape $\Delta_{\text{Plancherel}}$ of Plancherel-random Young diagrams, in x–y coordinates.

Let $\omega = (\omega_x, \omega_y) : [-\pi/2, \pi/2] \to \mathbb{R}^2$ be the planar curve defined by

$$\omega_x(\theta) = \left(\frac{2\theta}{\pi} + 1\right)\sin\theta + \frac{2}{\pi}\cos\theta,$$
$$\omega_y(\theta) = \left(\frac{2\theta}{\pi} - 1\right)\sin\theta + \frac{2}{\pi}\cos\theta.$$

It is straightforward to check that this is a parametric form of the Logan–Shepp–Vershik–Kerov limit shape Ω from (1.47), translated via (1.23) to the original x–y coordinate system whose axes are aligned with the Young diagram axes.

Next, given a Young diagram $\lambda = (\lambda_1, \ldots, \lambda_k)$, associate with λ the set $\text{set}_\lambda \subset \mathbb{R}^2$ defined by

$$\text{set}_\lambda = \bigcup_{\substack{1\le i\le k \\ 1\le j\le \lambda_i}} \big([i-1, i] \times [j-1, j]\big). \tag{1.57}$$

We will refer to set_λ as the **planar set** of λ.

Theorem 1.26 (Limit shape theorem for Plancherel-random partitions; strong version) *Define a set* $\Delta_{\text{Plancherel}} \subset \mathbb{R}^2$ *by*

$$\Delta_{\text{Plancherel}} = \Big\{ t \cdot \omega(\theta) \ : \ t \in [0,1],\ \theta \in [-\pi/2, \pi/2] \Big\}.$$

For each $n \geq 1$ *let* $\lambda^{(n)}$ *denote a random Young diagram chosen according to Plancherel measure of order* n*. Then the planar set* $\text{set}_{\lambda^{(n)}}$ *converges in probability to* $\Delta_{\text{Plancherel}}$ *as* $n \to \infty$*, in the following precise sense: for any* $0 < \epsilon < 1$*, we have that*

$$\mathbb{P}\left((1-\epsilon)\Delta_{\text{Plancherel}} \subset \frac{1}{\sqrt{n}} \text{set}_{\lambda^{(n)}} \subset (1+\epsilon)\Delta_{\text{Plancherel}} \right) \to 1 \text{ as } n \to \infty.$$

The connection of Theorem 1.26 to Theorem 1.22 together with the new information about the convergence of the scaled length $n^{-1/2}\lambda_1^{(n)}$ of the first row of $\lambda^{(n)}$ to 2 should be fairly obvious, so we leave it to readers to work out the (easy) details of the proof. The set $\Delta_{\text{Plancherel}}$ is shown in Fig. 1.15.

Exercises

Note on exercise difficulty levels. The unit of difficulty for exercises is the coffee cup ("☕"), with the difficulty scale ranging from ☕ (easy) to ☕☕☕☕☕ (research problem).

1.1 (☕☕) If $\Lambda = \lim_{n\to\infty} \ell_n/\sqrt{n}$ as in Theorem 1.6, show that the bounds $1 \leq \Lambda \leq e$ that follow from Lemmas 1.3 and 1.4 can be improved using elementary arguments, as follows:

   (a) Given a Poisson point process of unit intensity in $[0,\infty)\times[0,\infty)$, construct an increasing subset $(X_1,Y_1),(X_2,Y_2),(X_3,Y_3),\ldots$ of points by letting (X_1,Y_1) be the Poisson point that minimizes the coordinate sum $x+y$, and then by inductively letting (X_k,Y_k) be the Poisson point in $(X_{k-1},\infty)\times(Y_{k-1},\infty)$ that minimizes the coordinate sum $x+y$. Analyze the asymptotic behavior of this sequence, and deduce that $\Lambda \geq (8/\pi)^{1/2} \approx 1.595$.

   (b) In the proof of Lemma 1.4 observe that if $L(\sigma_n) \geq t$ then $X_{n,k} \geq \binom{t}{k}$, so the bound in (1.4) can be improved. Take $k \approx \alpha\sqrt{n}$ and $t \approx \beta\sqrt{n}$ and optimize over $\alpha < \beta$ to show that $\Lambda \leq 2.49$.

1.2 (☕) Verify that the random variables $-Y_{m,n}$ defined in (1.5) satisfy the conditions of Theorem A.3. (**Hint:** Use Lemma 1.4.)

1.3 (Steele [129]) The goal of this exercise is to prove the following explicit variance bound for $L(\sigma_n)$.

Theorem 1.27 *For some constant $C > 0$ and all $n \geq 1$, we have*

$$\mathrm{Var}(L(\sigma_n)) \leq C\sqrt{n}. \tag{1.58}$$

We make use of the following version of a general probabilistic concentration inequality due to Efron–Stein [36] and Steele [128].

Theorem 1.28 (Efron–Stein–Steele inequality) *Let $X_1, \ldots, X_n$ be independent random variables. Let $g : \mathbb{R}^n \to \mathbb{R}$ and $g_j : \mathbb{R}^{n-1} \to \mathbb{R}$, $(j = 1, \ldots, n)$, be measurable functions. Denote $Z = g(X_1, \ldots, X_n)$, and for any $1 \leq j \leq n$ denote $Z_j = g(X_1, \ldots, \hat{X}_j, \ldots, X_n)$, where $\hat{X}_j$ means that X_j is omitted from the list. Then we have*

$$\mathrm{Var}(Z) \leq \sum_{j=1}^{n} \mathbb{E}(Z - Z_j)^2. \tag{1.59}$$

See [20], Section 2, for the (easy) proof of Theorem 1.28.

(a) (☕☕) If $x_1, \ldots, x_n$ are distinct real numbers, denote by $L(x_1, \ldots, x_n)$ the maximal length of an increasing subsequence in the sequence $x_1, \ldots, x_n$. Take $X_1, \ldots, X_n$ to be i.i.d. random variables with the uniform distribution on $[0, 1]$. Let $Z = L(X_1, \ldots, X_n)$, and let $Z_j = L(X_1, \ldots, \hat{X}_j, \ldots, X_n)$ for $1 \leq j \leq n$. Note that, as discussed on p. 11, Z is equal in distribution to $L(\sigma_n)$.
Prove that the random variable $Z - Z_j$ takes on only the values 0 or 1, that is, it is an indicator random variable $Z - Z_j = \mathbf{1}_{E_j}$, and show that the event E_j is precisely the event that X_j participates in all maximal-length increasing subsequences in $X_1, \ldots, X_n$.

(b) (☕☕) Prove that $\sum_{j=1}^{n}(Z - Z_j)^2 = \sum_{j=1}^{n} \mathbf{1}_{E_j} \leq Z$.

(c) (☕) Apply Theorem 1.28, to conclude that $\mathrm{Var}(Z) \leq \mathbb{E}(Z)$.

(d) (☕) Finally, use (1.3) to deduce (1.58). (Note that C can be taken as any number greater than e if n is assumed large enough.)

1.4 (☕☕☕) Using Theorem 1.27, give an alternative proof of Theorem 1.6 that avoids the use of Kingman's subadditive ergodic theorem. Note that you will still need to use the geometric ideas discussed in Section 1.4 relating $L(\sigma_n)$ to the Poisson point process, and Fekete's subadditive lemma (Lemma A.1 in the Appendix).

1.5 (a) (☕☕☕) (Bollobás–Winkler [17]) Let $d \geq 2$. Given two points $\mathbf{x} = (x_1, \ldots, x_d)$, $\mathbf{x}' = (x'_1, \ldots, x'_d)$ in $\mathbb{R}^d$, denote $\mathbf{x} \preceq \mathbf{x}'$ if $x_j \leq x'_j$ for all j. A set of points in $\mathbb{R}^d$ is called a **chain** if any two of its elements are comparable in the partial order $\preceq$. For a set $A \subset \mathbb{R}^d$, denote by $L(A)$ the maximal size of a subset of A that is a chain.

Let $\mathbf{X}_1, \mathbf{X}_2, \ldots$ denote a sequence of independent and identically distributed random points chosen uniformly at random from the unit cube $[0,1]^d$, and for any n denote $A_n = \{\mathbf{X}_1, \ldots, \mathbf{X}_n\}$ (a random n-element subset of $[0,1]^d$). Prove the following generalization of Theorem 1.6: there exists a constant $0 < c_d < e$ such that we have the convergence in probability

$$n^{-1/d} L(A_n) \xrightarrow{P} c_d \text{ as } n \to \infty.$$

(b) (☕) Translate the above result into purely combinatorial language as a statement about random permutations. (For $d = 2$ it would be the claim from Theorem 1.6 that $L(\sigma_n)/\sqrt{n} \to \Lambda$ in probability as $n \to \infty$.)

(c) (☕☕☕☕☕) Find a formula (or even a good numerical estimate) for c_3.

1.6 (Lifschitz–Pittel [75]) Let X_n denote the total number of increasing subsequences (of any length) in the uniformly random permutation σ_n. For convenience, we include the empty subsequence of length 0, so that X_n can be written as $X_n = 1 + \sum_{k=1}^{n} X_{n,k}$ where the random variables $X_{n,k}$ are defined in the proof of Lemma 1.4 (p. 9).

(a) (☕☕☕) Prove the identities

$$\mathbb{E}X_n = \sum_{k=1}^{n} \frac{1}{k!}\binom{n}{k},$$

$$\mathbb{E}X_n^2 = \sum_{k,\ell \geq 0,\ k+\ell \leq n} \frac{4^\ell}{(k+\ell)!}\binom{n}{k+1}\binom{(k+1)/2+\ell-1}{\ell}.$$

(b) (☕☕☕) Use the above identities to prove the asymptotic estimates

$$\mathbb{E}X_n = (1+o(1))\frac{1}{2\sqrt{\pi e}\, n^{1/4}} \exp\left(2\sqrt{n}\right) \quad \text{as } n \to \infty,$$

$$\operatorname{Var} X_n = (1+o(1))\frac{\exp\left(2\sqrt{2+\sqrt{5}} \cdot \sqrt{n}\right)}{\sqrt{20\pi\left(2+\sqrt{5}\right)\exp\left(2+\sqrt{5}\right)}\, n^{1/4}} \quad \text{as } n \to \infty.$$

(c) (☕☕☕☕) Prove that there is a constant $\gamma > 0$ such that we have the convergence in probability

$$\frac{1}{\sqrt{n}} \log X_n \xrightarrow{P} \gamma \quad \text{as } n \to \infty.$$

(d) (☕☕☕☕☕) Find the value of γ.

1.7 (☕) Compute the Young diagram λ and Young tableaux (P, Q) obtained by applying the Robinson–Schensted algorithm to the permutation

$$\sigma = (13, 8, 3, 10, 2, 15, 7, 1, 5, 6, 11, 12, 14, 9, 4).$$

1.8 (☕) Apply the inverse Robinson–Schensted algorithm to the pair of Young tableaux

$P =$

1	4	6	10	11
2	5	7		
3	8	9		

$Q =$

1	3	5	8	9
2	4	10		
6	7	11		

to recover the associated permutation $\sigma \in S_{11}$.

1.9 (a) (☕☕☕) (Schensted [113]) Given a Young diagram λ, define an **increasing tableau of shape** λ to be a filling of the cells of λ with distinct numbers such that all rows and columns are arranged in increasing order. The insertion steps of the Robinson–Schensted algorithm, defined in Section 1.6, can be formalized as a mapping taking a number x and an increasing tableau P of shape λ whose entries are all distinct from x, and returning a new increasing tableau, denoted $R_x(P)$, obtained by the application to P of an insertion step with the number x as input. We call the operator R_x a **row-insertion operator**. By analogy, define now a **column-insertion operator** C_x that, when applied to the tableau P, will produce a new increasing tableau $C_x(P)$ by performing an insertion step in which the roles of rows and columns are reversed.

Prove that the operators R_x and C_y commute. That is, if x, y are distinct numbers and P is an increasing tableau with entries distinct from x and y, then we have that

$$C_y R_x P = R_x C_y P.$$

Hint: Divide into cases according to whether the maximum among x, y and the entries of P is x, y or one of the entries of P.

(b) (☕☕) Deduce the claim of Theorem 1.10(a) from part (a) above.

1.10 (☕☕☕) This exercise presents an outline of the steps needed to prove Theorem 1.10(b), based on the exposition of Knuth [71, Section 5.1.4]. Start with a definition: a **planar permutation** of order n is a two-line array of the form $\sigma = \begin{pmatrix} q_1 & q_2 & \cdots & q_n \\ p_1 & p_2 & \cdots & p_n \end{pmatrix}$, where $q_1 < \ldots < q_n$ and $p_1, \ldots, p_n$. We say that (q_i, p_i) is the ith element of σ. The inverse of σ is the planar permutation σ^{-1} obtained by switching the top and bottom rows of the array and sorting the columns to bring the top row to increasing order. Note that an ordinary permutation would correspond to the usual two-line notation for permutations, and in this case σ^{-1} corresponds to the usual notion of an inverse permutation. It is helpful to visualize σ simply as a set of points in the plane with distinct x- and y-coordinates, which emphasizes the symmetry between the q's and the p's. With such an interpretation, the operation of taking the inverse corresponds to reflection of the set of points along the diagonal $y = x$.

(a) Convince yourself that in this setting the Robinson–Schensted algorithm generalizes to a mapping that takes a planar permutation $\sigma = \begin{pmatrix} q_1 & \cdots & q_n \\ p_1 & \cdots & p_n \end{pmatrix}$ of order n and returns a triple (λ, P, Q) where $\lambda \in \mathcal{P}(n)$ and P, Q are increasing tableaux of shape λ (see Exercise 1.9 above for the definition) with the entries of P being $p_1, \ldots, p_n$ and the entries of Q being $q_1, \ldots, q_n$. Denote the sequence of intermediate insertion and recording tableaux obtained during the computation by $P^{(j)}, Q^{(j)}$, $j = 0, 1 \ldots, n$ (so that $P = P^{(n)}$ and $Q = Q^{(n)}$).

(b) For an integer $t \geq 1$, say that the element (q_i, p_i) **is in the class** t **of** σ if $P^{(i)}_{1,t} = p_i$ (where $P^{(i)}_{j,k}$ denotes the entry of $P^{(i)}$ in row i, column j), and denote this relation by $\gamma_\sigma(q_i, p_i) = t$.

As a warm-up problem, show that $\gamma_\sigma(q_i, p_i) = 1$ if and only if $p_i = \min(p_1, \ldots, p_i)$.

(c) If $\sigma' = \begin{pmatrix} q'_1 & \cdots & q'_m \\ p_1 & \cdots & p'_m \end{pmatrix}$ denotes σ with the columns corresponding to class 1 elements removed, show that for any t and i, $\gamma_{\sigma'}(q'_i, p'_i) = t$ if and only if $\gamma_\sigma(q'_i, p'_i) = t + 1$.

(d) Show that for any t, the class t elements can be labeled $(q_{i_1}, p_{i_1}), \ldots, (q_{i_k}, p_{i_k})$, where $q_{i_1} < \ldots < q_{i_k}$ and $p_{i_1} > \ldots > p_{i_k}$.

(e) With the notation of part (d) above for the class t elements, show that $P_{1,t} = p_{i_k}$ and $Q_{1,t} = q_{i_1}$, and that the planar permutation σ' associated with the "decapitated" tableaux P' and Q', obtained from P and Q respectively by deleting their first rows, is the union over all class numbers t of the columns $\begin{pmatrix} q_{i_2} & \cdots & q_{i_k} \\ p_{i_1} & \cdots & p_{i_{k-1}} \end{pmatrix}$.

(f) Show that for an element (q, p) of σ, $\gamma_\sigma(q, p) = t$ if and only if t is the largest number of indices $i_1, \ldots, i_t$ such that $p_{i_1} < \ldots < p_{i_t} =$ and $q_{i_1} < \ldots < q_{i_t} = q$. (That is, the class number of (q, p) is the maximal length of an increasing subsequence of σ that ends in (q, p).)

(g) Deduce from (f) that class number is symmetric in the q's and p's, in the sense that $\gamma_\sigma(q, p) = \gamma_{\sigma^{-1}}(p, q)$.

(h) Let P^{-1} and Q^{-1} denote the insertion and recording tableaux associated by the Robinson–Schensted algorithm with the inverse permutation σ^{-1}. Conclude from the observations in (d), the first part of (e) and (g) that for all class numbers t we have $P_{1,t} = Q^{-1}_{1,t}$ and $Q_{1,t} = P^{-1}_{1,t}$. That is, Theorem 1.10(b) is true at least for the first row of P and Q.

(i) Combine this with the second part of (e) to conclude by induction that the analogous statement is true for all the rows of P and Q and thus finish the proof of Theorem 1.10(b).

1.11 (a) (☕) A permutation $\sigma \in S_n$ is called an **involution** if $\sigma^2 = \text{id}$ (the identity permutation). Show that the number of Young tableaux with n cells is equal to the number of involutions in S_n.

(b) (☕☕) Denote the number of involutions in S_n by I_n. Show that

$$I_n = \sum_{k=0}^{\lfloor n/2 \rfloor} \frac{n!}{2^k k!(n-2k)!} \tag{1.60}$$

(c) (☕☕) Show that the exponential generating function of the sequence I_n is

$$\sum_{n=0}^{\infty} \frac{I_n x^n}{n!} = e^{x+x^2/2}.$$

(d) (☕) Show that $I_n = \mathbb{E}(X^n)$ where X is a random variable with a normal distribution $X \sim N(1, 1)$.

(e) (☕☕☕) (Chowla–Herstein–Moore [25]) Use (1.60) to show that I_n has the asymptotic behavior

$$I_n = (1 + o(1))\frac{1}{\sqrt{2}}\left(\frac{n}{e}\right)^{n/2} e^{\sqrt{n}-1/4} \quad \text{as } n \to \infty. \tag{1.61}$$

1.12 (☕☕☕) If λ, μ are two partitions such that the Young diagram of μ is contained in the Young diagram of λ, the difference between the two Young diagrams is called a skew Young diagram and denoted $\lambda \setminus \mu$. A skew Young tableau of shape $\lambda \setminus \mu$ is a filling on the integers $1, 2, \ldots, |\lambda \setminus \mu| = |\lambda| - |\mu|$ in the cells of the skew diagram $\lambda \setminus \mu$ following the same monotonicity rules as for an ordinary Young tableau, that is, the entries along each row and column are increasing.

Prove the fact, first proved by Aitken [1] (then rediscovered by Feit [39]; see also [71, Exercise 19, pp. 67, 609] and [125, Corollary 7.16.3, p. 344]), that the number $d_{\lambda\setminus\mu}$ of skew Young tableaux of shape $\lambda \setminus \mu$ is given by

$$d_{\lambda\setminus\mu} = |\lambda \setminus \mu|! \det_{i,j=1}^{k} \left(\frac{1}{(\lambda_i - i - \mu_j + j)!} \right)$$

where k is the number of parts of λ, μ_j is interpreted as 0 for j greater than the number of parts of μ and $1/r!$ is defined as 0 for $r < 0$.

1.13 (☕) Let $p(n)$ be the number of partitions of n. Show that there exists a constant $c > 0$ such that the inequality $p(n) > e^{c\sqrt{n}}$ holds for all $n \geq 1$.

1.14 (☕) For $n \in \mathbb{N}$, let $c(n)$ denote the number of **ordered partitions** (also called **compositions**) of n, that is, ways of expressing n as a sum of positive integers, where different orders are considered distinct representations. Prove that $c(n) = 2^{n-1}$, and deduce trivially that $p(n) \leq 2^{n-1}$.

1.15 A **strict partition** of n is a partition into parts that are distinct. Define

$$q(n, k) = \text{number of strict partitions of } n \text{ into } k \text{ parts},$$

$$q(n) = \sum_{k=1}^{\lfloor \sqrt{2n} \rfloor} q(n, k) = \text{number of strict partitions of } n,$$

$$Q(n) = \sum_{k=1}^{n} q(n).$$

(a) (☕☕) Prove that $q(n, k) \leq \frac{1}{k!}\binom{n+k-1}{k-1}$ for all n, k.

(b) (☕☕) Deduce that there exist constants $C_1, C_2 > 0$ such that the inequality $Q(n) \leq C_1 e^{C_2\sqrt{n}}$ holds for all $n \geq 1$.

(c) (☕☕) Prove that $p(n) \leq Q(n)^2$, and therefore by part (b) above $p(n)$ satisfies the upper bound $p(n) \leq C_1^2 e^{2C_2\sqrt{n}}$ for all $n \geq 1$.

Hint: Find a way to dissect the Young diagram of a partition of n into two diagrams encoding strict partitions.

1.16 (a) (☕) Define the generating function $F(z) = 1 + \sum_{n=1}^{\infty} p(n)z^n$. It is traditional to define $p(0) = 1$, so this can also be written as $F(z) = \sum_{n=0}^{\infty} p(n)z^n$. Deduce from the previous problem that the power series converges in the region $|z| < 1$. Prove that in this range **Euler's product formula** holds:

$$F(z) = \prod_{k=1}^{\infty} \frac{1}{1 - z^k}.$$

(b) (☕☕) Show that if $0 < x < 1$ then we have $\log F(x) \leq \frac{\pi^2}{6}\frac{x}{1-x}$. (You may need to use the fact that $\sum_{n=1}^{\infty} n^{-2} = \pi^2/6$; see Exercise 1.18.)

(c) (☕☕) Show that for real x satisfying $0 < x < 1$ we have $p(n) < x^{-n}F(x)$. Using the bound above for $F(x)$, find a value of x (as a function of n) that makes this a particularly good bound, and deduce that the bound $p(n) \leq e^{\pi\sqrt{2n/3}}$ holds for all $n \geq 1$.

1.17 (a) (☕) Prove that the integral (1.18) converges absolutely for any $f \in \mathcal{F}$.

(b) (☕) Prove that if $h\colon \mathbb{R} \to \mathbb{R}$ is a Lipschitz function with compact support then the integral

$$Q(h) = -\frac{1}{2}\int_{-\infty}^{\infty}\int_{-\infty}^{\infty} \log|s - t|\, h'(t)h'(s)\, dt\, ds$$

converges absolutely.

1.18 (☕☕) Verify (1.45) by reducing it, via integration by parts, a substitution and some additional massaging, to the integral evaluation

$$-\int_0^{\pi/2} \log(\sin(t))\,dt = \tfrac{1}{2}\pi\log(2), \tag{1.62}$$

which can then be verified by expanding the function $-\log\left(1-e^{2it}\right)$ in powers of e^{2it} and integrating termwise. As a nice corollary to the evaluation (1.62), obtain the famous identity $\sum_{n=1}^{\infty} n^{-2} = \pi^2/6$, first proved by Euler in 1735.

1.19 (a) (☕☕) Prove (1.50) and (1.52). Note that (1.52) is an improper integral.

(b) (☕☕) If you are familiar with the Euler gamma function, prove the following generalizations of (1.50) and (1.52):

$$\begin{aligned} \int_0^\infty e^{iu}u^{\alpha-1}\,du &= e^{\pi i\alpha/2}\Gamma(\alpha), \\ \int_0^\infty (e^{iu}-1)u^{\alpha-2}\,du &= \frac{-ie^{\pi i\alpha/2}\Gamma(\alpha)}{1-\alpha}, \end{aligned} \qquad (0<\alpha<1).$$

1.20 (☕☕) In the notation of the proof of Lemma 1.21, let $f_1(x) = f(x)e^{-x}$, $h_1(x) = h(x)e^{-x}$. Convince yourself that $h_1 \in L^1(\mathbb{R})$ (and therefore has a well-defined Fourier transform), then show, using a modified (and more rigorous) version of the computations in the proof, that $\hat{h}(s) = \hat{f}(s)$ for all $s \in \mathbb{R}$. Conclude that $h(x) = f(x)$ almost everywhere, as claimed in the proof.

1.21 (☕☕☕) (Romik [107]) For each $n \geq 1$, let σ_n be a uniformly random permutation in S_n, and let X_n denote the number of bumping operations that are performed when applying the Robinson–Schensted algorithm to σ_n. Show that as $n \to \infty$ we have with asymptotically high probability that $X_n = (1+o(1))\kappa n^{3/2}$ for some constant κ. Find a connection between the constant κ and the limit shape Ω of Plancherel-random Young diagrams, and use it to show that $\kappa = \frac{128}{27\pi^2}$.

1.22 (☕☕) A **Markov chain** (more precisely, discrete state-space Markov chain) is a sequence of random variables $X_1, X_2, \ldots$ taking values in some countable set $\mathcal{Z}$ (called the **state space** of the chain), that has the property that for any $n \geq 1$ and $x_1, \ldots, x_{n+1} \in \mathcal{Z}$, the relation

$$\mathbb{P}(X_{n+1} = x_{n+1} \,|\, X_1 = x_1, \ldots, X_n = x_n) = \mathbb{P}(X_{n+1} = x_{n+1} \,|\, X_n = x_n)$$

holds, provided the left-hand side makes sense, that is, the conditioning event has positive probability. The motto to remember for a Markov chain is "conditioned on the present, the future is independent of the past," that is, if we are given the value X_n of the chain at time n then the conditional probability distribution of the next value X_{n+1} (and, by induction, all subsequent values) is known and does not depend on the history of the sequence prior to time n. This conditional distribution is often given a notation such as

$$p_n(x, y) = \mathbb{P}(X_{n+1} = y \,|\, X_n = x) \qquad (x, y \in \mathcal{Z}),$$

and referred to as the **transition matrix**, or **transition kernel**, of the chain. (An especially common and well-understood class of Markov chain are the so-called **time-homogeneous** ones in which $p_n(\cdot, \cdot)$ does not depend on n, but time-inhomogeneous chains also appear frequently in many situations.) Prove that the Plancherel growth process defined in Section 1.19 is a Markov chain. Note that the transition kernel is given explicitly by the right-hand side of (1.56). In particular this means that one can randomly "grow" a Plancherel-random partition of order n, without using random permutations at all, by starting with the trivial diagram of order 1, and then repeatedly replacing the current diagram μ by a new diagram λ such that $\mu \nearrow \lambda$, where λ is sampled according to the transition probabilities given in (1.56). The next exercise shows an algorithm for efficiently sampling from the probability distribution given by these transition probabilities without actually having to compute these probabilities.

1.23 (☕☕☕) (Greene–Nijenhuis–Wilf [51]) Prove that the following "inverse hook walk" can be used to simulate the "growth" steps of the Plancherel growth step – that is, given the partition $\lambda^{(n-1)} = \mu$ in the $(n-1)$th step of the process, we can choose randomly a Young diagram $\lambda \vdash n$ such that $\mu \nearrow \lambda$ with the transition probabilities being given by (1.56). The inverse hook walk is defined as follows:

(a) Start from a cell with positive coordinates (i, j) lying *outside* the Young diagram of μ. It can be any cell as long as $i > \mu_1$ and $j > \mu'_1$, so that the inverse hook walk will have a positive probability of reaching any of the external corners of μ.

(b) Now perform an inverse hook walk, repeatedly replacing the current cell with a uniformly random cell from the **inverse hook** of the current cell. The inverse hook is defined as the set of cells lying to the left of the current cell, or above it, and being still outside the Young diagram of μ.

(c) The walk terminates when it reaches a cell $\mathbf{c}$ that is the only one in its inverse hook. These cells are exactly the external corners of μ. Let $\lambda = \mu \cup \{\mathbf{c}\}$.

2

The Baik–Deift–Johansson theorem

Chapter summary. Twenty years after the seminal works of Logan–Shepp and Vershik–Kerov, who showed that the maximal length $L(\sigma_n)$ of an increasing subsequence in a uniformly random permutation of order n is typically about $2\sqrt{n}$, Baik, Deift, and Johansson proved a remarkable theorem concerning the limiting distribution of the fluctuations of $L(\sigma_n)$ from this typical value. One particularly exciting aspect of the result is the appearance of the **Tracy–Widom distribution** from random matrix theory. In this chapter we will prove this celebrated result. To this end, we will study some elegant techniques such as the use of **determinantal point processes** –a particularly nice class of random sets – and some interesting facts about classical special functions, the **Bessel functions** and the **Airy function**.

2.1 The fluctuations of $L(\sigma_n)$ and the Tracy–Widom distribution

Having proved in the previous chapter that the first-order asymptotic behavior of $L(\sigma_n)$, the maximal length of an increasing subsequence in a uniformly random permutation of order n, is given by $(2 + o(1))\sqrt{n}$ (both in the typical case and on the average), the natural next step is to ask how far we can expect $L(\sigma_n)$ to fluctuate from this asymptotic value, that is, what exactly is hiding inside the "$o(1)$" term. Note that these fluctuations can be separated into two parts: first, the deterministic deviation $\ell_n - 2\sqrt{n}$ of the mean value of $L(\sigma_n)$ from $2\sqrt{n}$; and second, the random fluctuations $L(\sigma_n) - \ell_n$ of $L(\sigma_n)$ around its mean, whose size can be measured, for example, by looking at the standard deviation $(\mathrm{Var}(L(\sigma_n) - \ell_n))^{1/2}$ of this random variable. These two sources of fluctuation are conceptually distinct, in the sense that there seems to be no obvious a priori reason to assume that they scale similarly with n, although it turns out that in fact they do.

Although the question of the fluctuations of $L(\sigma_n)$ is of clear interest, for a while no further progress in this direction was made following the works of Logan–Shepp and Vershik–Kerov. Starting in the early 1990s, however, the problem started attracting a great deal of attention, and over several years various bounds on the size of the fluctuations were derived by several authors[1] using a variety of ingenious methods. These bounds were not sharp enough to provide an understanding of how the fluctuations scale as a function of n. However, some evidence pointed to the correct scale being of order $n^{1/6}$: this was put forward as a conjecture by Odlyzko and Rains [94], based on numerical evidence and a suggestive analytic bound from the Vershik–Kerov paper. The same conjecture was made independently by J. H. Kim [67], who in 1996 managed to prove that with high probability the positive part of the fluctuations $L(\sigma_n) - 2\sqrt{n}$ could not be of greater magnitude than $O(n^{1/6})$. Odlyzko and Rains's computational data also made it apparent that the distribution of $L(\sigma_n)$ does not become symmetric around its mean value as n grows large; thus, any naive hope of finding a limiting Gaussian distribution for $L(\sigma_n)$ seemed unwarranted.

The breakthrough came with the work of Jinho Baik, Percy A. Deift, and Kurt Johansson [11], who in the late 1990s found a remarkable limiting law for the fluctuations, thus not only settling the question of the scale of the fluctuations, which turned out to be indeed of order $n^{1/6}$, but also proving a much more precise result concerning their limiting distribution. The existence and precise nature of this result came as a surprise, as it revealed a deep structure and a connection between this problem in the combinatorics of random permutations and the theory of random matrices, integrable systems, and related topics in probability and mathematical physics.

To formulate Baik, Deift, and Johansson's result, we need to define the limiting distribution, called the **Tracy–Widom distribution** and denoted F_2. This distribution function, along with several closely related random processes and distribution functions, was studied by Craig A. Tracy and Harold Widom in the 1990s in connection with the asymptotic behavior of the largest eigenvalues of random matrices. To define it, first, let $\mathrm{Ai}(x)$ denote the **Airy function**, an important special function of mathematical analysis, defined by

$$\mathrm{Ai}(x) = \frac{1}{\pi}\int_0^\infty \cos\left(\tfrac{1}{3}t^3 + xt\right) dt, \qquad (x \in \mathbb{R}). \tag{2.1}$$

(It is not difficult to see that the integral converges for all real x as an improper integral, and defines a smooth function; other properties of $\mathrm{Ai}(x)$ are discussed in the following sections.) Next, we define the **Airy kernel** $\mathbf{A}\colon \mathbb{R}\times\mathbb{R} \to \mathbb{R}$, by

$$\mathbf{A}(x,y) = \begin{cases} \dfrac{\mathrm{Ai}(x)\,\mathrm{Ai}'(y) - \mathrm{Ai}'(x)\,\mathrm{Ai}(y)}{x-y} & \text{if } x \neq y, \\ \mathrm{Ai}'(x)^2 - x\,\mathrm{Ai}(x)^2 & \text{if } x = y. \end{cases} \tag{2.2}$$

(The values on the diagonal make $\mathbf{A}(\cdot,\cdot)$ continuous, by L'Hôpital's rule and the differential equation satisfied by $\mathrm{Ai}(x)$; see (2.8).) We now define $F_2 : \mathbb{R} \to \mathbb{R}$ by the formula

$$F_2(t) = 1 + \sum_{n=1}^\infty \frac{(-1)^n}{n!} \int_t^\infty \ldots \int_t^\infty \det_{i,j=1}^{n}\left(\mathbf{A}(x_i,x_j)\right) dx_1 \ldots dx_n. \tag{2.3}$$

Although the formula defining F_2 seems complicated at first sight, it has a more conceptual interpretation as a **Fredholm determinant** – a kind of determinant defined for certain linear operators acting on an infinite-dimensional vector space – which leads to the more compact notation $F_2(t) = \det\left(\mathbf{I} - \mathbf{A}|_{L^2(t,\infty)}\right)$. It is not difficult to show using standard facts about the Airy function that the integrals and infinite series in (2.3) converge (see Lemma 2.25 on p. 119), but the fact that the expression on the right-hand side defines a probability distribution function is nontrivial.

Theorem 2.1 *The function F_2 is a distribution function.*

We will prove Theorem 2.1 in Section 2.10, as part of a deeper analysis of F_2 and its properties. The main result of the analysis, which yields Theorem 2.1 as an easy corollary, is a remarkable structure theorem, due to Tracy and Widom, that represents F_2 in terms of the solution to a certain ordinary differential equation, the **Painlevé II** equation.

We are ready to state the celebrated Baik–Deift–Johansson theorem, whose proof is our main goal in this chapter.

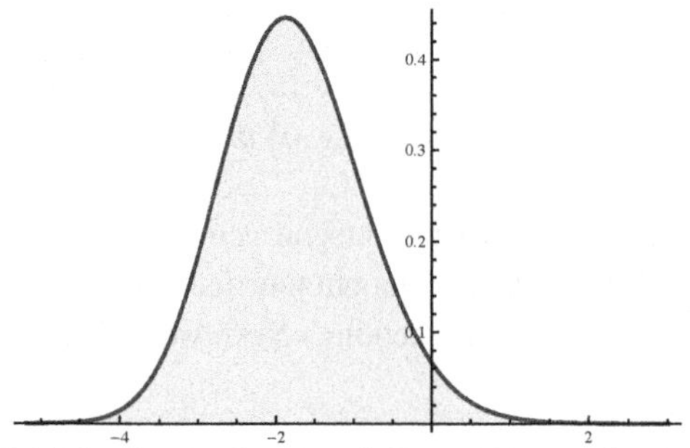

Figure 2.1 The density function $f_2(t) = F_2'(t)$ associated with the Tracy–Widom distribution.

Theorem 2.2 (The Baik–Deift–Johansson theorem) *For each $n \geq 1$, let σ_n denote a uniformly random permutation of order n. Then for any $x \in \mathbb{R}$ we have that*

$$\mathbb{P}\left(\frac{L(\sigma_n) - 2\sqrt{n}}{n^{1/6}} \leq x\right) \to F_2(x) \quad \text{as } n \to \infty. \tag{2.4}$$

That is, the scaled fluctuations $n^{-1/6}(L(\sigma_n) - 2\sqrt{n})$ converge in distribution to F_2.

Fig. 2.1 shows the graph of the probability density function $f_2(t) = F_2'(t)$.[2]

2.2 The Airy ensemble

The Tracy–Widom distribution originally arose as the limiting distribution of the largest eigenvalue of a random matrix chosen from the so-called **Gaussian Unitary Ensemble** (or **GUE**), a natural model in random matrix theory for a random Hermitian matrix (see box on p. 84). Tracy and Widom also studied the distributions of the second-largest eigenvalue, third-largest, and so on. Collectively the largest eigenvalues of a GUE random matrix converge after scaling to a random point process on $\mathbb{R}$ called the **Airy ensemble**. Baik, Deift and Johansson recognized that Theorem 2.2 can be generalized to include quantities analogous to the largest eigenvalues of a GUE matrix other than the maximal one. The correct object to consider turns out to be random partitions chosen according to Plancherel measure.

Recall that the length $\lambda_1^{(n)}$ of the first row of such a Plancherel-random partition $\lambda^{(n)}$ is equal in distribution to $L(\sigma_n)$. Baik et al. proved also [12] that the length $\lambda_2^{(n)}$ of the second row of $\lambda^{(n)}$ converges in distribution after scaling to the second-largest element of the Airy ensemble, and conjectured that more generally, for each $k \geq 1$, the joint distribution of the lengths of the first k rows of $\lambda^{(n)}$ converges after scaling to the largest k elements in the Airy ensemble. This conjecture was given three different proofs soon afterwards, one by Andrei Okounkov [95],[3] and two other roughly equivalent ones found independently by Alexei Borodin, Grigori Olshanski, and Andrei Okounkov [19] and by Kurt Johansson [63].

In this chapter, our primary goals will be to prove Theorems 2.1 and 2.2. However, we will aim higher and actually end up proving the stronger result on convergence of the scaled lengths of the first rows of a Plancherel-random partition to the Airy ensemble. This will require a bit more background and preparatory work, but in this way we will get a better picture of the mathematical structure that underlies these remarkable asymptotic phenomena. In particular, a key element in the proof is the use of the concept of a **determinantal point process**. These processes (which are a type of random point process, another concept we will define more properly) have appeared in recent years in connection with an increasing number of problems in combinatorics, probability, and statistical physics, so it is worthwhile to add them to our arsenal of mathematical tools.

To state the precise result, we need some additional definitions. The Airy ensemble is defined in terms of the Airy kernel, defined in (2.2), according to the following general recipe. Define a **random point process** on $\mathbb{R}$ to be any random locally finite subset X of $\mathbb{R}$. (There is a formal way to define such objects using standard measure-theoretic concepts, but for our purposes it will be enough to consider X as a kind of random variable taking values in the set of subsets of $\mathbb{R}$, such that one may ask questions of the form "what is the probability that exactly k points of X fall in an interval $I \subseteq \mathbb{R}$?" for an arbitrary nonnegative integer k and subinterval I of $\mathbb{R}$.) For such a random point process, for each $n \in \mathbb{N}$ we define its **n-point correlation function** $\rho_X^{(n)} : \mathbb{R}^n \to [0, \infty)$ by

$$\rho_X^{(n)}(x_1, \ldots, x_n) = \lim_{\epsilon \downarrow 0} \left[(2\epsilon)^{-n} \, \mathbb{P} \left(\bigcap_{j=1}^{n} \left\{ \left| X \cap [x_j - \epsilon, x_j + \epsilon] \right| = 1 \right\} \right) \right], \quad (2.5)$$

Random matrix theory and the Gaussian Unitary Ensemble

The mapping from a square matrix to its eigenvalues is an interesting function. **Random matrix theory** studies the effects that this function has on matrices chosen at random in some simple or natural way. That is, given a random matrix whose entries have some given joint distribution, what can we say about the joint distribution of the eigenvalues, and about their asymptotic behavior? Usually the measure on the entries of the matrix is a fairly simple one, for example, they are taken as i.i.d. samples from some distribution. If one is interested in matrices from one of the well-behaved families of matrices, such as Hermitian or symmetric matrices, a standard trick is to condition the matrix of i.i.d. samples to lie in the family; e.g., in the case of symmetric matrices this means that only the entries on the main diagonal and above it are taken as i.i.d. samples, and then the values of the entries below the diagonal are dictated by the symmetry condition.

A particularly nice and well-behaved random matrix model is the **Gaussian Unitary Ensemble**, or **GUE**. In this case, one takes the entries on the main diagonal to be i.i.d. $N(0,1)$ (standard Gaussian) random variables; independently the entries above the diagonal are i.i.d. complex Gaussian random variables of type $N(0,\frac{1}{2}) + iN(0,\frac{1}{2})$ (i.e., each of the real and imaginary part is an $N(0,\frac{1}{2})$ r.v., and they are independent); and each entry below the diagonal is the complex conjugate of the corresponding entry in the reflected position above the diagonal. This results in a random Hermitian matrix, with the nice combination of properties that its entries are independent (modulo the Hermitian constraint) and its distribution is invariant under the action of the unitary group by conjugation.

It can be shown that the vector of real eigenvalues $(\xi_1,\ldots,\xi_n)$ (considered with a randomized order) of a random GUE matrix of order n has joint density function

$$f_n(x_1,\ldots,x_n) = \frac{1}{Z_n}\prod_{1\le i<j\le n}(x_i - x_j)^2 \exp\left(-\frac{1}{2}\sum_{i=1}^n x_i^2\right),$$

where $Z_n = (2\pi)^{n/2}\prod_{k=1}^{n-1}(k!)$. An analysis of this explicit formula, and other techniques, can be used to reveal much information about the asymptotic behavior of the GUE eigenvalues. See [4] for a good introduction to this field.

if the limit exists. The correlation function $\rho_X^{(n)}$ is similar in nature to a probability density in that it measures the relative likelihood of finding a point of the process simultaneously in the vicinity of each of the points $x_1,\ldots,x_n$. The 1-point correlation function $\rho_X^{(1)}$ (or sometimes the associated measure $\rho_X^{(1)}(x)\,dx$) is called the **intensity** of the process. If these limits exist almost everywhere for all $n \ge 1$, then, under some mild technical conditions, it is known that the correlation functions characterize the distribution of the random set X (see [121]). In this case we say that X is an **absolutely continuous** point process.

An absolutely continuous **determinantal point process** is a special kind of absolutely continuous point process for which the correlation functions can be expressed as determinants of a certain form. More precisely, an absolutely continuous point process X is called determinantal if there is a function $\mathbf{K}\colon \mathbb{R} \times \mathbb{R} \to \mathbb{R}$ (called the **correlation kernel** of the process) such that for all $n \ge 1$ we have

$$\rho_X^{(n)}(x_1, \ldots, x_n) = \det_{i,j=1}^{n} \Big(\mathbf{K}(x_i, x_j)\Big), \qquad (x_1, \ldots, x_n \in \mathbb{R}). \tag{2.6}$$

The **Airy ensemble** is the determinantal process X_{Airy} whose correlation kernel is the Airy kernel $\mathbf{A}(x, y)$. Of course, one must become convinced that such an object exists and is unique—that is, that the determinants of the form (2.6) in the case when $\mathbf{K}(x, y) = \mathbf{A}(x, y)$ are indeed the n-point correlation functions of a unique random point process; we discuss this in more detail in Section 2.11. In particular, it can be shown that the Airy ensemble almost surely has a maximal element. Therefore, for notational convenience, rather than consider it as a set of unlabeled elements, we label its (random) elements in decreasing order starting with the largest one, as follows:

$$X_{\text{Airy}} = \{\zeta_1 > \zeta_2 > \zeta_3 > \ldots\}. \tag{2.7}$$

With this preparation, we are ready to state the strengthened version of the Baik–Deift–Johansson theorem that was proved by Borodin, Okounkov, Olshanski, and Johansson.

Theorem 2.3 (Edge statistics for Plancherel measure[4]) *For each $n \ge 1$, let $\lambda^{(n)}$ denote a random partition chosen according to Plancherel measure of order n, let $\lambda_j^{(n)}$ denote the length of its jth row, and denote $\bar{\lambda}_j^{(n)} = n^{-1/6}(\lambda_j^{(n)} - 2\sqrt{n})$. Then for each $k \ge 1$, we have the convergence in distribution*

$$(\bar{\lambda}_1^{(n)}, \ldots, \bar{\lambda}_k^{(n)}) \xrightarrow{d} (\zeta_1, \ldots, \zeta_k) \ \text{as } n \to \infty.$$

Our approach to the proofs of Theorems 2.2 and 2.3 follows those of Borodin–Okounkov–Olshanski [19] and Johansson [63]. The Airy function

The Airy function

The Airy function was introduced to mathematics by Sir George Biddell Airy, an English 19th century astronomer and mathematician, who is known among other things for establishing in 1851 the prime meridian (the 0° longitude line), based at the Royal Observatory in Greenwich, England. He discovered the function while analyzing the mathematics of the phenomenon of rainbows; the formula he derived and published in an 1838 paper, describing approximately the oscillations of the intensity of light of a given wavelength near the caustic (the curve in the rainbow where the light of that wavelength appears), is essentially the square of the Airy function, $\mathrm{Ai}(x)^2$, in an appropriate coordinate system. The Airy function has many other applications in mathematical analysis, the theory of differential equations, optics, quantum physics, and other fields. Its fundamental property is the relation

$$\mathrm{Ai}''(x) = x\,\mathrm{Ai}(x). \tag{2.8}$$

That is, $\mathrm{Ai}(x)$ is a solution of the second-order ordinary differential equation $y''(x) = x\,y(x)$, known as the **Airy differential equation** (see Exercise 2.15).

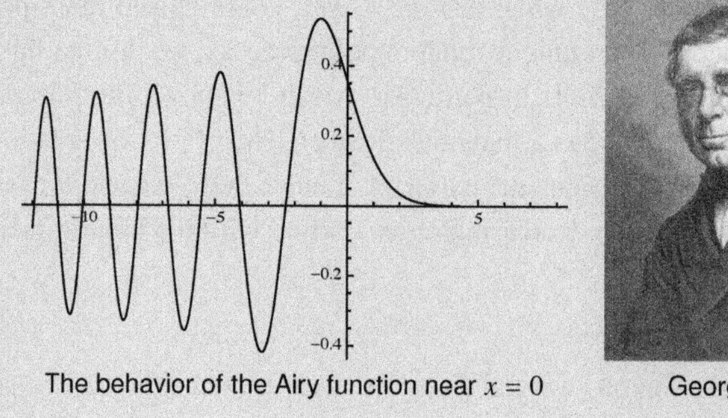

The behavior of the Airy function near $x = 0$

George Airy

and kernel will play a crucial role in the analysis. We discuss some of the properties of these functions in the following sections. See the box for some additional background on the history of the Airy function and its importance in mathematics.

2.3 Poissonized Plancherel measure

Theorems 2.2 and 2.3 are asymptotic results, but the first key element in their proof is an exact identity relating Plancherel measure to determinantal point processes. To formulate this identity, we need to first replace the family of finite-order Plancherel measures with an averaged-out version of them called **Poissonized Plancherel measure**, which we define as follows. Let $\theta > 0$ be a continuous positive parameter (roughly, it will have the same role as that played by the discrete parameter n). Denote the set $\cup_{n=0}^{\infty}\mathcal{P}(n)$ of all integer partitions (including the trivial empty partition of order 0) by $\mathcal{P}^*$. The Poissonized Plancherel measure with parameter θ is a measure P_θ on the set $\mathcal{P}^*$, assigning to a partition λ the probability

$$P_\theta(\lambda) = e^{-\theta}\frac{\theta^{|\lambda|}d_\lambda^2}{(|\lambda|!)^2}.$$

To verify that this is indeed a probability measure, observe that

$$P_\theta(\mathcal{P}^*) = \sum_{n=0}^{\infty} P_\theta(\mathcal{P}(n)) = \sum_{n=0}^{\infty} e^{-\theta}\frac{\theta^n}{n!}\left(\sum_{\lambda\vdash n}\frac{d_\lambda^2}{n!}\right) = \sum_{n=0}^{\infty} e^{-\theta}\frac{\theta^n}{n!} = 1.$$

This computation also illustrates the fact, immediate from the definition, that P_θ is simply a "Plancherel measure with a Poisson-random n" (or, in more technical probabilistic language, it is "a mixture of the Plancherel measures of all orders with Poisson weights $e^{-\theta}\theta^n/n!$"). That is, we can sample a random partition $\lambda^{\langle\theta\rangle}$ with distribution P_θ in a two-step experiment, by first choosing a random variable N with the Poisson distribution $\mathrm{Poi}(\theta)$, and then, conditioned on the event that $N = n$, choosing $\lambda^{\langle\theta\rangle}$ to be a Plancherel-random partition of order n. This idea also ties in nicely with Hammersley's approach to the study of the maximal increasing subsequence length discussed in Section 1.4 (see Exercise 2.1).

Note also that when θ is large, the random order N of the partition is with high probability close to θ (since N has mean θ and variance θ). So, at least intuitively it seems plausible that many reasonably natural asymptotic results we might prove for the Poissonized Plancherel-random partition $\lambda^{\langle\theta\rangle}$ will also hold true for the original Plancherel-random $\lambda^{(n)}$. Indeed, for our context we will be able to make such a deduction using a nice trick of "de-Poissonization"; see Section 2.9.

It now turns out that the Poissonized Plancherel-random partition $\lambda^{\langle\theta\rangle}$ can be encoded in terms of a determinantal point process, using the concept of **Frobenius coordinates**. Let $\lambda \in \mathcal{P}(n)$ be a partition. Define numbers $p_1, \ldots, p_d$ by

$$p_j = \lambda_j - j,$$

where d is the largest $j \geq 1$ for which $\lambda_j - j$ is nonnegative. Graphically, p_j is the number of cells in the jth row of the Young diagram of λ to the right of the main diagonal. Note that the p_j form a decreasing sequence. Similarly, define another decreasing sequence of numbers $q_1 > \ldots > q_d \geq 0$, where q_j is the number of cells in the jth *column* of λ below the diagonal, formally given by

$$q_j = \lambda'_j - j.$$

The fact that there are equally many q_j as p_j is easy to see (the number d is the size of the so-called **Durfee square** of λ, that is, the maximal k such that a square of dimensions $k \times k$ leaning against the top-left corner of the Young diagram fits inside it). Together, the p_j and q_j are called the Frobenius coordinates of λ. It is customary to write the vector of Frobenius coordinates in the form $(p_1, \ldots p_d \mid q_1, \ldots, q_d)$. Clearly the Frobenius coordinates determine the partition λ (see also Exercise 2.13).

For our purposes, it will be more convenient to use the following variant of the Frobenius coordinates. Denote $\mathbb{Z}' = \mathbb{Z} - \frac{1}{2}$. The set of **modified Frobenius coordinates** is defined by

$$\mathrm{Fr}(\lambda) = \left\{p_1 + \tfrac{1}{2}, \ldots, p_d + \tfrac{1}{2}, -q_1 - \tfrac{1}{2}, \ldots, -q_d - \tfrac{1}{2}\right\} \subset \mathbb{Z}'. \tag{2.9}$$

One rationale for adding $\frac{1}{2}$ to each of the p_j and q_j to get half-integer values is that this has the geometric interpretation of including the Young diagram cells lying on the main diagonal, each of which is broken up into two equal halves, one being assigned to p_j and the other to q_j. Thus for example we have the identity $|\lambda| = \sum_{j=1}^d \left(p_j + \frac{1}{2}\right) + \sum_{j=1}^d \left(q_j + \frac{1}{2}\right)$, which will play a small role later on. Fig. 2.2 is an illustration of the Frobenius coordinates for the diagram $\lambda = (4, 4, 3, 1, 1)$, whose Frobenius coordinates are $(3, 2, 0 \mid 4, 1, 0)$.

The hook-length formula (Theorem 1.12), which played a crucial role in our analysis of the asymptotic behavior of Plancherel-random partitions in the previous chapter, will be essential here as well. What we need is a different version of it that expresses d_λ, the number of standard Young tableaux of shape λ, as a determinant involving the Frobenius coordinates. We derive it as a corollary to two other auxiliary representations for d_λ.

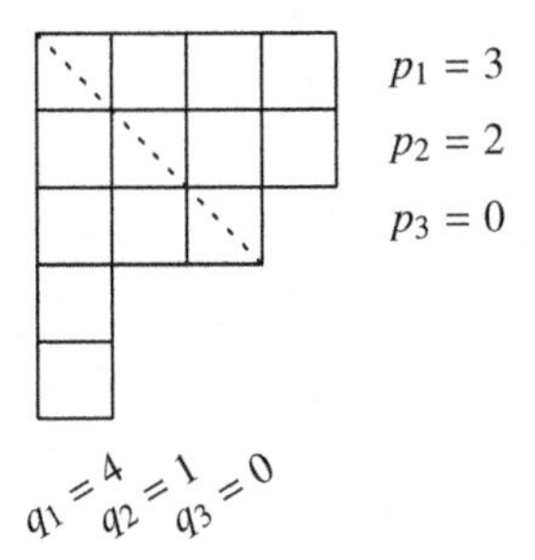

Figure 2.2 A Young diagram and its Frobenius coordinates.

Lemma 2.4 *If $\lambda \in \mathcal{P}(n)$ has m parts, then, with the convention that $\lambda_i = 0$ if $i > m$, for any $k \geq m$ we have*

$$\frac{d_\lambda}{|\lambda|!} = \frac{\prod_{1\le i<j\le k}(\lambda_i - \lambda_j + j - i)}{\prod_{1\le i\le k}(\lambda_i + k - i)!}. \tag{2.10}$$

Proof First note that it is enough to prove the claim for $k = m$, since it is easy to see that the right-hand side of (2.10) stays constant for $k \geq m$. Therefore, by the hook-length formula, we need to prove that

$$\prod_{(i,j)\in\lambda} h_\lambda(i,j) = \frac{\prod_{1\le i\le m}(\lambda_i + m - i)!}{\prod_{1\le i<j\le m}(\lambda_i - \lambda_j + j - i)}. \tag{2.11}$$

We claim that for any $1 \leq i \leq m$, the equality

$$\prod_{j=1}^{\lambda_i} h_\lambda(i,j) = \frac{(\lambda_i + m - i)!}{\prod_{i<j\le m}(\lambda_i - \lambda_j + j - i)} \tag{2.12}$$

holds; this clearly implies (2.11) by multiplying over $i = 1, \ldots, m$. To see why (2.12) is true, note that the largest hook-length in the ith row of λ is $\lambda_i + m - i$, so that the left-hand side of (2.12) is equal to $(\lambda_i + m - i)!$ divided by the product of the hook-lengths that are *missing* from that row. It is not difficult to see that these missing hook-lengths consist exactly of the numbers $(\lambda_i+m-i)-\lambda_m, (\lambda_i+m-i)-(\lambda_{m-1}+1), \ldots, (\lambda_i+m-i)-(\lambda_{i+1}+m-j-1)$. This list accounts exactly for the denominator in (2.12). □

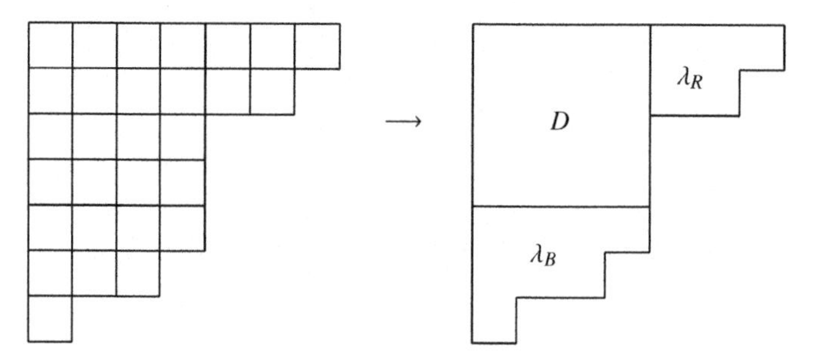

Figure 2.3 A Young diagram is broken up into its Durfee square and two smaller "right" and "bottom" Young diagrams.

Lemma 2.5 *If $\lambda = (p_1, \ldots, p_d \mid q_1, \ldots, q_d)$ in Frobenius coordinates, then we have*

$$\frac{d_\lambda}{|\lambda|!} = \frac{\prod_{1\le i,j\le d}(p_i - p_j)(q_i - q_j)}{\prod_{1\le i,j\le d}(p_i + q_j + 1)\prod_{i=1}^{d}(p_i!q_i!)}. \tag{2.13}$$

Proof Divide the Young diagram of λ into three shapes: the Durfee square D of size $d\times d$ leaning against the top-left corner, and the smaller Young diagrams λ_{R} and λ_{B} leaning against D from the right and bottom, respectively (Fig. 2.3).

Now compute the product of reciprocals of the hook numbers

$$\frac{d_\lambda}{|\lambda|!} = \prod_{(i,j)\in\lambda} h_\lambda(i,j)^{-1}$$

separately over the cells of the three shapes D, λ_{R} and λ_{B}, as follows. The contribution from D is exactly

$$\prod_{1\le i,j\le d}(p_i + q_j + 1)^{-1},$$

since for $1 \le i, j \le d$ we have that $h_\lambda(i, j) = p_i + q_j + 1$. The contribution from λ_{R} can be computed by applying (2.10) to that partition, whose parts satisfy $(\lambda_{\mathrm{R}})_i = \lambda_i - d$, and whose hook-lengths are the same as the corresponding hook-lengths for λ. This gives

$$\prod_{(i,j)\in\lambda_{\mathrm{R}}} h_\lambda(i,j)^{-1} = \frac{\prod_{1\le i<j\le d}(p_i - p_j)}{\prod_{i=1}^{d} p_i!}.$$

In a similar way, the contribution from λ_B is given by the analogous expression

$$\frac{\prod_{1\le i<j\le d}(q_i - q_j)}{\prod_{i=1}^{d} q_i!},$$

and combining these products gives the claim. □

Next, we need a classical determinant identity from linear algebra. A proof idea is suggested in Exercise 2.2.

Lemma 2.6 (The Cauchy determinant identity)

$$\det_{i,j=1}^{n}\left(\frac{1}{x_i + y_j}\right) = \frac{\prod_{1\le i<j\le n}(x_j - x_i)(y_j - y_i)}{\prod_{1\le i,j\le n}(x_i + y_j)}. \tag{2.14}$$

By combining Lemmas 2.5 and 2.6, we get the following determinantal form of the hook-length formula in Frobenius coordinates.

Corollary 2.7 *We have*

$$\frac{d_\lambda}{|\lambda|!} = \det_{i,j=1}^{d}\left(\frac{1}{(p_i + q_j + 1)p_i!q_j!}\right). \tag{2.15}$$

The identity (2.15) has useful implications for Poissonized Plancherel measure. We will show later that when the Young diagram is encoded using the modified Frobenius coordinates, the resulting process is a type of determinantal point process – a discrete version of the "absolutely continuous" determinantal point processes described in the previous section. The relevant background on determinantal point processes is presented in the next section. For now, we can formulate a result that already illustrates this idea to some extent. Define a function $\mathbf{L}_\theta \colon \mathbb{Z}' \times \mathbb{Z}' \to \mathbb{R}$ by

$$\mathbf{L}_\theta(x, y) = \begin{cases} 0 & \text{if } xy > 0, \\ \dfrac{1}{x - y} \cdot \dfrac{\theta^{(|x|+|y|)/2}}{\left(|x| - \frac{1}{2}\right)!\left(|y| - \frac{1}{2}\right)!} & \text{if } xy < 0. \end{cases} \tag{2.16}$$

Proposition 2.8 *Let $\lambda \in \mathcal{P}^*$. Let* $\mathrm{Fr}(\lambda) = \{x_1, x_2, \ldots, x_s\}$ *be the modified Frobenius coordinates of λ. Then we have*

$$P_\theta(\lambda) = e^{-\theta} \det_{i,j=1}^{s}\left(\mathbf{L}_\theta(x_i, x_j)\right). \tag{2.17}$$

Proof Let $(p_1, \ldots, p_d \mid q_1, \ldots, q_d)$ be the usual Frobenius coordinates of λ, related to the modified coordinates by

$$\{x_1, \ldots, x_s\} = \left\{p_1 + \tfrac{1}{2}, \ldots, p_d + \tfrac{1}{2}, -q_1 - \tfrac{1}{2}, \ldots, -q_d - \tfrac{1}{2}\right\}$$

(so in particular $s = 2d$). By Corollary 2.7 we have

$$P_\theta(\lambda) = e^{-\theta} \frac{\theta^{|\lambda|} d_\lambda^2}{(|\lambda|!)^2} = e^{-\theta} \theta^{|\lambda|} \left(\det_{i,j=1}^{d} \left(\frac{1}{(p_i + q_j + 1) p_i! q_j!} \right) \right)^2,$$

so we need to verify the identity

$$\theta^{|\lambda|} \left(\det_{i,j=1}^{d} \left(\frac{1}{(p_i + q_j + 1) p_i! q_j!} \right) \right)^2 = \det_{i,j=1}^{s} \left(\mathbf{L}_\theta(x_i, x_j) \right).$$

Since $\mathbf{L}_\theta(x, y) = 0$ if $\operatorname{sgn}(x) = \operatorname{sgn}(y)$, we see that the matrix $(\mathbf{L}_\theta(x_i, x_j))_{i,j}$ can be written as the block matrix

$$\left(\mathbf{L}_\theta(x_i, x_j) \right)_{i,j=1}^{s} = \left(\begin{array}{c|c} 0 & \left(-\mathbf{L}_\theta\left(p_i + \tfrac{1}{2}, -q_j - \tfrac{1}{2}\right) \right)_{i,j=1}^{d} \\ \hline \left(\mathbf{L}_\theta\left(p_j + \tfrac{1}{2}, -q_i - \tfrac{1}{2}\right) \right)_{i,j=1}^{d} & 0 \end{array} \right)$$

Therefore we can evaluate its determinant, making use of the fact that $|\lambda| = \sum_{i=1}^{d} (p_i + q_i + 1)$, as follows:

$$\begin{aligned} \det_{i,j=1}^{s} \left(\mathbf{L}_\theta(x_i, x_j) \right) &= (-1)^{d^2+d} \left(\det \left(\mathbf{L}_\theta \left(p_i + \tfrac{1}{2}, -q_j - \tfrac{1}{2} \right) \right)_{i,j=1}^{d} \right)^2 \\ &= \theta^{\sum_{i=1}^{d} \left(p_i + \frac{1}{2}\right) + \sum_{j=1}^{d} \left(q_j + \frac{1}{2}\right)} \left(\det_{i,j=1}^{d} \left(\frac{1}{(p_i + q_j + 1) p_i! q_j!} \right) \right)^2 \\ &= \theta^{|\lambda|} \left(\det_{i,j=1}^{d} \left(\frac{1}{(p_i + q_j + 1) p_i! q_j!} \right) \right)^2, \end{aligned}$$

which proves the claim. □

2.4 Discrete determinantal point processes

The goal of this section is to define and describe some of the basic properties of determinantal point processes whose elements lie in some countable set. The subject can be treated in much greater generality.[5] We start with the even simpler case of a process with a finite set of possible values, where the subject at its most basic level reduces to some elementary linear algebra.

Let Ω be a finite set. We consider square matrices with real entries whose rows and columns are indexed by elements of Ω. In contrast to the usual linear algebra notation, we denote such matrices as functions $\mathbf{M}\colon \Omega\times\Omega \to \mathbb{R}$, so that $\mathbf{M}(x, y)$ denotes the entry lying in the row with index x and the column with index y, and use the word **kernel** instead of matrix. Kernels will be denoted with boldface letters, for example, $\mathbf{M}, \mathbf{L}, \mathbf{K}$, and so on. We will also think of kernels as linear operators, acting on the space $\ell^2(\Omega)$ of complex-valued functions on Ω (which are identified with column vectors) in the usual way by matrix multiplication. The notation $\mathbf{ML}$ denotes the product of the kernels $\mathbf{M}$ and $\mathbf{L}$, that is, the usual product of matrices, equivalent to composition of linear operators. We use $\mathbf{I}$ to denote the identity matrix/operator acting either on the space $\ell^2(\Omega)$ or on an appropriate subspace of $\ell^2(\Omega)$ that will be clear from the context.

A **random point process in** Ω (or simply **point process**) is a random subset X of Ω, that is, a random variable taking values in the set of subsets of Ω.[6] We refer to Ω as the **underlying space** of X. Given the point process X, define its **correlation function** by

$$\rho_X(A) = \mathbb{P}(A \subseteq X), \qquad (A \subseteq \Omega). \tag{2.18}$$

(As a point of terminology, it is sometimes customary to define for each $n \geq 1$, in analogy with (2.5), the function

$$\rho_X^{(n)}(x_1, \ldots, x_n) = \mathbb{P}(x_1, \ldots, x_n \in X) = \rho_X(\{x_1, \ldots, x_n\}) \quad (x_1, \ldots, x_n \in \Omega),$$

referred to as the n**-point correlation function** of the process, or the correlation function of order n; thus the mapping $A \mapsto \rho_X(A)$ combines the correlation functions of all orders.)

We say that X is determinantal if there exists a kernel $\mathbf{K}\colon \Omega \times \Omega \to \mathbb{R}$, called the **correlation kernel of** X, such that for any $A = \{x_1, \ldots, x_s\} \subseteq \Omega$, we have that

$$\rho_X(A) = \det_{i,j=1}^{s} \Big(\mathbf{K}(x_i, x_j)\Big), \tag{2.19}$$

where we adopt the convention that the determinant of an empty matrix is 1. Note that the right-hand side does not depend on the labeling of the elements of A. We denote by $\mathbf{K}_A$ the submatrix $(\mathbf{K}(x, y))_{x,y\in A}$. In the language of linear operators, if we think of $\mathbf{K}$ as a linear operator on $\ell^2(\Omega)$, then $\mathbf{K}_A$ corresponds to its restriction to the copy of $\ell^2(A)$ embedded in $\ell^2(\Omega)$. With this notation, the determinant on the right-hand side of (2.19) can be written simply as $\det(\mathbf{K}_A)$.

In practice, the distribution of a finite point process is often given not in terms of its correlation function but in terms of a formula for the probability of individual configurations. We say that the point process X **has a determinantal configuration kernel** if there is a normalization constant $Z > 0$ and a kernel $\mathbf{L} \colon \Omega \times \Omega \to \mathbb{R}$, called the **configuration kernel of** X, such that for any subset $A = \{x_1, \ldots, x_s\} \subseteq \Omega$, we have

$$\mathbb{P}(X = A) = \frac{1}{Z} \det_{i,j=1}^{s} \left(\mathbf{L}(x_i, x_j)\right) = \frac{1}{Z} \det(\mathbf{L}_A). \tag{2.20}$$

By summing (2.20) over all subsets of Ω, we find that the normalization constant must satisfy

$$Z = \sum_{A \subseteq \Omega} \det(\mathbf{L}_A) = \det(\mathbf{I} + \mathbf{L}) \tag{2.21}$$

(the second equality is a trivial, though perhaps not widely known, observation about determinants). In particular, this implies that $\det(\mathbf{I} + \mathbf{L})$ must be nonzero, so $\mathbf{I} + \mathbf{L}$ is invertible.

It turns out that if a point process has a determinantal configuration kernel, then it is determinantal (the converse is not true – see Exercise 2.3), and there is a simple relationship between the correlation and configuration kernels. This property is one of the key features that makes the study of determinantal point processes so fruitful; its precise formulation is as follows.

Proposition 2.9 *If a point process X has a determinantal configuration kernel $\mathbf{L}$, then it is determinantal and its correlation kernel is given by*

$$\mathbf{K} = \mathbf{L}(\mathbf{I} + \mathbf{L})^{-1}. \tag{2.22}$$

Proof Let $\mathbf{K}$ be given by (2.22), and let $(t_x)_{x \in \Omega}$ denote a family of indeterminate variables indexed by the elements of Ω. We will prove the identity

$$\sum_{A \subseteq \Omega} \rho_X(A) \prod_{x \in A} t_x = \sum_{A \subseteq \Omega} \det(\mathbf{K}_A) \prod_{x \in A} t_x, \tag{2.23}$$

equating two polynomials in the set of indeterminates $(t_x)_{x \in \Omega}$. This will prove the claim by comparing coefficients. To prove (2.23), observe that by the definition of $\rho_X(A)$, the left-hand side can be rewritten as

$$\sum_{B \subseteq \Omega} \mathbb{P}(X = B) \prod_{x \in B} (1 + t_x) = \sum_{B \subseteq \Omega} \frac{1}{Z} \det(\mathbf{L}_B) \prod_{x \in B} (1 + t_x). \tag{2.24}$$

Defining a square matrix $\mathbf{D}$ whose rows and columns are indexed by elements of Ω by $\mathbf{D}(x,x) = t_x$ and $\mathbf{D}(x,y) = 0$ for $x \neq y$ (i.e., $\mathbf{D}$ is the diagonal matrix whose entries are the indeterminates t_x), and denoting $\mathbf{D}' = \mathbf{I} + \mathbf{D}$, we see that the right-hand side of (2.24) is equal to

$$\frac{1}{Z}\sum_{B\subseteq\Omega} \det\left((\mathbf{D}'\mathbf{L})_B\right) = \frac{\det(\mathbf{I}+\mathbf{D}'\mathbf{L})}{\det(\mathbf{I}+\mathbf{L})} = \frac{\det(\mathbf{I}+\mathbf{L}+\mathbf{DL})}{\det(\mathbf{I}+\mathbf{L})}. \tag{2.25}$$

Since $\mathbf{K} = \mathbf{L}(\mathbf{I}+\mathbf{L})^{-1}$, we have that $(\mathbf{I}+\mathbf{DK})(\mathbf{I}+\mathbf{L}) = \mathbf{I}+\mathbf{L}+\mathbf{DL}$, so, by the multiplicativity of the determinant, the last expression in (2.25) is equal to $\det(\mathbf{I}+\mathbf{DK})$. This in turn can be expanded to give the right-hand side of (2.23), proving the claim. □

Note that if a configuration kernel $\mathbf{L}$ exists, the kernels $\mathbf{K}$ and $\mathbf{L}$ contain exactly the same information about the process X, since (2.22) can be inverted to give $\mathbf{L} = \mathbf{K}(\mathbf{I}-\mathbf{K})^{-1}$. However, in practice $\mathbf{K}$ encodes the probabilities of some interesting events associated with X in a more accessible way. The relation (2.19) is one example of this fact, and we give several more examples. In Propositions 2.10, 2.11 and 2.12, X will denote a determinantal point process with correlation kernel $\mathbf{K}$. The following formula is in some sense dual to (2.19) (see Exercises 2.3 and 2.5).

Proposition 2.10 *Let $A \subseteq \Omega$, and denote $A^c = \Omega \setminus A$. We have*

$$\mathbb{P}(X \subseteq A) = \det(\mathbf{I} - \mathbf{K}_{A^c}).$$

Proof By the inclusion–exclusion principle, we have that

$$\begin{aligned}
\mathbb{P}(X \subseteq A) &= 1 - \mathbb{P}\left(\cup_{x\in A^c}\{x \in X\}\right) \\
&= 1 - \sum_{x\in A^c}\mathbb{P}(x \in X) + \sum_{\{x,y\}\subseteq A^c}\mathbb{P}(\{x,y\}\subseteq X) - \dots \\
&= 1 + \sum_{k=1}^{|\Omega|}(-1)^k\left(\sum_{B\subseteq A^c,\,|B|=k}\det(\mathbf{K}_B)\right) \\
&= 1 + \sum_{\emptyset\subsetneq B\subseteq A^c}(-1)^{|B|}\det(\mathbf{K}_B) = \det(\mathbf{I}-\mathbf{K}_{A^c}).
\end{aligned}$$
□

More generally, we have the following formula for the distribution of the number of points of X occurring inside a set $A \subseteq \Omega$.

Proposition 2.11 *For any $A \subseteq \Omega$ and $n \geq 0$, we have that*

$$\mathbb{P}(|X \cap A| = n) = \frac{(-1)^n}{n!} \frac{d^n}{dx^n}\Big|_{x=1} \det(\mathbf{I} - x\mathbf{K}_A). \tag{2.26}$$

Proof Letting $s = -x$, we have

$$\begin{aligned}\det(\mathbf{I} + s\mathbf{K}_A) &= \sum_{B \subseteq A} \det(\mathbf{K}_B) s^{|B|} = \sum_{B \subseteq A} \mathbb{P}(B \subseteq X \cap A) s^{|B|} \\ &= \sum_{E \subseteq A} \mathbb{P}(X \cap A = E)(1+s)^{|E|} = \sum_{n=0}^{|A|} \mathbb{P}(|X \cap A| = n)(1+s)^n.\end{aligned} \tag{2.27}$$

So, we have shown that $\det(\mathbf{I} - x\mathbf{K}_A) = \sum_{n=0}^{|A|} \mathbb{P}(|X \cap A| = n)(1-x)^n$, which implies (2.26). □

Continuing in even greater generality, the joint distribution of the numbers of points of X falling in several disjoint sets can also be expressed in terms of $\mathbf{K}$. First, we introduce a small bit of additional notation: if $A \subset \Omega$, let $\mathbf{P}_A \colon \Omega \times \Omega \to \mathbb{R}$ denote the kernel given by

$$\mathbf{P}_A(x, y) = \begin{cases} 1 & \text{if } x = y \in A, \\ 0 & \text{otherwise.} \end{cases} \tag{2.28}$$

That is, $\mathbf{P}_A$ is a projection operator to the subspace of $\ell^2(\Omega)$ spanned by the coordinates in A.

The following result is an easy generalization of Proposition 2.11 and its proof is left to the reader (Exercise 2.6).

Proposition 2.12 *Let $k \geq 1$. For any disjoint subsets $A_1, \ldots, A_k \subseteq \Omega$ and integers $n_1, \ldots, n_k \geq 0$, we have*

$$\begin{aligned}&\mathbb{P}\Big(|X \cap A_1| = n_1, \ldots, |X \cap A_k| = n_k\Big) \\ &\quad = \frac{(-1)^N}{N!} \frac{\partial^{n_1+\ldots+n_k}}{\partial x_1^{n_1} \ldots \partial x_k^{n_k}}\Big|_{x_1=\ldots=x_k=1} \det\left(\mathbf{I} - \sum_{j=1}^{k} x_j \mathbf{P}_{A_j} \cdot \mathbf{K} \cdot \mathbf{P}_{A_1 \cup \ldots \cup A_k}\right),\end{aligned}$$

where $N = n_1 + \ldots + n_k$.

This concludes our summary of the basic facts regarding determinantal point processes in the simplest case of a finite underlying space. Next, we consider determinantal point processes that are random subsets of a *countable* set Ω. In this case a kernel $\mathbf{M} \colon \Omega \times \Omega \to \mathbb{R}$ can be thought of as an infinite matrix. As before, we say that X is a **random point process in** Ω if X is a random subset of Ω (i.e., a random variable taking values in $\{0,1\}^{\Omega}$, or equivalently a collection of random variables $(Y_x)_{x\in\Omega}$ where $Y_x = 1$ if and only if $x \in X$). If X is a point process, we define the correlation function $\rho_X(\cdot)$ of X by $\rho_X(A) = \mathbb{P}(A \subseteq X)$ as before (where A is any finite subset of Ω), and say that X is determinantal if there exists a kernel $\mathbf{K}\colon \Omega\times\Omega \to \mathbb{R}$ such that (2.19) holds for any finite subset $A = \{x_1, \ldots, x_s\} \subset \Omega$. As before, we say that X has a determinantal configuration kernel if there is a normalization constant $Z > 0$ and a kernel $\mathbf{L}\colon \Omega \times \Omega \to \mathbb{R}$ such that (2.20) holds for any finite subset $A = \{x_1, \ldots, x_s\} \subset \Omega$, except that in this defintion we make the added requirement that X be almost surely a finite set, so that we can still express the normalization constant as $Z = \sum_{A\subset\Omega,\, |A|<\infty} \det(\mathbf{L}_A)$.

Fortunately for us, much of the theory presented in the foregoing discussion remains valid, but some care must be taken when interpreting some of the standard linear-algebraic concepts in this infinite-dimensional setting. In particular, for the second equality in (2.21) to make sense (where the summation is now understood to range over finite subsets of Ω), one must have a well-defined notion of the determinant of an infinite matrix. One possible solution is to take this equality as the *definition* of the determinant of $\mathbf{I} + \mathbf{L}$, and to keep the notation $\det(\mathbf{I} + \mathbf{L})$ for this new determinant concept. This may seem a bit contrived, but it turns out that it is in fact quite natural and leads to the elegant and extremely useful theory of **Fredholm determinants**. The full theory of these determinants is subtle and relies on deep ideas from operator theory, but we present here a minimal discussion of Fredholm determinants that is adequate for our relatively limited needs.[7]

One elementary result we need is Hadamard's inequality, a basic inequality from linear algebra.

Lemma 2.13 (Hadamard's inequality) (a) *If $M = (m_{i,j})_{i,j=1}^n$ is a Hermitian positive-semidefinite matrix, then*

$$\det(M) \leq \prod_{i=1}^{n} m_{i,i}. \tag{2.29}$$

(b) *If M is any $n \times n$ square matrix whose column vectors are denoted*

$v_1, \dots, v_n$, *then*

$$|\det(M)| \le \prod_{j=1}^{n} \|v_j\|_2 \le n^{n/2} \prod_{j=1}^{n} \|v_j\|_\infty. \tag{2.30}$$

Proof (a) For $i = 1, \dots, n$, let e_i denote the ith standard basis vector in $\mathbb{C}^n$. Note that $m_{i,i} = e_i^\top M e_i \ge 0$. In particular, if M is singular then (2.29) holds, since $\det(M) = 0$. Next, assume that M is nonsingular. In this case M is positive-definite and $m_{i,i} > 0$ for $i = 1, \dots, n$. Let D be the diagonal matrix with entries $1/\sqrt{m_{i,i}}$, $i = 1, \dots, n$. Let $M' = DMD$. Then M' is also Hermitian and positive-definite (in particular, it is diagonalizable and has positive eigenvalues), and its diagonal entries are $m'_{i,i} = 1$. Let $\lambda_1, \dots, \lambda_n$ be the eigenvalues of M'. We have $\sum_i \lambda_i = \operatorname{tr} M' = \sum_i m'_{i,i} = n$. It follows that

$$\det(M') = \prod_{i=1}^{n} \lambda_i \le \left(\frac{1}{n}\sum_{i=1}^{n} \lambda_i\right)^n = \left(\frac{1}{n}\operatorname{tr} M'\right)^n = 1,$$

On the other hand, $\det(M') = \det(D)^2 \det(M) = \det(M)/\prod_i m_{i,i}$, so we get (2.29).

(b) Defining $P = M^* M$, we have that P is Hermitian and positive-semidefinite, and its diagonal entries are $p_{i,i} = \|v_i\|_2^2$. The inequality (2.29) applied to P now gives

$$|\det(M)|^2 = \det(P) \le \prod_{i=1}^{n} p_{i,i} = \prod_{i=1}^{n} \|v_i\|_2^2,$$

which gives (2.30). □

Fix some ordering $\omega_1, \omega_2, \dots$ of the elements of Ω. We consider the class $\mathcal{E}$ of kernels $\mathbf{T}\colon \Omega \times \Omega \to \mathbb{C}$ (we shall see that there is some advantage to considering complex-valued kernels) whose entries satisfy the bound

$$|\mathbf{T}(\omega_i, \omega_j)| \le C e^{-c(i+j)} \tag{2.31}$$

for some constants $C, c > 0$ (which may depend on $\mathbf{T}$ but not on i, j). If $\mathbf{T}$ is in the class $\mathcal{E}$, we say it has **exponential decay**. Trivially, kernels in $\mathcal{E}$ are bounded linear operators on the Hilbert space $\ell^2(\Omega)$. If $\mathbf{S}, \mathbf{T} \in \mathcal{E}$, it is immediate to check that the operators $\mathbf{S} + \mathbf{T}$, $\mathbf{ST}$ and $\lambda\mathbf{S}$ (where λ is an arbitrary complex number) also have exponential decay.

For a kernel $\mathbf{T} \in \mathcal{E}$, we define the **Fredholm determinant of I + T** by

$$\det(\mathbf{I}+\mathbf{T}) = \sum_{A \subset X, |A|<\infty} \det(\mathbf{T}_A), \tag{2.32}$$

provided that the sum converges absolutely, which we claim it does. This can be written equivalently as

$$\begin{aligned}
\det(\mathbf{I}+\mathbf{T}) &= \sum_{k=0}^{\infty} \sum_{A \subset X, |A|=k} \det(T_A) \\
&= \sum_{k=0}^{\infty} \frac{1}{k!} \left[\sum_{x_1 \in \Omega} \sum_{x_2 \in \Omega} \cdots \sum_{x_k \in \Omega} \det_{i,j=1}^{k} \Big(\mathbf{T}(x_i, x_j) \Big) \right] \\
&= \sum_{k=0}^{\infty} \frac{1}{k!} \left[\sum_{m_1=1}^{\infty} \cdots \sum_{m_k=1}^{\infty} \det_{i,j=1}^{k} \Big(\mathbf{T}(\omega_{m_i}, \omega_{m_j}) \Big) \right]. \qquad (2.33)
\end{aligned}$$

To see that for a kernel with exponential decay the Fredholm determinant is well-defined, note that by (2.31) and Hadamard's inequality (2.30) we have

$$\begin{aligned}
\left| \det_{i,j=1}^{k} \Big(\mathbf{T}(\omega_{m_i}, \omega_{m_j}) \Big) \right| &= e^{-2c \sum_{j=1}^{n} m_j} \left| \det_{i,j=1}^{k} \Big(e^{c(m_i+m_j)} \mathbf{T}(x_i, x_j) \Big) \right| \\
&\le k^{k/2} C^k e^{-2c \sum_{j=1}^{k} m_j},
\end{aligned}$$

so that

$$\sum_{k=0}^{\infty} \frac{1}{k!} \left[\sum_{m_1=1}^{\infty} \cdots \sum_{m_k=1}^{\infty} \left| \det_{i,j=1}^{k} \Big(\mathbf{T}(\omega_{m_i}, \omega_{m_j}) \Big) \right| \right] \le \sum_{k=0}^{\infty} \frac{C^k k^{k/2}}{k!} \left(\sum_{m=1}^{\infty} e^{-2cm} \right)^k < \infty$$

which shows that we indeed have absolute convergence.

If $\mathbf{T}$ is a kernel with no nonzero entries outside of the first n rows and columns, then the sum on the right-hand side of (2.32) is a finite sum, and by standard linear algebra it coincides with the ordinary matrix determinant $\det_{i,j=1}^{n}(\delta_{ij} + \mathbf{T}(\omega_i, \omega_j))$ (where δ_{ij} is the Kronecker delta symbol). This shows that the Fredholm determinant is a natural extension of the familiar determinant from linear algebra. Note that some authors define the Fredholm determinant to be the quantity that in our notation would be written as $\det(\mathbf{I} - \mathbf{T})$, and, to add further potential for confusion, may refer to this quantity (inconsistently) as the "Fredholm determinant of **T**" or "Fredholm determinant associated with **T**."

In the proof of Proposition 2.9, we used the multiplicativity property of ordinary determinants. We will need the analogue of this property for Fredholm determinants.

Lemma 2.14 (Multiplicativity of Fredholm determinants) *If* $\mathbf{S}, \mathbf{T} \in \mathcal{E}$ *then we have*

$$\det(\mathbf{I} + \mathbf{S} + \mathbf{T} + \mathbf{ST}) = \det(\mathbf{I} + \mathbf{S}) \det(\mathbf{I} + \mathbf{T}). \tag{2.34}$$

Proof For each $n \geq 1$ denote by $\mathbf{P}_n$ the projection operator $\mathbf{P}_{\{\omega_1,\ldots,\omega_n\}}$ (in the notation of (2.28)). Then $\mathbf{P}_n\mathbf{S}\mathbf{P}_n$ is the matrix obtained from $\mathbf{S}$ by zeroing out all entries except in the first n rows and columns. Denote

$$\mathcal{E}_n = \{\mathbf{S} \in \mathcal{E} : \mathbf{S} = \mathbf{P}_n\mathbf{S}\mathbf{P}_n\}.$$

We start by noting that if $\mathbf{S}, \mathbf{T} \in \mathcal{E}_m$ for some $m \geq 1$ then (2.34) holds by the multiplicativity of ordinary matrix determinants.

Next, assume that $\mathbf{S} \in \mathcal{E}_m$, $\mathbf{T} \in \mathcal{E}$. Denote $\mathbf{T}_n = \mathbf{P}_n\mathbf{T}\mathbf{P}_n$. We have

$$\det(\mathbf{I} + \mathbf{T}_n) = \sum_{A \subseteq \{\omega_1,\ldots,\omega_n\}} \det(\mathbf{T}_A) \xrightarrow[n\to\infty]{} \det(\mathbf{I} + \mathbf{T}),$$

and similarly

$$\begin{aligned} \det(\mathbf{I}+\mathbf{S} + \mathbf{T}_n + \mathbf{S}\mathbf{T}_n) &= \sum_{A \subseteq \{\omega_1,\ldots,\omega_n\}} \det((\mathbf{S} + \mathbf{T}_n + \mathbf{S}\mathbf{T}_n)_A) \\ &= \sum_{A \subseteq \{\omega_1,\ldots,\omega_n\}} \det((\mathbf{S} + \mathbf{T} + \mathbf{ST})_A) \xrightarrow[n\to\infty]{} \det(\mathbf{I} + \mathbf{S} + \mathbf{T} + \mathbf{ST}) \end{aligned}$$

where the second equality is valid for $n \geq m$. Combining these facts with the observation that (again due to the multiplicativity of finite matrix determinants)

$$\det(\mathbf{I} + \mathbf{S} + \mathbf{T}_n + \mathbf{S}\mathbf{T}_n) = \det(\mathbf{I} + \mathbf{S}) \det(\mathbf{I} + \mathbf{T}_n)$$

gives (2.34).

Finally, to prove (2.34) in the general case $\mathbf{S}, \mathbf{T} \in \mathcal{E}$, denote $\mathbf{S}_m = \mathbf{P}_m\mathbf{S}\mathbf{P}_m$. Then $\mathbf{S}_m \in \mathcal{E}_m$ so by the case proved above we have

$$\det(\mathbf{I} + \mathbf{S}_m + \mathbf{T} + \mathbf{S}_m\mathbf{T}) = \det(\mathbf{I} + \mathbf{S}_m) \det(\mathbf{I} + \mathbf{T}) \xrightarrow[n\to\infty]{} \det(\mathbf{I} + \mathbf{S}) \det(\mathbf{I} + \mathbf{T}).$$

On the other hand, we can write

$$\begin{aligned}
&\det(\mathbf{I} + \mathbf{S}_m + \mathbf{T} + \mathbf{S}_m\mathbf{T}) \\
&\quad = \sum_{A\subset\mathbb{N},\ |A|<\infty} \det((\mathbf{S}_m + \mathbf{T} + \mathbf{S}_m\mathbf{T})_A) \\
&\quad = \sum_{A\subseteq\{1,\dots,m\}} \det((\mathbf{S}_m + \mathbf{T} + \mathbf{S}_m\mathbf{T})_A) + \sum_{\substack{A\subseteq\mathbb{N},\ |A|<\infty \\ A\not\subseteq\{1,\dots,m\}}} \det((\mathbf{S}_m + \mathbf{T} + \mathbf{S}_m\mathbf{T})_A) \\
&\quad = \sum_{A\subseteq\{1,\dots,m\}} \det((\mathbf{S} + \mathbf{T} + \mathbf{S}\mathbf{T})_A) + \sum_{\substack{A\subseteq\mathbb{N},\ |A|<\infty \\ A\not\subseteq\{1,\dots,m\}}} \det((\mathbf{S}_m + \mathbf{T} + \mathbf{S}_m\mathbf{T})_A).
\end{aligned}$$

In the last expression, the first sum converges to $\det(\mathbf{I} + \mathbf{S} + \mathbf{T} + \mathbf{ST})$ as $m \to \infty$, so if we can show that the second sum tends to 0 the result will follow. Indeed, the kernels $\mathbf{S}_m + \mathbf{T} + \mathbf{S}_m\mathbf{T}$ have exponential decay uniformly in m, that is, there exist constants $C, c > 0$ such that for any $i, j \geq 1$, the (ω_i, ω_j)-entry of this kernel is bounded in absolute value by $Ce^{-c(i+j)}$. It therefore follows using Hadamard's inequality as before that

$$\begin{aligned}
&\sum_{\substack{A\subseteq\mathbb{N},\ |A|<\infty \\ A\not\subseteq\{1,\dots,m\}}} |\det((\mathbf{S}_m + \mathbf{T} + \mathbf{S}_m\mathbf{T})_A)| \\
&\qquad \leq \sum_{k=1}^{\infty} \frac{k^{k/2}}{k!} k \sum_{j_1=m+1}^{\infty} \sum_{j_2,\dots,j_k=1}^{\infty} C^k e^{-2c\sum_{d=1}^{k} j_d} \\
&\qquad = \sum_{k=1}^{\infty} \frac{k^{k/2+1}}{k!} C^k \left(\sum_{j=1}^{\infty} e^{-2cj}\right)^{k-1} \cdot \sum_{j=m+1}^{\infty} e^{-2cj}.
\end{aligned}$$

The summation over k is a convergent sum, and it is multiplied by an expression that tends to 0 as $m \to \infty$, as claimed. □

We can now extend Proposition 2.9 to the countable setting.

Proposition 2.15 *If X is a point process in a countable set Ω with a determinantal configuration kernel $\mathbf{L}\colon \Omega \times \Omega \to \mathbb{R}$, $\mathbf{L}$ has exponential decay, and $\mathbf{I} + \mathbf{L}$ is an invertible operator in $\ell^2(\Omega)$, then X is determinantal and its correlation kernel is given by* (2.22).

Before starting the proof, note that the assumption that $\mathbf{I} + \mathbf{L}$ is invertible is unnecessary, since it can be shown that (as with ordinary determinants) that already follows from the fact that $Z = \det(\mathbf{I} + \mathbf{L})$ is nonzero. However, we did not prove such an implication in the context of Fredholm determinants, and for our purposes the extra assumption will not be an impediment.

Proof We follow the same reasoning that was used in the proof of Proposition 2.9, except that rather than take $(t_x)_{x\in\Omega}$ to be a set of indeterminates, we let $t_x = 1$ for $x \in E$ and $t_x = 0$ for $x \in \Omega \setminus E$, where E is an arbitrary finite subset of Ω. All the steps of the proof remain valid, with the only modification being that the summation in (2.24) is taken over all *finite* subsets $B \subset \Omega$. In that case the algebraic manipulations of the proof all involve sums that are easily seen to be absolutely convergent, and we can use Lemma 2.14 in the last step (with $\mathbf{S} = \mathbf{DK}$, a kernel with only finitely many nonzero entries, and $\mathbf{T} = \mathbf{L}$) where previously the multiplicativity of the determinant was used. This establishes that (2.23) holds for the substitution $t_x = \chi_E(x)$, or in other words that

$$\sum_{A\subseteq E} \rho_X(A) = \sum_{A\subseteq E} \det(\mathbf{K}_A)$$

for any finite $E \subset \Omega$. It is easy to see that this implies that $\rho_X(E) = \det(\mathbf{K}_E)$ for all finite $E \subset \Omega$, by induction on $|E|$. □

Propositions 2.10, 2.11, and 2.12 also have analogues that hold in the case of determinantal processes on a countable space. Note that if $\mathbf{T} \in \mathcal{E}$ then we have

$$\det(\mathbf{I} + x\mathbf{T}) = \sum_{A\subset X,\, |A|<\infty} x^{|A|} \det(T_A),$$

so the fact that this sum converges absolutely for any complex x implies that $\det(\mathbf{I} + x\mathbf{T})$ is an entire function of x.

Proposition 2.16 *If X is a determinantal point process in a countable set Ω with correlation kernel $\mathbf{K}$, and $\mathbf{K}$ has exponential decay, then for any $A \subseteq \Omega$ we have the identity of entire functions*

$$\det(\mathbf{I} - x\mathbf{K}_A) = \sum_{n=0}^{\infty} \mathbb{P}(|X \cap A| = n)(1-x)^n, \qquad (x \in \mathbb{C}).$$

Consequently, we have for all $n \geq 0$ that

$$\mathbb{P}(|X \cap A| = n) = \frac{(-1)^n}{n!} \frac{d^n}{dx^n}\Big|_{x=1} \det(\mathbf{I} - x\mathbf{K}_A).$$

Proof The proof consists simply of going through the computation in Proposition 2.11 carefully and checking that all the steps are meaningful and valid in the present context. □

The following analogue of Proposition 2.12 is proved using similar reasoning to that used in the proof above, starting from the proof of Proposition 2.12 that you will construct in Exercise 2.6. The details are omitted.

Proposition 2.17 *Let $k \geq 1$, and let $A_1, \ldots, A_k \subseteq \Omega$ be disjoint. Under the assumptions of Proposition 2.16 above, the expression*

$$\det\left(\mathbf{I} - \sum_{j=1}^{k} x_j \mathbf{P}_{A_j} \cdot \mathbf{K} \cdot \mathbf{P}_{A_1 \cup \ldots \cup A_k}\right),$$

defines an entire function of the complex variables $x_1, \ldots, x_n$, and we have the identity

$$\det\left(\mathbf{I} - \sum_{j=1}^{k} x_j \mathbf{P}_{A_j} \cdot \mathbf{K} \cdot \mathbf{P}_{A_1 \cup \ldots \cup A_k}\right)$$
$$= \sum_{n_1, \ldots, n_k \geq 0} \mathbb{P}\left(\bigcap_{j=1}^{k} \{|X \cap A_j| = n_j\}\right)(1 - x_1)^{n_1} \ldots (1 - x_k)^{n_k}.$$

Consequently, for any integers $n_1, \ldots, n_k \geq 0$ we have that

$$\mathbb{P}\Big(|X \cap A_1| = n_1, \ \ldots, \ |X \cap A_k| = n_k\Big)$$
$$= \frac{(-1)^N}{N!} \frac{\partial^{n_1 + \ldots + n_k}}{\partial x_1^{n_1} \ldots \partial x_k^{n_k}}\bigg|_{x_1 = \ldots = x_k = 1} \det\left(\mathbf{I} - \sum_{j=1}^{k} x_j \mathbf{P}_{A_j} \cdot \mathbf{K} \cdot \mathbf{P}_{A_1 \cup \ldots \cup A_k}\right),$$

where $N = n_1 + \ldots + n_k$.

2.5 The discrete Bessel kernel

We now consider Proposition 2.8 again in light of the discussion in the previous section. The set $\{x_1, \ldots, x_s\}$ on the right-hand side of (2.17) has equally many positive and negative elements, since it is the set of modified Frobenius coordinates of a Young diagram. It is easy to see that for any set $\{x_1, \ldots, x_s\} \subset \mathbb{Z}'$ without this property, the determinant in (2.17) will be equal to 0. It follows that the set of modified Frobenius coordinates $\mathrm{Fr}(\lambda^{\langle\theta\rangle})$ of a random Young diagram $\lambda^{\langle\theta\rangle}$ chosen according to Poissonized

Plancherel measure with parameter θ forms a point process in $\mathbb{Z}'$ with determinantal configuration kernel $\mathbf{L}_\theta(\cdot,\cdot)$ given by (2.16). Note that $\mathbf{L}_\theta$ is trivially seen to have exponential decay (with respect to the ordering of the elements of $\mathbb{Z}'$ according to increasing distance from 0). In particular, the Fredholm determinant $\det(\mathbf{I}+\mathbf{L})$ is well defined, and, by (2.17) and the preceding comments, is given by

$$\det(\mathbf{I}+\mathbf{L}_\theta) = \sum_{A\subset\mathbb{Z}',\,|A|<\infty} \det((\mathbf{L}_\theta)_A) = e^\theta.$$

So, considering the general theory we developed in the previous section, we see that to gain better insight into the behavior of this point process, we need to identify the corresponding kernel $\mathbf{K}_\theta = \mathbf{L}_\theta(\mathbf{I}+\mathbf{L}_\theta)^{-1}$ whose determinants will describe the correlation function of this process, in accordance with Proposition 2.15. This family of kernels was found by Borodin, Okounkov, and Olshanski, and a variant of it was found independently by Johansson. It is defined in terms of Bessel functions $J_\alpha(z)$, so we begin by recalling the definition of these classical special functions; see the box on p. 106 for additional background.

Let $\Gamma(t)$ denote the Euler gamma function (for readers unfamiliar with this function, its definition and basic properties are reviewed in Exercise 2.7). Let $\alpha\in\mathbb{R}$. The **Bessel function of order** α is the analytic function of a complex variable z defined by the power series expansion

$$J_\alpha(z) = \left(\frac{z}{2}\right)^\alpha \sum_{m=0}^\infty \frac{(-1)^m (z/2)^{2m}}{m!\,\Gamma(m+\alpha+1)}, \qquad (z\in\mathbb{C}). \tag{2.35}$$

We will care only about values of z that are positive real numbers, but in fact it is easy to see that with the standard interpretation of the power function z^α, $J_\alpha(z)$ is analytic on the complex plane with a branch cut at the negative real line.

Recall that for $n\in\mathbb{N}$ we have $\Gamma(n)=(n-1)!$, so in the case when $\alpha=n$ is an integer we can write (2.35) as

$$J_n(z) = \left(\frac{z}{2}\right)^n \sum_{m=\max(-n,0)}^\infty \frac{(-1)^m (z/2)^{2m}}{m!\,(m+n)!}, \qquad (z\in\mathbb{C}). \tag{2.36}$$

It follows in particular that we have the identity

$$J_{-n}(z) = (-1)^n J_n(z), \qquad (z\in\mathbb{C}, n\in\mathbb{Z}), \tag{2.37}$$

relating the Bessel functions of positive and negative integer orders.

Note that $J_\alpha(z)$ is in fact a function of two variables, α and z, although because of how it often arises in applications it is more common to think of it as a one-parameter family $(J_\alpha(z))_{\alpha\in\mathbb{R}}$ of functions of one variable. For our purposes actually the dependence on α will be equally important to the dependence on z. We denote $J'_\alpha(z) = \frac{\partial}{\partial z} J_\alpha(z)$ and $\dot{J}_\alpha(z) = \frac{\partial}{\partial \alpha} J_\alpha(z)$ (the analytic dependence of $J_\alpha(z)$ on α is also not difficult to establish). As a further notational shortcut, in what follows we denote $J_x = J_x(2\sqrt{\theta})$, and similarly $J'_x = J'_x(2\sqrt{\theta})$, $\dot{J}_x = \dot{J}_x(2\sqrt{\theta})$.

Define the **discrete Bessel kernel** with parameter $\theta > 0$ as the function $\mathbf{J}_\theta \colon \mathbb{Z}\times\mathbb{Z} \to \mathbb{R}$ given by

$$\mathbf{J}_\theta(s,t) = \begin{cases} \sqrt{\theta}\,\dfrac{J_s J_{t+1} - J_{s+1} J_t}{s-t} & \text{if } s \neq t, \\ \sqrt{\theta}\left(\dot{J}_s J_{s+1} - J_s \dot{J}_{s+1}\right) & \text{if } s = t. \end{cases} \tag{2.38}$$

It will be convenient to consider $\mathbf{J}_\theta(\cdot,\cdot)$ to be defined also for noninteger real arguments. With such a definition $\mathbf{J}_\theta$ becomes a continuous function on $\mathbb{R}\times\mathbb{R}$, by L'Hôpital's rule.

Next, we define another variant of the discrete Bessel kernel, denoted $\mathbf{K}_\theta \colon \mathbb{Z}'\times\mathbb{Z}' \to \mathbb{R}$, by

$$\mathbf{K}_\theta(x,y) = \operatorname{sgn}(x)^{x-\frac{1}{2}} \operatorname{sgn}(y)^{y+\frac{1}{2}} \mathbf{J}_\theta(x - \tfrac{1}{2}, y - \tfrac{1}{2}). \tag{2.39}$$

A short computation using (2.37) gives the more explicit expression

$$\mathbf{K}_\theta(x,y) = \begin{cases} \sqrt{\theta}\,\dfrac{J_{|x|-\frac{1}{2}} J_{|y|+\frac{1}{2}} - J_{|x|+\frac{1}{2}} J_{|y|-\frac{1}{2}}}{|x| - |y|} & \text{if } xy > 0,\ x \neq y, \\ \sqrt{\theta}\operatorname{sgn}(x)\left(\dot{J}_{|x|-\frac{1}{2}} J_{|x|+\frac{1}{2}} - J_{|x|-\frac{1}{2}} \dot{J}_{|x|+\frac{1}{2}}\right) & \text{if } x = y, \\ \sqrt{\theta}\,\dfrac{J_{|x|-\frac{1}{2}} J_{|y|-\frac{1}{2}} + J_{|x|+\frac{1}{2}} J_{|y|+\frac{1}{2}}}{x - y} & \text{if } xy < 0. \end{cases} \tag{2.40}$$

The following claim is easy, and we will verify it shortly after recalling some standard facts about the Bessel functions.

Lemma 2.18 *For each $\theta > 0$, the kernel $\mathbf{K}_\theta$ has exponential decay.*

We are now ready to state one of the most important results of this chapter.

The Bessel functions

The family of Bessel functions $J_\alpha(z)$ was defined in the 18th century by Daniel Bernoulli – son of Johann Bernoulli, mentioned in the previous chapter in connection with the brachistochrone problem – and generalized and studied systematically by the German 19th century astronomer and mathematician Friedrich Wilhelm Bessel (who is also remembered today as the first person to accurately estimate the distance of Earth to a star other than the Sun).

The Bessel functions are solutions of the second-order linear differential equation

$$x^2y'' + xy' + (x^2 - \alpha^2)y = 0, \tag{2.41}$$

and are more precisely called **Bessel functions of the first kind**, since in the literature there are actually several families of Bessel functions, denoted $J_\alpha(z)$, $Y_\alpha(z)$, $I_\alpha(z)$, and $K_\alpha(z)$. Bessel functions appear prominently in several places in mathematical physics and the theories of ordinary and partial differential equations. In particular, they arise perhaps most naturally in the study of several of the well-known second-order linear partial differential equations (the Laplace equation, the wave equation and the heat equation) on a cylindrical domain. For more information, see [22], [73], [145].

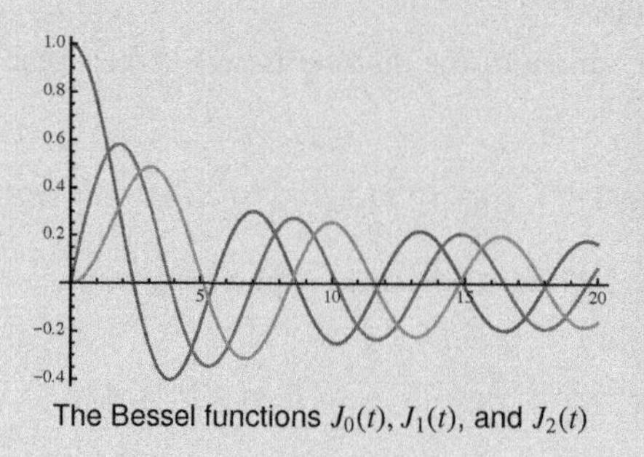

The Bessel functions $J_0(t)$, $J_1(t)$, and $J_2(t)$

Friedrich Bessel

Theorem 2.19 *For each $\theta > 0$, we have the operator relation*

$$\mathbf{K}_\theta = \mathbf{L}_\theta(\mathbf{I} + \mathbf{L}_\theta)^{-1}. \tag{2.42}$$

Consequently, the point process $\mathrm{Fr}(\lambda^{\langle\theta\rangle})$ *is determinantal with correlation kernel* $\mathbf{K}_\theta$ *and its correlation function is given by*

$$\rho_{\mathrm{Fr}(\lambda^{\langle\theta\rangle})}(A) = \det_{x,y\in A}\Big(\mathbf{K}_\theta(x,y)\Big), \qquad (A \subset \mathbb{Z}',\ |A| < \infty). \tag{2.43}$$

The proof of Theorem 2.19 is a somewhat involved computation that reduces to well-known identities satisfied by the Bessel functions. We recall the relevant facts we need from the theory of the Bessel functions in the following lemma.

Lemma 2.20 *The Bessel functions $J_\alpha(z)$ satisfy the following relations:*

$$\frac{\partial}{\partial z}J_\alpha(2z) = -2J_{\alpha+1}(2z) + \frac{\alpha}{z}J_\alpha(2z) = 2J_{\alpha-1}(2z) - \frac{\alpha}{z}J_\alpha(2z), \tag{2.44}$$

$$J_{\alpha+1}(2z) = \frac{\alpha}{z}J_\alpha(2z) - J_{\alpha-1}(2z), \tag{2.45}$$

$$J_\alpha(2z) = (1+o(1))\frac{z^\alpha}{\Gamma(\alpha+1)} \quad \textit{as } \alpha \to \infty \textit{ with } z > 0 \textit{ fixed}, \tag{2.46}$$

$$\dot{J}_\alpha(2z) = O\left(\frac{z^\alpha}{\Gamma(\alpha+1)}\right) \textit{as } \alpha \to \infty \textit{ with } z > 0 \textit{ fixed}, \tag{2.47}$$

$$\frac{\sin(\pi\alpha)}{\pi z} = J_\alpha(2z)J_{1-\alpha}(2z) + J_{-\alpha}(2z)J_{\alpha-1}(2z), \tag{2.48}$$

$$J_\alpha(2z) = \frac{1}{\Gamma(\alpha)}\sum_{m=0}^{\infty}\frac{1}{m+\alpha}\frac{z^{m+\alpha}}{m!}J_m(2z), \qquad (z \neq 0), \tag{2.49}$$

$$J_{\alpha-1}(2z) = \frac{z^{\alpha-1}}{\Gamma(\alpha)} - \frac{1}{\Gamma(\alpha)}\sum_{m=0}^{\infty}\frac{1}{m+\alpha}\frac{z^{m+\alpha}}{m!}J_{m+1}(2z), \qquad (z \neq 0), \tag{2.50}$$

Proof The equality of the left-hand side of (2.44) to the other two expressions in that equation follows easily by equating coefficients of z in the power series expansions of the respective expressions. Equating these two expressions for $\frac{\partial}{\partial z}J_\alpha(2z)$ yields the recurrence relation (2.45). Relation (2.46) is simply the claim that the first summand in the infinite series defining $J_\alpha(2z)$ is asymptotically the dominant term when z is a fixed positive number and $\alpha \to \infty$; this follows from the fact that

$$\Gamma(m+\alpha+1) = (\alpha+1)(\alpha+2)\dots(\alpha+m)\Gamma(\alpha+1) \geq \alpha^m\Gamma(\alpha+1),$$

so the contribution to $J_\alpha(2z)$ from the summands corresponding to values $m \geq 1$ in (2.35) is bounded by the initial term corresponding to $m = 0$, multiplied by $\sum_{m=1}^\infty (z^2/\alpha)^m = \frac{z^2}{\alpha - z^2}$. Similarly, to investigate the asymptotics of $\dot{J}_\alpha(2z)$, differentiate the series in (2.35) termwise with respect to α (which is easily justified due to the rapid convergence of both the original series and of the series of derivatives), to obtain that

$$\begin{aligned}\dot{J}_\alpha(2z) &= \log z J_\alpha(2z) - z^\alpha \sum_{m=0}^{\infty} \frac{\Gamma'(\alpha+m+1)z^{2m}}{m!\Gamma(\alpha+m+1)^2} \\ &= \log z J_\alpha(2z) - z^\alpha \sum_{m=0}^{\infty} \frac{z^{2m}}{m!\Gamma(\alpha+m+1)}\psi(\alpha+m+1),\end{aligned}$$

where $\psi(x) = \Gamma'(x)/\Gamma(x)$ is the logarithmic derivative of the gamma function. Now recall that $\psi(x) = \log x + O(1)$ as $x \to \infty$ (Exercise 2.7), to conclude easily (using similar reasoning to that used above) that the term $\log z J_\alpha(2z)$ and the first term corresponding to $m = 0$ in the infinite sum, both growing as $O(z^\alpha/\Gamma(\alpha+1))$ as a function of α, are the dominant contributions in the limit as $\alpha \to \infty$, which proves (2.47).

Next, identity (2.48) can be seen using (2.44) to be equivalent to

$$-\frac{\sin(\pi\alpha)}{\pi z} = J_\alpha(2z)J'_{-\alpha}(2z) - J'_{-\alpha}(2z)J_{\alpha-1}(2z) \tag{2.51}$$

A proof of this is outlined in Exercise 2.8.

Next, to verify (2.49), rewrite the right-hand side by expanding each term $J_m(2z)$ into a power series, to get

$$\begin{aligned}&\frac{z^\alpha}{\Gamma(\alpha)} \sum_{m=0}^{\infty}\sum_{k=0}^{\infty} \frac{(-1)^k}{(m+\alpha)k!m!(m+k)!} z^{2(m+k)} \\ &\qquad = \frac{z^\alpha}{\Gamma(\alpha)} \sum_{n=0}^{\infty} \left(\sum_{m=0}^{n} \frac{(-1)^{n+m}}{(m+\alpha)m!(n-m)!n!} \right) z^{2n}.\end{aligned}$$

Comparing this to the power series expansion of the left-hand side of (2.49), we see that it is enough to prove for each $n \geq 0$ the finite summation identity

$$\sum_{m=0}^{n} \frac{(-1)^m}{m+\alpha}\binom{n}{m} = \frac{n!\,\Gamma(\alpha)}{\Gamma(\alpha+n+1)} = \frac{n!}{\alpha(\alpha+1)\ldots(\alpha+n)}. \tag{2.52}$$

This is not difficult to prove and is left to the reader (Exercise 2.9).

Finally, to prove (2.50), first prove the much simpler identity

$$\sum_{n=0}^{\infty} \frac{z^n}{n!} J_n(2z) = 1, \tag{2.53}$$

(Exercise 2.10). Next, multiply the right-hand side of (2.50) by $\Gamma(\alpha)$ and rewrite it as

$$
\begin{aligned}
z^{\alpha-1} - &\sum_{m=0} \frac{1}{m+\alpha} \frac{z^{m+\alpha}}{m!} J_{m+1}(2z) \\
&= z^{\alpha-1} - z^{\alpha-1} \sum_{n=1}^{\infty} \frac{n}{n+(\alpha-1)} \frac{z^n}{n!} J_n(2z) \\
&= z^{\alpha-1} - z^{\alpha-1} \sum_{n=1}^{\infty} \left(1 - (\alpha-1)\frac{1}{n+(\alpha-1)}\right) \frac{z^n}{n!} J_n(2z) \\
&= z^{\alpha-1} \left(1 - \sum_{n=0}^{\infty} \frac{z^n}{n!} J_n(2z) + J_0(2z)\right) \\
&\quad + (\alpha-1) z^{\alpha-1} \left(\sum_{n=0}^{\infty} \frac{1}{n+(\alpha-1)} \frac{z^n}{n!} J_n(2z) - \frac{1}{\alpha-1} J_0(2z)\right).
\end{aligned}
$$

Applying (2.49) and (2.53), we get (2.50). □

Proof of Lemma 2.18 As with the case of $\mathbf{L}_\theta$, the claim that $\mathbf{K}_\theta$ has exponential decay refers to an ordering of the elements of $\mathbb{Z}'$ according to increasing distance from 0, so we need to show that

$$|\mathbf{K}_\theta(x, y)| \le C e^{-c(|x|+|y|)}$$

for all $x, y \in \mathbb{Z}'$ and constants $C, c > 0$ (that may depend on θ). This is easy to check using the formulas for $\mathbf{L}_\theta$ and $\mathbf{K}_\theta$, the asymptotic relations (2.46), (2.47) and the rapid growth of the gamma function. □

Proof of Theorem 2.19 Throughout the proof, we denote $z = \sqrt{\theta}$ and for simplicity omit the subscript θ from various quantities, for example, writing $\mathbf{K}$ instead of $\mathbf{K}_\theta$ and $\mathbf{L}$ in place of $\mathbf{L}_\theta$. Our goal is to prove that $\mathbf{K} = \mathbf{L}(\mathbf{I} + \mathbf{L})^{-1}$, or, equivalently that

$$\mathbf{K} + \mathbf{KL} - \mathbf{L} = \mathbf{0} = \mathbf{K} + \mathbf{LK} - \mathbf{L}. \tag{2.54}$$

It is not difficult to check that these two equations are equivalent (see Exercise 2.14; note that for infinite-dimensional operators the relation $\mathbf{ST} = \mathbf{I}$ does not generally imply that $\mathbf{TS} = \mathbf{I}$, but here one can use the symmetries of $\mathbf{L}$ and $\mathbf{K}$), so we prove the first one. When $z = 0$ (i.e., $\theta = 0$), $\mathbf{K} = \mathbf{L} = \mathbf{0}$ (the zero kernel), so this holds. Therefore, it is enough to verify that

$$(\mathbf{K} + \mathbf{KL} - \mathbf{L})' = \mathbf{K}' + \mathbf{K}'\mathbf{L} + \mathbf{KL}' - \mathbf{L}' = \mathbf{0}, \tag{2.55}$$

where a prime denotes differentiation with respect to z. (The concept of differentiating an operator may be slightly confusing, so some clarification is in order. Here, we think of **K** and **L** as matrices, and differentiation acts on each entry; it is easy to convince oneself that the rule for differentiating a product of matrices is $(\mathbf{KL})' = \mathbf{KL}' + \mathbf{K}'\mathbf{L}$, in analogy with the usual rule for differentiating a product of functions.)

We therefore have to compute each of the quantities appearing in (2.55). This is a straightforward but somewhat tedious computation, relying heavily on Lemma 2.20. We include it for completeness, but the impatient reader should not feel too bad about skipping ahead. For **L**, we have immediately that

$$\mathbf{L}'(x,y) = \begin{cases} 0 & \text{if } xy > 0, \\ \operatorname{sgn}(x) \dfrac{z^{|x|+|y|-1}}{\left(|x|-\frac{1}{2}\right)!\left(|y|-\frac{1}{2}\right)!} & \text{if } xy < 0. \end{cases} \tag{2.56}$$

Next, for **K**, consider first the case when $xy > 0$. Making use of the relations (2.44) (and recalling also that J_α is shorthand for $J_\alpha(2\sqrt{\theta}) = J_\alpha(2z)$), we have that

$$\begin{aligned}
\mathbf{K}'(x,y) &= \frac{d}{dz}\left(z \frac{J_{|x|-\frac{1}{2}} J_{|y|+\frac{1}{2}}}{|x|-|y|} \right) = \frac{J_{|x|-\frac{1}{2}} J_{|y|+\frac{1}{2}}}{|x|-|y|} \\
&\quad + \frac{z}{|x|-|y|} \left[\left(-2J_{|x|+\frac{1}{2}} + \frac{|x|-\frac{1}{2}}{z} J_{|x|-\frac{1}{2}} \right) J_{|y|+\frac{1}{2}} \right. \\
&\quad + J_{|x|-\frac{1}{2}} \left(2J_{|y|-\frac{1}{2}} - \frac{|y|+\frac{1}{2}}{z} J_{|y|+\frac{1}{2}} \right) - \left(2J_{|x|-\frac{1}{2}} - \frac{|x|+\frac{1}{2}}{z} J_{|x|+\frac{1}{2}} \right) J_{|y|-\frac{1}{2}} \\
&\quad \left. - J_{|x|+\frac{1}{2}} \left(-2J_{|y|+\frac{1}{2}} + \frac{|y|-\frac{1}{2}}{z} J_{|y|-\frac{1}{2}} \right) \right] \\
&= J_{|x|-\frac{1}{2}} J_{|y|+\frac{1}{2}} + J_{|x|+\frac{1}{2}} J_{|y|-\frac{1}{2}}.
\end{aligned} \tag{2.57}$$

In the case when $xy < 0$, using a similar computation one may verify that

$$\mathbf{K}'(x,y) = \operatorname{sgn}(x) \left(J_{|x|-\frac{1}{2}} J_{|y|-\frac{1}{2}} - J_{|x|+\frac{1}{2}} J_{|y|+\frac{1}{2}} \right). \tag{2.58}$$

Next, using (2.57) and (2.58) we proceed to evaluate $\mathbf{K}'\mathbf{L}$, a matrix whose entries are given by

$$(\mathbf{K}'\mathbf{L})(x,y) = \sum_{t \in \mathbb{Z}'} \mathbf{K}'(x,t)\mathbf{L}(t,y).$$

Again we divide into cases according to whether x and y have the same sign. Assume that $xy > 0$, then

$$
\begin{aligned}
(\mathbf{K}'\mathbf{L})(x,y) &= \sum_{t\in\mathbb{Z}',\, tx>0} \frac{\operatorname{sgn}(x)}{t-y} \frac{z^{|t|+|y|}}{\left(|t|-\frac{1}{2}\right)!\left(|y|-\frac{1}{2}\right)} \left(J_{|x|-\frac{1}{2}} J_{|t|-\frac{1}{2}} - J_{|x|+\frac{1}{2}} J_{|t|+\frac{1}{2}}\right) \\
&= -\frac{z^{|y|+\frac{1}{2}}}{\left(|y|-\frac{1}{2}\right)!} \left(J_{|x|-\frac{1}{2}} \sum_{m=0}^{\infty} \frac{1}{m+|y|+\frac{1}{2}} \frac{z^m}{m!} J_m \right. \\
&\qquad \left. - J_{|x|+\frac{1}{2}} \sum_{m=0}^{\infty} \frac{1}{m+|y|+\frac{1}{2}} \frac{z^m}{m!} J_{m+1} \right).
\end{aligned}
$$

By (2.49) and (2.50), this is equal to

$$
(\mathbf{K}'\mathbf{L})(x,y) = \frac{z^{|y|-\frac{1}{2}}}{\left(|y|-\frac{1}{2}\right)!} J_{|x|+\frac{1}{2}} - J_{|x|-\frac{1}{2}} J_{|y|+\frac{1}{2}} - J_{|x|+\frac{1}{2}} J_{|y|-\frac{1}{2}}. \tag{2.59}
$$

Assume now that $xy < 0$; in this case we can write in a similar fashion that

$$
\begin{aligned}
(\mathbf{K}'\mathbf{L})(x,y) &= \sum_{t\in\mathbb{Z}',\, tx>0} \frac{1}{t-y} \frac{z^{|t|+|y|}}{\left(|t|-\frac{1}{2}\right)!\left(|y|-\frac{1}{2}\right)!} \left(J_{|x|-\frac{1}{2}} J_{|t|+\frac{1}{2}} + J_{|x|+\frac{1}{2}} J_{|t|-\frac{1}{2}}\right) \\
&= \frac{z^{|y|+\frac{1}{2}}}{\left(|y|-\frac{1}{2}\right)!} \left(J_{|x|-\frac{1}{2}} \sum_{m=0}^{\infty} \frac{\operatorname{sgn}(x)}{m+|y|+\frac{1}{2}} \frac{z^m}{m!} J_{m+1} \right. \\
&\qquad \left. + J_{|x|+\frac{1}{2}} \sum_{m=0}^{\infty} \frac{\operatorname{sgn}(x)}{m+|y|+\frac{1}{2}} \frac{z^m}{m!} J_m \right) \\
&= \operatorname{sgn}(x) \left(J_{|x|+\frac{1}{2}} J_{|y|+\frac{1}{2}} - J_{|x|-\frac{1}{2}} J_{|y|-\frac{1}{2}} + \frac{z^{|y|-\frac{1}{2}}}{\left(|y|-\frac{1}{2}\right)!} J_{|x|-\frac{1}{2}} \right).
\end{aligned} \tag{2.60}
$$

Finally, the last quantity to compute in (2.55) is $\mathbf{K}\mathbf{L}'$. Again, if $xy > 0$, we have

$$
\begin{aligned}
(\mathbf{KL}')(x,y) &= -\sum_{t:tx<0} \frac{\operatorname{sgn}(x)}{x-t} \frac{z^{|t|+|y|}}{\left(|t|-\frac{1}{2}\right)!\left(|y|-\frac{1}{2}\right)!}\left(J_{|x|-\frac{1}{2}}J_{|t|-\frac{1}{2}} + J_{|x|+\frac{1}{2}}J_{|t|+\frac{1}{2}}\right) \\
&= -\operatorname{sgn}(x)\frac{z^{|y|+\frac{1}{2}}}{\left(|y|-\frac{1}{2}\right)!}\left(J_{|x|-\frac{1}{2}}\sum_{m=0}^{\infty}\frac{\operatorname{sgn}(x)}{m+|x|+\frac{1}{2}}\frac{z^m}{m!}J_m \right. \\
&\qquad\qquad \left. +J_{|x|+\frac{1}{2}}\sum_{m=0}^{\infty}\frac{\operatorname{sgn}(x)}{m+|x|+\frac{1}{2}}\frac{z^m}{m!}J_{m+1}\right) \\
&= -\frac{z^{|y|+\frac{1}{2}}}{\left(|y|-\frac{1}{2}\right)!}\left[J_{|x|-\frac{1}{2}}\frac{\left(|x|-\frac{1}{2}\right)!}{z^{|x|+\frac{1}{2}}}J_{|x|+\frac{1}{2}} \right. \\
&\qquad\qquad \left. +J_{|x|+\frac{1}{2}}\left(\frac{1}{z}-\frac{\left(|x|-\frac{1}{2}\right)!}{z^{|x|+\frac{1}{2}}}J_{|x|-\frac{1}{2}}\right)\right] \\
&= -\frac{z^{|y|+\frac{1}{2}}}{\left(|y|-\frac{1}{2}\right)!}J_{|x|+\frac{1}{2}}. \qquad (2.61)
\end{aligned}
$$

In the other case when $xy < 0$, we have

$$
(\mathbf{KL}')(x,y) = \sum_{tx>0} \frac{\operatorname{sgn}(x)}{|x|-|t|} \frac{z^{|t|+|y|}}{\left(|t|-\frac{1}{2}\right)!\left(|y|-\frac{1}{2}\right)!}\left(J_{|x|-\frac{1}{2}}J_{|t|+\frac{1}{2}} + J_{|x|+\frac{1}{2}}J_{|t|-\frac{1}{2}}\right),
$$

with the understanding that for the term corresponding to $t = x$, the singularity that appears is resolved by replacing $\frac{1}{|x|-|t|}(J_{|x|-\frac{1}{2}}J_{|t|+\frac{1}{2}} + J_{|x|+\frac{1}{2}}J_{|t|-\frac{1}{2}})$ with the limit of this expression as $t \to x$. This is justified by the fact that $\mathbf{K}(x,x) = \lim_{t\to x}\mathbf{K}(x,t)$, a consequence of the continuity of $\mathbf{J}_\theta(\cdot,\cdot)$ that was noted earlier. In other words, we can write $(\mathbf{KL}')(x,y)$ as a limit of perturbed values of the form

$$
\begin{aligned}
(\mathbf{KL}')(x,y) &= -\operatorname{sgn}(x)\frac{z^{|y|+\frac{1}{2}}}{\left(|y|-\frac{1}{2}\right)!} \\
&\quad\times \lim_{\epsilon\to 0}\left(J_{|x|-\frac{1}{2}}\sum_{k=0}^{\infty}\frac{1}{k-\left(|x|-\frac{1}{2}\right)+\epsilon}\frac{z^k}{k!}J_{k+1}\right. \\
&\qquad\qquad \left. -J_{|x|+\frac{1}{2}}\sum_{k=0}^{\infty}\frac{1}{k-\left(|x|-\frac{1}{2}\right)+\epsilon}\frac{z^k}{k!}J_k\right).
\end{aligned}
$$

This can now be evaluated using (2.49) and (2.50), to give

$$
\begin{aligned}
(\mathbf{KL}')(x,y) \\
&= -\operatorname{sgn}(x)\frac{z^{|y|+\frac{1}{2}}}{\left(|y|-\frac{1}{2}\right)!}\lim_{\epsilon\to 0}\Bigg[J_{|x|-\frac{1}{2}}\left(\frac{1}{z}-\Gamma\left(\epsilon-|x|+\tfrac{1}{2}\right)z^{|x|-\frac{1}{2}-\epsilon}J_{-|x|-\frac{1}{2}}\right) \\
&\qquad\qquad -J_{|x|+\frac{1}{2}}\Gamma\left(\epsilon-|x|+\tfrac{1}{2}\right)z^{|x|-\frac{1}{2}-\epsilon}J_{-|x|+\frac{1}{2}}\Bigg] \\
&= -\operatorname{sgn}(x)\frac{z^{|y|+\frac{1}{2}}}{\left(|y|-\frac{1}{2}\right)!}\left[\frac{1}{z}J_{|x|-\frac{1}{2}}-\lim_{\alpha\to|x|-\frac{1}{2}}\Big(\Gamma(-\alpha)\left(J_\alpha J_{-\alpha-1}+J_{-\alpha}J_{\alpha+1}\right)\Big)\right].
\end{aligned}
$$

To evaluate the limit, note that, by (2.48), we have

$$
\begin{aligned}
\Gamma(-\alpha)&\left(J_\alpha J_{-\alpha-1}+J_{-\alpha}J_{\alpha+1}\right) \\
&= \Gamma(-\alpha)\frac{1}{\pi z}\sin(-\pi\alpha) \\
&= \frac{\Gamma(-\alpha)}{z\Gamma(-\alpha)\Gamma(\alpha+1)} = \frac{1}{\Gamma(\alpha+1)} \to \frac{1}{\left(|x|-\frac{1}{2}\right)!z} \quad \text{as } \alpha\to|x|-\tfrac{1}{2}.
\end{aligned}
$$

(Actually, the full power of (2.48) isn't really used here, but rather the fact that the right-hand side of (2.48) is equal to $(\pi z\Gamma(\alpha)\Gamma(1-\alpha))^{-1}$, which can be proved without reference to the identity $\Gamma(\alpha)\Gamma(1-\alpha)=\pi\sin(\pi\alpha)^{-1}$; see Exercise 2.8.) So we get finally that

$$
(\mathbf{KL}')(x,y) = -\operatorname{sgn}(x)\frac{z^{|y|-\frac{1}{2}}}{\left(|y|-\frac{1}{2}\right)!}\left(J_{|x|-\frac{1}{2}}-\frac{z^{|x|-\frac{1}{2}}}{\left(|x|-\frac{1}{2}\right)!}\right). \tag{2.62}
$$

To conclude the proof, it is now straightforward to combine the results of our computation above, namely, (2.56), (2.59), (2.60), (2.61), and (2.62), and verify that (2.55) holds. □

It is worth taking a short pause at this point to look back at the results we proved and appreciate their significance. Theorem 2.19, and in particular (2.43), is the main exact (as opposed to asymptotic) result that, when combined with the tools that we developed for working with determinantal point processes, enables us to launch an attack on Theorems 2.2 and 2.3. The next part of the proof, which we undertake in the next few sections, is a detailed asymptotic analysis of the quantities appearing in these exact results.

We conclude this section by proving an infinite sum representation of the kernel $\mathbf{J}_\theta$ that will be useful later on. Although the "original" discrete Bessel kernel $\mathbf{J}_\theta$ defined in (2.38) played no role until now except in being used to define its (somewhat less elegant) variant $\mathbf{K}_\theta$, it will be more convenient to work with for the purpose of the asymptotic analysis. The kernel $\mathbf{J}_\theta$ also has a more direct combinatorial significance, which is not needed for our purposes but is discussed in Exercise 2.13.

Proposition 2.21 *For $s, t \in \mathbb{R}$ we have*

$$\mathbf{J}_\theta(s,t) = \sum_{m=1}^{\infty} J_{s+m} J_{t+m}. \tag{2.63}$$

Proof It is easy to check (with the help of (2.47)) that the right-hand side of (2.63) is, like the left-hand side, continuous on the diagonal $s = t$; so it is enough to prove the identity under the assumption $s \neq t$. Using the recurrence (2.45), we have

$$\mathbf{J}_\theta(s+1,t+1) - \mathbf{J}_\theta(s,t) \tag{2.64}$$

$$\begin{aligned} &= \frac{z}{s-t}\left(J_{s+1}J_{t+2} - J_{s+2}J_{t+1}\right) - \frac{z}{s-t}\left(J_s J_{t+1} - J_{s+1}J_t\right) \\ &= \frac{z}{s-t}\Bigg[J_{s+1}\left(\frac{t+1}{z}J_{t+1} - J_t\right) - \left(\frac{s+1}{z}J_{s+1} - J_s\right)J_{t+1} \\ &\qquad\qquad - J_s J_{t+1} + J_{s+1}J_t\Bigg] = -J_{s+1}J_{t+1}. \end{aligned} \tag{2.65}$$

Consequently, we have that

$$\begin{aligned} \mathbf{J}_\theta(s,t) &= J_{s+1}J_{t+1} + \mathbf{J}_\theta(s+1,t+1) \\ &= J_{s+1}J_{t+1} + J_{s+2}J_{t+2} + \mathbf{J}_\theta(s+2,t+2) \\ &= \ldots = \sum_{m=1}^{N} J_{s+m}J_{t+m} + \mathbf{J}_\theta(s+N,t+N) \quad (\text{for any } N \geq 1). \end{aligned} \tag{2.66}$$

The asymptotics of the term $\mathbf{J}_\theta(s+N, t+N)$ as $N \to \infty$ can be determined using (2.46), which gives that

$$\begin{aligned}
\mathbf{J}_\theta(s+N, t+N) &= \frac{z}{s-t}\left(J_{s+N}J_{t+N+1} - J_{s+N+1}J_{t+N}\right) \\
&= \frac{z^{s+t+2N+2}}{s-t}\left(\frac{1+o(1)}{(s+N)!(t+N+1)!} - \frac{1+o(1)}{(s+N+1)!(t+N)!}\right) \\
&= (1+o(1))\frac{z^{s+t+2N+2}}{(s+N+1)!(t+N+1)!}.
\end{aligned}$$

This last quantity tends to 0 as $N \to \infty$, so we get (2.63). □

2.6 Asymptotics for the Airy function and kernel

In what follows, we will need some results on the asymptotic behavior of the Airy function $\mathrm{Ai}(x)$, its derivative $\mathrm{Ai}'(x)$, and the Airy kernel $\mathbf{A}(x, y)$ for large positive arguments. Recall that we defined the Airy function in terms of the integral representation (2.1); this representation involves a conditionally convergent integral that is somewhat difficult to work with, so we start by proving another more convenient representation.

Lemma 2.22 *The integral on the right-hand side of* (2.1) *converges as an improper integral for any* $x \in \mathbb{R}$. *The function* $\mathrm{Ai}(x)$ *it defines has the equivalent representation*

$$\mathrm{Ai}(x) = \frac{1}{2\pi}\int_{-\infty}^{\infty} \exp\left(\frac{1}{3}(a-it)^3 + x(-a+it)\right) dt \tag{2.67}$$

where $i = \sqrt{-1}$ *and* $a > 0$ *is an arbitrary positive number. Furthermore,* $\mathrm{Ai}(x)$ *is a smooth function and for any* $j \geq 1$ *its* j*th derivative* $\mathrm{Ai}^{(j)}(x)$ *is given by*

$$\mathrm{Ai}^{(j)}(x) = \frac{1}{2\pi}\int_{-\infty}^{\infty} (-a+it)^j \exp\left(\frac{1}{3}(a-it)^3 + x(-a+it)\right) dt. \tag{2.68}$$

Proof Considering $x \in \mathbb{R}$ as a fixed number, define an analytic function

$f(z) = \exp\left(xz - \frac{1}{3}z^3\right)$. For any $M > 0$ we have that

$$\begin{aligned}
&\int_0^M \cos\left(\tfrac{1}{3}t^3 + xt\right) dt \\
&= \frac{1}{2}\int_0^M \exp\left(-\tfrac{1}{3}(it)^3 + x(it)\right) dt + \frac{1}{2}\int_0^M \exp\left(-\tfrac{1}{3}(-it)^3 + x(-it)\right) dt \\
&= \frac{1}{2i}\int_{[-iM,iM]} f(z)\, dz,
\end{aligned}$$

where the last integral is a complex contour integral and where $[\xi,\zeta]$ denotes the line segment connecting two complex numbers ξ and ζ. Thus, the right-hand side of (2.1) can be rewritten as

$$\begin{aligned}
&\frac{1}{\pi}\int_0^\infty \cos\left(\tfrac{1}{3}t^3 + xt\right) dt \\
&\qquad = \frac{1}{2\pi i}\int_{i(-\infty)}^{i\infty} f(z)\, dz := \lim_{M\to\infty} \frac{1}{2\pi i}\int_{[-iM,iM]} f(z)\, dz, \qquad (2.69)
\end{aligned}$$

provided that the limit exists. It is also immediate to see that the right-hand side of (2.67) can be similarly written as the contour integral

$$\begin{aligned}
&\frac{1}{2\pi}\int_{-\infty}^\infty \exp\left(\frac{1}{3}(a - it)^3 + x(-a + it)\right) dt \\
&\qquad = \frac{1}{2\pi i}\int_{-a-i\infty}^{-a+i\infty} f(z)\, dz := \lim_{M\to\infty} \frac{1}{2\pi i}\int_{[-a-iM,-a+iM]} f(z)\, dz, \qquad (2.70)
\end{aligned}$$

again provided that the limit exists. But note that if $z = s + it$ then

$$|f(z)| = \exp \operatorname{Re}\left(xz - \tfrac{1}{3}z^3\right) = \exp\left(xs - \tfrac{1}{3}s^3 + st^2\right), \qquad (2.71)$$

which in the case $s = -a < 0$ immediately implies that the integral in (2.70) converges absolutely. Furthermore, by applying Cauchy's theorem to the contour integral of $f(z)$ over the rectangular contour with corners

$-a \pm iM, \pm iM$, we see that

$$\left| \int_{[-iM,iM]} f(z)\,dz - \int_{[-a-iM,-a+iM]} f(z)\,dz \right|$$
$$\leq \int_{-a}^{0} |f(s-iM)|\,ds + \int_{-a}^{0} |f(s+iM)|\,ds$$
$$= 2\int_{-a}^{0} \exp\left[s\left(x - \tfrac{1}{3}s^2 + M^2\right)\right]\,ds$$
$$\leq 2\int_{-\infty}^{0} \exp\left[s\left(x + M^2\right)\right]\,ds = \frac{1}{M^2+x} \xrightarrow[M\to\infty]{} 0.$$

This implies that the limit in (2.69) exists and is equal to (2.70). Finally, since it can be seen using (2.71) that for any $j \geq 1$ the partial derivative

$$\frac{\partial^j}{\partial x^j} \exp\left(xz - \tfrac{1}{3}z^3\right) = z^j \exp\left(xz - \tfrac{1}{3}z^3\right)$$

is absolutely integrable on the contour $\{\operatorname{Im} z = -a\}$, the dominated convergence theorem justifies differentiation under the integral sign in (2.67), proving (2.68). □

Lemma 2.23 (Asymptotics of the Airy function and its derivative) *The Airy function* $\operatorname{Ai}(x)$ *and its derivative* $\operatorname{Ai}'(x)$ *have the following asymptotics as* $x \to \infty$*:*

$$\operatorname{Ai}(x) = (1+o(1))\frac{1}{2\sqrt{\pi}} x^{-1/4} e^{-\frac{2}{3}x^{3/2}}, \tag{2.72}$$

$$\operatorname{Ai}'(x) = (1+o(1))\frac{1}{2\sqrt{\pi}} x^{1/4} e^{-\frac{2}{3}x^{3/2}}. \tag{2.73}$$

Proof Since we are interested in large values of x, assume $x > 0$. Start by rewriting the integral (2.67) (interpreted as a contour integral as in (2.69)) in a way that highlights the way the integrand scales as x grows; by making the substitution $z = \sqrt{x}\,w$ we see that

$$\begin{aligned}
\operatorname{Ai}(x) &= \frac{1}{2\pi i}\int_{-a-i\infty}^{-a+i\infty} \exp\left(xz - \tfrac{1}{3}z^3\right)\,dz \\
&= \frac{\sqrt{x}}{2\pi i}\int_{-ax^{-1/2}-i\infty}^{-ax^{-1/2}+i\infty} \exp\left(x^{3/2}w - \tfrac{1}{3}x^{3/2}w^3\right)\,dw \\
&= \frac{\sqrt{x}}{2\pi i}\int_{-1-i\infty}^{-1+i\infty} \exp\left(x^{3/2}g(w)\right)\,dw,
\end{aligned} \tag{2.74}$$

where in the last step we define $g(w) = w - \frac{1}{3}w^3$ and choose the value of a

(which up to now was allowed to be an arbitrary positive number) to be $a = \sqrt{x}$. The reason this choice makes sense is that $g(w)$ has a critical point at $w = -1$, and choosing the contour of integration so as to pass through this point results in an integral that can be easily estimated for large x; indeed, $g(w)$ can be expanded in a (finite) power series around the critical point as

$$g(w) = -\tfrac{2}{3} + (w+1)^2 - \tfrac{1}{3}(w+1)^3.$$

Substituting this into (2.74) results in the improved (for asymptotic purposes) representation

$$\begin{aligned}\operatorname{Ai}(x) &= \frac{\sqrt{x}}{2\pi} \exp\left(-\tfrac{2}{3}x^{3/2}\right) \int_{-\infty}^{\infty} \exp\left(x^{3/2}((it)^2 - \tfrac{1}{3}(it)^3)\right) dt \\ &= \frac{\sqrt{x}}{2\pi} \exp\left(-\tfrac{2}{3}x^{3/2}\right) x^{-3/4} \int_{-\infty}^{\infty} \exp\left(-u^2 + \tfrac{1}{3}ix^{-3/4}u^3\right) du.\end{aligned}$$

By the dominated convergence theorem, the last integral converges as $x \to \infty$ to $\int_{-\infty}^{\infty} e^{-u^2}\,du = \sqrt{\pi}$, so we get that the Airy function behaves asymptotically as

$$(1 + o(1)) \frac{1}{2\sqrt{\pi}} x^{-1/4},$$

as claimed in (2.72). The proof of (2.73) is similar and is left to the reader (Exercise 2.16). □

Lemma 2.24 *For some constant $C > 0$ the bound*

$$\mathbf{A}(x, y) \le C \exp^{-(x^{3/2} + y^{3/2})}. \tag{2.75}$$

holds for all $x, y > 0$.

Proof Since $\mathbf{A}(\cdot,\cdot)$ is continuous, it is enough to verify (2.75) for $x \neq y$. If $|x - y| > 1$ then this follows immediately from the definition of $\mathbf{A}(x, y)$ together with the relations (2.72) and (2.73). If $|x - y| \le 1$, use the mean-value theorem to write

$$\begin{aligned}\mathbf{A}(x, y) &= \operatorname{Ai}(x)\left(\frac{\operatorname{Ai}'(y) - \operatorname{Ai}'(x)}{x - y}\right) - \operatorname{Ai}'(x)\left(\frac{\operatorname{Ai}(y) - \operatorname{Ai}(x)}{x - y}\right) \\ &= \operatorname{Ai}(x)\operatorname{Ai}''(\tilde{x}) - \operatorname{Ai}'(x)\operatorname{Ai}'(\bar{x}),\end{aligned}$$

where $\tilde{x} = \alpha x + (1 - \alpha)y$, $\bar{x} = \beta x + (1 - \beta)y$ for some $0 \le \alpha, \beta \le 1$. Because of the Airy differential equation (2.8), this can be written as

$$\tilde{x}\,\mathrm{Ai}(x)\,\mathrm{Ai}(\tilde{x}) - \mathrm{Ai}'(x)\,\mathrm{Ai}'(\tilde{x}).$$

Now again use (2.72) and (2.73) to conclude that (2.75) holds for a suitable constant C. □

As a first application of (2.75), we can now justify an important claim we made at the beginning of the chapter.

Lemma 2.25 *The series of integrals defining $F_2(t)$ in* (2.3) *converges absolutely.*

Proof Note that for any fixed $t \in \mathbb{R}$ the bound (2.75) actually holds, possibly with a larger constant C depending on t, for all $x, y \geq t$. This implies that for $x_1, \ldots, x_n \geq t$ we have (using also (2.30)) that

$$\begin{aligned} \left| \det_{i,j=1}^{n} \left(\mathbf{A}(x_i, x_j) \right) \right| &= \exp\left(-2 \sum_{j=1}^{n} x_j^{3/2} \right) \left| \det_{i,j=1}^{n} \left(e^{x_i^{3/2} + x_j^{3/2}} \mathbf{A}(x_i, x_j) \right) \right| \\ &\leq n^{n/2} C^n \exp\left(-2 \sum_{j=1}^{n} x_j^{3/2} \right), \end{aligned} \tag{2.76}$$

and therefore that, for some $C_1 > 0$ (again depending on t), we have

$$\left| \int_t^\infty \ldots \int_t^\infty \det_{i,j=1}^{n} \left(\mathbf{A}(x_i, x_j) \right) dx_1 \ldots dx_n \right| \leq n^{n/2} C_1^n$$

for all $n \geq 1$. Since $n! \geq (n/e)^n$, that proves absolute convergence of the series on the right-hand side of (2.3). □

2.7 Asymptotics for $\mathbf{J}_\theta$

The goal of this section is to prove the following asymptotic result, which will be another key ingredient in the proof of Theorem 2.3.

Theorem 2.26 (Asymptotics for the discrete Bessel kernel)

(a) *Let $z = \sqrt{\theta}$ as before. For $x, y \in \mathbb{R}$ we have*

$$z^{1/3} \mathbf{J}_\theta(2z + xz^{1/3}, 2z + yz^{1/3}) \to \mathbf{A}(x, y) \qquad \text{as } z \to \infty, \tag{2.77}$$

where the converegence is uniform as x, y range over a compact subset of $\mathbb{R}$.

(b) *There exist constants $Z, C, c > 0$ such that for all $x, y > 0$ and $z > Z$ we have*

$$z^{1/3}\left|\mathbf{J}_\theta(2z + xz^{1/3}, 2z + yz^{1/3})\right| \le Ce^{-c(x+y)}. \tag{2.78}$$

For the proof, we rely on the following version of a classical result, sometimes referred to as **Nicholson's approximation** (although its rigorous formulation is due to G. N. Watson), relating the Airy function to the asymptotics of a Bessel function $J_\alpha(z)$ where the order α is close to the argument z. The proof of this asymptotic result requires an elaborate study of integral representations of the Bessel functions, so we do not include it here. See [19, Section 4] and [145] for the details.[8]

Theorem 2.27 (Nicholson's approximation for Bessel functions)

(a) *For $x \in \mathbb{R}$, we have*

$$z^{1/3} J_{2z+xz^{1/3}}(2z) \to \operatorname{Ai}(x) \qquad \text{as } z \to \infty,$$

uniformly as x ranges over a compact subset of $\mathbb{R}$.

(b) *There exist constants $Z, C, c > 0$ such that for all $z > Z$ and $x > 0$ we have*

$$z^{1/3}\left|J_{2z+xz^{1/3}}(2z)\right| \le Ce^{-cx}.$$

We also need the following integral representation of the Airy kernel,[9] analogous to the representation (2.63) for the kernel $\mathbf{J}_\theta$.

Lemma 2.28 *For any $x, y \in \mathbb{R}$ we have*

$$\mathbf{A}(x, y) = \int_0^\infty \operatorname{Ai}(x + t) \operatorname{Ai}(y + t)\, dt. \tag{2.79}$$

Proof By continuity arguments it is clearly enough to prove the claim when $x \neq y$. In this case, using the Airy differential equation $\operatorname{Ai}''(u) =$

$u\,\mathrm{Ai}(u)$ we have

$$\begin{aligned}
\frac{\partial}{\partial t}\mathbf{A}(x+t,y+t) &= \frac{1}{x-y}\frac{\partial}{\partial t}\Big(\mathrm{Ai}(x+t)\,\mathrm{Ai}'(y+t) - \mathrm{Ai}'(x+t)\,\mathrm{Ai}(y+t)\Big) \\
&= \frac{1}{x-y}\Big(\mathrm{Ai}(x+t)\,\mathrm{Ai}''(y+t) - \mathrm{Ai}''(x+t)\,\mathrm{Ai}(y+t)\Big) \\
&= \frac{1}{x-y}\Big((y+t)\,\mathrm{Ai}(x+t)\,\mathrm{Ai}(y+t) - (x+t)\,\mathrm{Ai}(x+t)\,\mathrm{Ai}(+t)\Big) \\
&= -\,\mathrm{Ai}(x+t)\,\mathrm{Ai}(y+t), \qquad (2.80)
\end{aligned}$$

(compare with (2.65)). Therefore we can write

$$\begin{aligned}
\mathbf{A}(x,y) &= \mathbf{A}(x+T,y+T) - \int_0^T \frac{\partial}{\partial t}\mathbf{A}(x+t,y+t)\,dt \\
&= \mathbf{A}(x+T,y+T) + \int_0^T \mathrm{Ai}(x+t)\,\mathrm{Ai}(y+t)\,dt,
\end{aligned}$$

and letting $T \to \infty$ (in analogy with the limit argument $N \to \infty$ applied to (2.66)) gives the claim, using (2.75). □

Proof of Theorem 2.26 Throughout the proof we denote $\tilde{x} = 2z + xz^{1/3}$ and $\tilde{y} = 2z + yz^{1/3}$. We first prove (2.78). Let Z, C, c be the constants in Theorem 2.27(b). Then we have

$$\begin{aligned}
z^{1/3}\Big|\mathbf{J}_\theta(\tilde{x},\tilde{y})\Big| &= \left|z^{-1/3}\sum_{m=1}^{\infty}\left(z^{1/3}J_{\tilde{x}+m}\right)\left(z^{1/3}J_{\tilde{y}+m}\right)\right| \\
&\le z^{-1/3}\sum_{m=1}^{\infty}\left(Ce^{-c(x+mz^{-1/3})}\right)\left(Ce^{-c(y+mz^{-1/3})}\right) \\
&= C^2 e^{-c(x+y)} z^{-1/3}\sum_{m=1}^{\infty} q^m, \qquad (2.81)
\end{aligned}$$

where $q = \exp(-2cz^{-1/3})$. Since $q = 1 - 2cz^{-1/3} + O(z^{-2/3})$, we have that $\sum_{m=1}^{\infty} q^m = q/(1-q) = (2c)^{-1}z^{1/3} + O(1)$, and hence (2.78) follows from (2.81) after relabelling C.

Next, we prove (2.77). Fix a large number $T > 0$ whose value will be specified later, and denote $\tilde{T} = \lfloor T z^{1/3} \rfloor$. We have

$$\begin{aligned}
&\left| z^{1/3} \mathbf{J}_\theta(\tilde{x}, \tilde{y}) - \mathbf{A}(x, y) \right| \\
&\qquad = \left| z^{-1/3} \sum_{m=1}^{\infty} \left(z^{1/3} J_{\tilde{x}+m} \right) \left(z^{1/3} J_{\tilde{y}+m} \right) - \int_0^\infty \operatorname{Ai}(x+t) \operatorname{Ai}(y+t)\, dt \right| \\
&\qquad \le E_1 + E_2 + E_3 + E_4, \qquad\qquad (2.82)
\end{aligned}$$

where E_1, E_2, E_3 and E_4 are quantities defined by

$$\begin{aligned}
E_1 &= \left| \int_T^\infty \operatorname{Ai}(x+t) \operatorname{Ai}(y+t)\, dt \right| = \left| \mathbf{A}(x+T, y+T) \right|, \\
E_2 &= z^{1/3} \left| \sum_{m=\tilde{T}+1}^{\infty} J_{\tilde{x}+m} J_{\tilde{y}+m} \right| = z^{1/3} \left| \mathbf{J}_\theta(\tilde{x}+\tilde{T}, \tilde{y}+\tilde{T}) \right|, \\
E_3 &= \left| z^{-1/3} \sum_{m=1}^{\tilde{T}} \operatorname{Ai}(x+mz^{-1/3}) \operatorname{Ai}(y+mz^{-1/3}) - \int_0^T \operatorname{Ai}(x+t) \operatorname{Ai}(y+T)\, dt \right|, \\
E_4 &= \left| z^{-1/3} \sum_{m=1}^{\tilde{T}} \left(z^{1/3} J_{\tilde{x}+m} \right) \left(z^{1/3} J_{\tilde{y}+m} \right) \right. \\
&\qquad \left. - z^{-1/3} \sum_{m=1}^{\tilde{T}} \operatorname{Ai}(x+mz^{-1/3}) \operatorname{Ai}(y+mz^{-1/3}) \right|.
\end{aligned}$$

Each of the E_i can now be bounded in a fairly straightforward manner. For E_1, by (2.75) we have

$$E_1 \le C \exp\left(-(x+T)^{3/2} - (y+T)^{3/2} \right),$$

which, under the assumption that $z \to \infty$ as x, y range over a compact set, can be made arbitrarily small by picking T to be a sufficiently large number. For E_2 we can use (2.78) to get the bound

$$E_2 \le C \exp\left[-c\,(x+y+2T) \right]$$

which has similar properties. Next, the sum in E_3 is a Riemann sum for the integral $\int_0^T \operatorname{Ai}(x+t) \operatorname{Ai}(y+t)\, dt$, so $E_3 \to 0$ as $z \to \infty$ with T fixed, and it is easy to see that the convergence is uniform as x, y range over a compact set. Finally, Theorem 2.27(i) implies that $E_4 \to 0$ as $z \to \infty$ with T fixed, again with the required uniformity.

Combining the preceding estimates we see that given any $\epsilon > 0$, by fixing T to be a large enough number we get a bound for the left-hand side of (2.82) of the form $\epsilon + E_3 + E_4$, where $E_3 + E_4 \to 0$ as $z \to \infty$ uniformly on compacts in x and y. This proves (2.77). □

2.8 The Tracy–Widom limit law for Poissonized Plancherel measure

In this section we prove the following version of Theorem 2.2 for the length of the first row of a random Young diagram chosen according to Poissonized Plancherel measure.

Theorem 2.29 *For each $\theta > 0$, let $\lambda^{\langle\theta\rangle}$ denote a random partition chosen according to Poissonized Plancherel measure with parameter θ, let $\lambda_1^{\langle\theta\rangle}$ denote the length of its first row. Then for any $t \in \mathbb{R}$ we have*

$$\mathbb{P}\left(\lambda_1^{\langle\theta\rangle} \le 2\sqrt{\theta} + t\theta^{1/6}\right) \to F_2(t) \ \text{ as } \theta \to \infty, \tag{2.83}$$

where F_2 is the Tracy–Widom distribution defined in (2.3). *That is, the family of random variables $\theta^{-1/6}(\lambda_1^{\langle\theta\rangle} - 2\sqrt{\theta})$ converges in distribution to F_2 as $\theta \to \infty$.*

Recall that we still do not know the fact, claimed in Theorem 2.1, that F_2 is a distribution function. What we prove here is the relation (2.83). The separate claim about convergence in distribution will follow once we prove Theorem 2.1 in Section 2.10, using unrelated methods.

Note that $\lambda_1^{\langle\theta\rangle}$ is equal in distribution to $L(\sigma_N)$, the maximal length of an increasing subsequence in a permutation σ_N that was chosen in a two-step experiment, where we first choose a Poisson random variable $N \sim \text{Poi}(\theta)$, and then draw a uniformly random permutation of order N. Thus, the statement of Theorem 2.29 comes very close to explaining Theorem 2.2, and indeed in the next section we will show how Theorem 2.2 can be deduced from it using a relatively simple "de-Poissonization" trick. Similarly, later we will formulate and prove an analogue of Theorem 2.3 for Poissonized Plancherel measure, and deduce Theorem 2.3 from it using de-Poissonization.

To start our attack on Theorem 2.29, observe that, by Theorem 2.19 and Proposition 2.16, the probability on the left-hand side of (2.83) can be written as a Fredholm determinant. More precisely, fixing $t \in \mathbb{R}$ and letting $(A_\theta)_{\theta>0}$ be the family of subintervals of $\mathbb{R}$ given by $A_\theta = (2\sqrt{\theta} + t\theta^{1/6}, \infty)$, we have that

$$\begin{aligned}
&\mathbb{P}\left(\lambda_1^{\langle\theta\rangle} \le 2\sqrt{n} + tn^{1/6}\right) = \mathbb{P}\left(\mathrm{Fr}(\lambda_1^{\langle\theta\rangle}) \subseteq \mathbb{Z}' \setminus A_\theta\right) \\
&= \det\left(\mathbf{I} - (\mathbf{K}_\theta)_{\mathbb{Z}'\cap A_\theta}\right) \\
&= 1 + \sum_{n=1}^{\infty} \frac{(-1)^n}{n!} \sum_{x_1 \in \mathbb{Z}'\cap A_\theta} \ldots \sum_{x_n \in \mathbb{Z}'\cap A_\theta} \det_{i,j=1}^{n} \left(\mathbf{K}_\theta(x_i, x_j)\right). \qquad (2.84)
\end{aligned}$$

We want to prove that this expression converges to $F_2(t)$, whose definition in (2.3) is finally starting to make sense. It is convenient to replace $\mathbf{K}_\theta(x, y)$ by its simpler variant $\mathbf{J}_\theta(x, y)$, noting that by (2.39), we have $\mathbf{K}_\theta(x, y) = \mathbf{J}_\theta(x - \frac{1}{2}, y - \frac{1}{2})$ for positive x, y (which is the range we are considering, since, for any fixed t, $A_\theta \subset (0, \infty)$ if θ is large enough). So the Fredholm determinant in (2.84) can be replaced by

$$\det\left(\mathbf{I} - (\mathbf{J}_\theta)_{\mathbb{Z}\cap A_\theta}\right) = 1 + \sum_{n=1}^{\infty} \frac{(-1)^n}{n!} \sum_{x_1 \in \mathbb{Z}\cap A_\theta} \ldots \sum_{x_n \in \mathbb{Z}\cap A_\theta} \det_{i,j=1}^{n} \left(\mathbf{J}_\theta(x_i, x_j)\right), \qquad (2.85)$$

with negligible asymptotic effect. Furthermore, by Theorem 2.26, the family of kernels $\mathbf{J}_\theta(\cdot, \cdot)$ converges to the Airy kernel $\mathbf{A}(\cdot, \cdot)$, after appropriate scaling, as $\theta \to \infty$. So, it seems intuitively plausible that the n-dimensional sums appearing in the Fredholm determinant in (2.85) should behave like Riemann sums that converge to the n-dimensional integrals in the definition of $F_2(t)$, and that the infinite series of such convergent Riemann-like sums should also converge to the corresponding infinite series of integrals. Indeed, we shall soon confirm this intuition rigorously by making careful use of the asymptotic bounds and estimates that were derived in the previous section.

The following continuity estimate for determinants will be useful.

Lemma 2.30 *For a square matrix $M = (m_{i,j})_{i,j=1}^n$ of real numbers, denote $\|M\|_\infty = \max_{1\le i,j\le n} |m_{i,j}|$. If $M = (m_{i,j})_{i,j=1}^n$, $M' = (m'_{i,j})_{i,j=1}^n$ are two square matrices of order n, then we have*

$$\left|\det(M) - \det(M')\right| \le n^{1+n/2} \max(\|M\|_\infty, \|M'\|_\infty)^{n-1} \|M - M'\|_\infty. \qquad (2.86)$$

Proof Denote by r_j and r'_j the jth column vector of M and M', respectively. Define a sequence of matrices $A_1, A_2, \ldots, A_n$ by writing each A_j as a list of column vectors, namely

$$\begin{aligned}
A_1 &= (\, r_1 - r_1' \quad r_2 \quad r_3 \quad \ldots \quad r_n \,),\\
A_2 &= (\, r_1' \quad r_2 - r_2' \quad r_3 \quad \ldots \quad r_n \,),\\
A_3 &= (\, r_1' \quad r_2' \quad r_3 - r_3' \quad \ldots \quad r_n \,),\\
&\vdots\\
A_n &= (\, r_1' \quad r_2' \quad r_3' \quad \ldots \quad r_n - r_n' \,).
\end{aligned}$$

By the additivity property of the determinant with respect to columns, we see that

$$\det(M) - \det(M') = \sum_{j=1}^{n} \det(A_j),$$

since the right-hand side becomes a telescopic sum. By Hadamard's inequality (2.30), for each $1 \le j \le n$ we have

$$\begin{aligned}
\left|\det(A_j)\right| &\le n^{n/2} \|r_j - r_j'\|_\infty \prod_{1\le i\le n,\ i\ne j} \max(\|r_j\|_\infty, \|r_j'\|_\infty)\\
&\le n^{n/2} \|M - M'\|_\infty \max(\|M\|_\infty, \|M'\|_\infty)^{n-1},
\end{aligned}$$

and this yields (2.86) on summing over $1 \le j \le n$. □

Proof of Theorem 2.29 Fix $t \in \mathbb{R}$. Denote $\tilde{t} = 2\sqrt{\theta} + t\theta^{1/6}$, and for any numbers $v < w$ with $v \in \mathbb{R}$ and $w \in \mathbb{R} \cup \{\infty\}$, denote

$$j_n(v, w) = \sum_{v<x_1<w} \cdots \sum_{v<x_n<w} \det_{i,j=1}^{n} \Big(\mathbf{J}_\theta(x_i, x_j)\Big), \tag{2.87}$$

$$a_n(v, w) = \int_v^w \cdots \int_v^w \det_{i,j=1}^{n} \Big(\mathbf{A}_\theta(x_i, x_j)\Big)\, dx_1 \ldots dx_n,$$

where the sums in (2.87) are over integer values, and for convenience we suppress the dependence on θ of the quantities $\tilde{t}$, $j_n(v, w)$ and $a_n(v, w)$. From the discussion so far, we see that the theorem will follow if we show that $\sum_{n=1}^{\infty} \frac{1}{n!} |j_n(\tilde{t}, \infty) - a_n(t, \infty)| \to 0$ as $\theta \to \infty$. As a first step, fixing $n \ge 1$, we will derive a bound for $|j_n(\tilde{t}, \infty) - a_n(t, \infty)|$, being careful to keep track of the dependence on n of the various quantities.

The computation is structurally similar to the one in the proof of Theorem 2.26 in the previous section. Fix a number $T > \max(t, 0)$ that will be specified later, and denote $\tilde{T} = 2\sqrt{\theta} + T\theta^{1/6}$. We then have that

$$
\begin{aligned}
&|j_n(\tilde{t},\infty) - a_n(t,\infty)| \\
&\qquad \le |a_n(T,\infty)| + |j_n(\tilde{T},\infty)| + |j_n(\tilde{t},\tilde{T}) - a_n(t,T)|. \qquad (2.88)
\end{aligned}
$$

For the first summand in this expression, (2.75) gives using Hadamard's inequality that

$$
\begin{aligned}
|a_n(T,\infty)| &\le \int_T^\infty \cdots \int_T^\infty C^n e^{-2\sum_{j=1}^n x_i^{3/2}} \det_{i,j=1}^n \left(e^{x_i^{3/2}+x_j^{3/2}} \mathbf{A}(x_i,x_j) \right) dx_1 \ldots dx_n \\
&\le \left(C \int_T^\infty e^{-2x^{3/2}}\, dx \right)^n n^{n/2} \le C^n e^{-nT} n^{n/2}. \qquad (2.89)
\end{aligned}
$$

Next, for $|j_n(\tilde{T},\infty)|$, using (2.78) gives in a similar fashion that

$$
\begin{aligned}
|j_n(\tilde{T},\infty)| &\le \sum_{x_1>\tilde{T}} \cdots \sum_{x_n>\tilde{T}} \theta^{-n/6} C^n n^{n/2} \exp\left(-2c\theta^{-1/6} \sum_{j=1}^n (x_j - 2\sqrt{\theta}) \right) \\
&\le C^n n^{n/2} \left(\theta^{-1/6} \sum_{m \ge T\theta^{1/6}} e^{-2c\theta^{-1/6} m} \right),
\end{aligned}
$$

and it can be readily checked that as θ grows large this is also bounded by an expression of the form $C_2^n e^{-2cnT} n^{n/2}$ for some constant $C_2 > 0$.

Next, to bound $|j_n(\tilde{t},\tilde{T}) - a_n(t,T)|$, which is where the approximation of an integral by a sum comes into play, decompose this quantity further by writing

$$
|j_n(\tilde{t},\tilde{T}) - a_n(t,T)| \le |j_n(\tilde{t},\tilde{T}) - r_n(t,T)| + |r_n(t,T) - a_n(t,T)|, \qquad (2.90)
$$

where we denote

$$
r_n(t,T) = \theta^{-n/6} \sum_{\tilde{t}<x_1<\tilde{T}} \cdots \sum_{\tilde{t}<x_n<\tilde{T}} \det_{i,j=1}^n \left(\mathbf{A}\left(\frac{x_i - 2\sqrt{\theta}}{\theta^{1/6}}, \frac{x_j - 2\sqrt{\theta}}{\theta^{1/6}} \right) \right).
$$

Note that $r_n(t,T)$ is a true Riemann sum for the n-dimensional integral $a_n(t,T)$, where on each axis the interval $[t,T]$ is partitioned into intervals of length $\theta^{-1/6}$. Thus, we have that

$$
\begin{aligned}
&|r_n(t,T) - a_n(t,T)| \le (T-t)^n \\
&\qquad \times \max\left\{ |D(\mathbf{x}) - D(\mathbf{y})| \ :\ \mathbf{x},\mathbf{y} \in [t,T]^n, \|\mathbf{x}-\mathbf{y}\|_2 < \sqrt{n}\,\theta^{-1/6} \right\},
\end{aligned}
$$

where we use the notation $D(\mathbf{x}) = \det_{i,j=1}^n \left(\mathbf{A}(x_i,x_j) \right)$. This can be made more explicit by using (2.86), which translates this into the bound

$$|r_n(t,T) - a_n(t,T)| \le M^{n-1}\,(T-t)^n\,n^{1+n/2}\left(\sqrt{n}\,\theta^{-1/6}\right)\Delta \tag{2.91}$$

where $M = \max_{x,y\in[t,T]} |\mathbf{A}(x,y)|$ and $\Delta = \max_{x,y\in[t,T]} |\nabla\mathbf{A}(x,y)|$. The important thing to note is that M and Δ depend on t and T but not on n.

The last term to bound is $|j_n(\tilde{t},\tilde{T}) - r_n(t,T)|$. This is where the convergence of the Bessel kernel to the Airy kernel comes in; Theorem 2.26 gives

$$|j_n(\tilde{t},\tilde{T}) - r_n(t,T)| \le C^n(T-t)^n n^{n/2+1}\epsilon(\theta), \tag{2.92}$$

where $\epsilon(\theta) \to 0$ as $\theta \to \infty$ (and does not depend on n).

Combining the bounds we derived in (2.88)–(2.92), and tidying up by combining the various constants, we get the bound

$$\begin{aligned} &|j_n(\tilde{t},\infty) - a_n(t,\infty)| \\ &\qquad \le C(t)^n e^{-cnT} + n^{n/2+3}M(t,T)^n\theta^{-1/6} + M(t,T)^n\epsilon(\theta), \end{aligned} \tag{2.93}$$

where $\epsilon(\theta) \to 0$ as $\theta \to \infty$, $C(t) > 0$ depends on t but not on T, n or θ, and $M(t,T) > 0$ depends on t and T but not on n or θ.

Now multiply (2.93) by $1/n!$ and sum over $n \ge 1$, to get

$$\sum_{n=1}^{\infty}\frac{1}{n!}|j_n(\tilde{t},\infty) - a_n(t,\infty)| \le \left(\exp\left(C(t)e^{-cT}\right) - 1\right) + Q(t,T)(\theta^{-1/6} + \epsilon(\theta)),$$

for some $Q(t,T) > 0$. Choosing $T > 0$ to be a sufficiently large number, the first summand can be made arbitrarily small – smaller than some $\delta > 0$, say. Letting θ go to ∞ then shows that

$$\limsup_{\theta\to\infty}\sum_{n=1}^{\infty}\frac{1}{n!}|j_n(\tilde{t},\infty) - a_n(t,\infty)| \le \delta.$$

Since δ was arbitrary, this proves our claim that

$$\sum_{n=1}^{\infty}\frac{1}{n!}|j_n(\tilde{t},\infty) - a_n(t,\infty)| \to 0 \quad \text{as } \theta \to \infty,$$

and therefore finishes the proof. □

2.9 A de-Poissonization lemma

For each $\theta > 0$, let N_θ denote a random variable with the Poisson distribution $\mathrm{Poi}(\theta)$. Recall that N_θ has mean θ and variance θ. Therefore we have

that

$$f(\theta) := \mathbb{P}\left(|N_\theta - \theta| \geq 3\sqrt{\theta\log\theta}\right) \to 0 \ \text{ as } \theta \to \infty, \tag{2.94}$$

since, for example, by Chebyshev's inequality this probability is bounded by $(9\log\theta)^{-1}$ (the rate of convergence is immaterial for our purposes, but in case you care, Exercise 2.19 suggests an improved bound).

To deduce Theorem 2.2 from Theorem 2.29, we use the following simple de-Poissonization lemma due to Johansson [63].

Lemma 2.31 *Let $P_1 \geq P_2 \geq P_3 \geq \ldots$ be a nonincreasing sequence of numbers in $[0,1]$. Denote*

$$\theta_n = n - 4\sqrt{n\log n}, \quad \phi_n = n + 4\sqrt{n\log n}.$$

Then for any $n \geq 1$ we have that

$$e^{-\phi_n}\sum_{m=0}^{\infty}\frac{\phi_n^m}{m!}P_m - f(\phi_n) \leq P_n \leq e^{-\theta_n}\sum_{m=0}^{\infty}\frac{\theta_n^m}{m!}P_m + f(\theta_n). \tag{2.95}$$

Proof It is easy to check that $n < \phi_n - 3\sqrt{\phi_n\log\phi_n}$. Therefore we have that

$$\begin{aligned} e^{-\phi_n}\sum_{m=0}^{\infty}\frac{\phi_n^m}{m!}P_m &= e^{-\phi_n}\sum_{0\leq m\leq n}\frac{\phi_n^m}{m!}P_m + e^{-\phi_n}\sum_{m>n}\frac{\phi_n^m}{m!}P_m \\ &\leq e^{-\phi_n}\sum_{0\leq m\leq n}\frac{\phi_n^m}{m!}\cdot 1 + \left(e^{-\phi_n}\sum_{m>n}\frac{\phi_n^m}{m!}\right)P_n \\ &\leq f(\phi_n) + \left(e^{-\phi_n}\sum_{m\geq 0}\frac{\phi_n^m}{m!}\right)P_n = f(\phi_n) + P_n, \end{aligned}$$

which proves the first inequality in the chain (2.95). Similarly, for the second inequality, observe that $n > \theta_n + 3\sqrt{\theta_n\log\theta_n}$, and that therefore we can write

$$\begin{aligned} e^{-\theta_n}\sum_{m=0}^{\infty}\frac{\theta_n^m}{m!}P_m &= e^{-\theta_n}\sum_{0\leq m<n}\frac{\theta_n^m}{m!}P_m + e^{-\theta_n}\sum_{m\geq n}\frac{\theta_n^m}{m!}P_m \\ &\geq e^{-\theta_n}\sum_{0\leq m\leq n}\frac{\theta_n^m}{m!}P_n \\ &\geq \left(e^{-\theta_n}\sum_{m\geq 0}\frac{\theta_n^m}{m!} - f(\theta_n)\right)P_n = P_n - P_nf(\theta_n) \\ &\geq P_n - f(\theta_n), \end{aligned}$$

which proves the claim. □

Lemma 2.32 *Let σ_n denote for each $n \geq 1$ a uniformly random permutation in S_n. For any $t \in \mathbb{R}$ we have*

$$\mathbb{P}(L(\sigma_n) \leq t) \geq \mathbb{P}(L(\sigma_{n+1}) \leq t). \tag{2.96}$$

Proof One can obtain σ_{n+1} by starting with σ_n and inserting the value $n + 1$ at a uniformly random position with respect to the existing values $\sigma_n(1), \ldots, \sigma_n(n)$. Such an insertion can only create longer increasing subsequences than the ones already present in σ_n, which means that under this coupling of σ_n and σ_{n+1} we have that $L(\sigma_{n+1}) \geq L(\sigma_n)$. This implies the event inclusion $\{L(\sigma_{n+1}) \leq t\} \subseteq \{L(\sigma_n) \leq t\}$, which immediately gives (2.96). □

Proof of Theorem 2.2 As with the analogous claim we showed for Poissonized Plancherel measure, we prove (2.4), which will prove the theorem modulo Theorem 2.1, which is proved separately in the next section.

Let $\lambda^{(n)}$ denote as before a random Young diagram chosen according to Plancherel measure of order n, and let $\lambda^{\langle\theta\rangle}$ denote a random Young diagram chosen according to Poissonized Plancherel measure with parameter θ. For $t \in \mathbb{R}$, denote $P_n(t) = \mathbb{P}\left(\lambda_1^{(n)} \leq t\right)$ (the distribution function of $\lambda_1^{(n)}$), and denote $Q_\theta(t) = \mathbb{P}\left(\lambda_1^{\langle\theta\rangle} \leq t\right)$ (the distribution function of $\lambda_1^{\langle\theta\rangle}$). By the definition of Poissonized Plancherel measure as a Poisson mixture of ordinary Plancherel measures, we have

$$Q_\theta(t) = e^{-\theta} \sum_{n=0}^{\infty} \frac{\theta^n}{n!} P_n(t). \tag{2.97}$$

By Lemma 2.32, for fixed t the sequence $P_1(t), P_2(t), \ldots$ is nonincreasing. Therefore we can deduce from Lemma 2.31 that

$$Q_{\phi_n}(\tilde{t}) - f(\phi_n) \leq P_n(\tilde{t}) \leq Q_{\theta_n}(\tilde{t}) + f(\theta_n), \tag{2.98}$$

where we denote $\tilde{t} = 2\sqrt{n} + tn^{1/6}$. But by Theorem 2.47, both of the terms Q_{θ_n} and Q_{ϕ_n} sandwiching the P_n-term in (2.98) converge to $F_2(t)$ as $n \to \infty$ (one can check easily that the fact that n is used for the scaling of the arguments rather than θ_n and ϕ_n does not matter). So, we get the same limit for $P_n(\tilde{t})$, and the theorem is proved. □

2.10 An analysis of F_2

Having proved (2.4), the main claim of Theorem 2.2, we now take a closer look at the function F_2, since no treatment of this important result concerning the limiting fluctuations of $L(\sigma_n)$ would be complete without a deeper understanding of the function describing these limiting fluctuations and its properties. First and most importantly, we need to prove the assertion from Theorem 2.1 that F_2 is a distribution function; without this knowledge, the convergence in Theorem 2.2 is not equivalent to convergence in distribution of the scaled fluctuations $n^{-1/6}(L(\sigma_n) - 2\sqrt{n})$, and in particular this leaves room for the possibility that the unscaled fluctuations $L(\sigma_n) - 2\sqrt{n}$ are of an order of magnitude bigger than $n^{1/6}$. Note that it is easy to see from the definition of F_2 that it is a smooth function, and that $F_2(t) \to 1$ as $t \to \infty$ (see Lemma 2.34). It also follows from Theorem 2.2 that $0 \leq F_2(t) \leq 1$ for all t and that F_2 is a nondecreasing function of t, but for these facts too it would be preferable to have a direct and shorter proof that does not rely on the complicated algebraic, analytic, and combinatorial investigations that went into proving Theorem 2.2. The missing fact required to deduce that F_2 is a distribution function is that $F_2(t) \to 0$ as $t \to -\infty$.

In addition to this immediate and clearly formulated goal, the analysis is also driven by a more vague desire to better understand how to fit this new function F_2 into the taxonomy of the various special functions of mathematical analysis. Can it be expressed in some simple way in terms of known functions? (Here, "simple" is used as a subjective term that is not precisely defined. For example, F_2 is defined in terms of the Airy function, but the definition involves an infinite summation of multidimensional integrals, so one can reasonably argue it is not very simple – in particular, from a computational standpoint.) This question is of theoretical and also of practical significance, since, for example, it has a bearing on the important problem of evaluating F_2 numerically to high precision.

To answer these questions, we prove the following result, due to Tracy and Widom [136], that represents F_2 in terms of a certain special function that arises in the theory of differential equations, the **Painlevé II transcendent** (see box on the opposite page for some background related to this terminology). This representation implies Theorem 2.1 as an immediate corollary.

The Painlevé transcendents

The theory of Painlevé transcendents was developed around the year 1900 by Paul Painlevé, a French mathematician and human rights activist, who later went on to other notable achievements, entering his country's politics and going on to hold various ministerial positions in the French government including two short terms as prime minister of France in 1917 and 1925. The problem leading to the Painlevé transcendents concerned the classification of solutions to second-order differential equations of the form $y'' = R(y, y', t)$, where R is a rational function. Painlevé showed that the elements of a natural class of solutions of this equation, consisting of those solutions whose only movable singularities are poles, could each be transformed into one of fifty different canonical forms. He was further able to show that 44 of these canonical forms consisted of functions that could be expressed in terms of known special functions, leaving six whose solutions required the introduction of new functions. These six "fundamentally new" solutions became known as the **Painlevé transcendents**, and have since made occasional appearances in various areas of analysis, mathematical physics and other fields. The equations giving rise to them are known as the **Painlevé I–VI equations**.

$$(P_I) \quad y''=6y^2+x$$
$$(P_{II}) \quad y''=2y^3+xy+\alpha$$
$$(P_{III}) \quad y''=\frac{1}{y}(y')^2-\frac{1}{x}y'+\frac{\alpha y^2+\beta}{x}+\gamma y^3+\frac{\delta}{y}$$
$$(P_{IV}) \quad y''=\frac{1}{2y}(y')^2+\frac{3}{2}y^3+4xy^2+2(x^2-\alpha)y+\frac{\beta}{y}$$
$$(P_V) \quad y''=\left(\frac{1}{2y}+\frac{1}{y-1}\right)(y')^2-\frac{1}{x}y'+\frac{(y-1)^2}{x^2}\left(\alpha y+\frac{\beta}{y}\right)+\frac{\gamma y}{x}+\frac{\delta y(y+1)}{y-1}$$
$$(P_{VI}) \quad y''=\frac{1}{2}\left(\frac{1}{y}+\frac{1}{y-1}+\frac{1}{y-x}\right)(y')^2-\left(\frac{1}{x}+\frac{1}{x-1}+\frac{1}{y-x}\right)y'+\frac{y(y-1)(y-x)}{x^2(x-1)^2}\left(\alpha+\frac{\beta x}{y^2}+\frac{\gamma(x-1)}{(y-1)^2}+\frac{\delta x(x-1)}{(y-x)^2}\right)$$

The six Painlevé equations

Paul Painlevé

Theorem 2.33 *Let $q\colon \mathbb{R} \to \mathbb{R}$ be the unique solution of the Painlevé II equation*

$$q''(t) = tq(t) + 2q(t)^3 \tag{2.99}$$

that has the asymptotic behavior

$$q(t) = (1 + o(1))\,\mathrm{Ai}(t) \qquad \text{as } t \to \infty. \tag{2.100}$$

Then we have the representation

$$F_2(t) = \exp\left(-\int_t^\infty (x-t)q(x)^2\,dx\right), \qquad (t \in \mathbb{R}). \tag{2.101}$$

Note that the theorem contains the statement that a function $q(t)$ satisfying (2.99) and (2.100) is unique. We shall define $q(t)$ using an explicit expression and prove that it satisfies these conditions, but the fact that they determine it uniquely is a known result from the theory of Painlevé equations (see [57]) that we will not prove. Furthermore, it should be noted that the general form of the Painlevé II equation is

$$y''(x) = 2y(x)^3 + xy(x) + \alpha,$$

where α is an arbitrary constant, so that (2.99) represents the special case $\alpha = 0$ of the equation.

Proof of Theorem 2.1 Write (2.101) as $F_2(t) = \exp\left(-\int_{-\infty}^\infty U(x,t)q(x)^2\,dx\right)$, where $U(x,t) = (x-t)\mathbf{1}_{[t,\infty)}(x)$. The fact that $U(x,t) \ge 0$ implies that $0 \le F_2(t) \le 1$ for all t. The fact that $U(x,t)$ is nonincreasing in t implies that $F_2(t)$ is nondecreasing. To check that $F_2(t) \to 0$ as $t \to -\infty$, observe that for $t < 0$ we have that $F_2(t) \le \exp\left(-|t|\int_0^\infty q(x)^2\,dx\right)$. By (2.100), $\int_0^\infty q(x)^2\,dx > 0$, so the result follows. □

The analysis required for the proof of Theorem 2.33 will consist of a sequence of claims about a set of auxiliary functions that we now define. Our exposition follows that of [4]. Introduce the notation

$$\det_{\mathbf{A}}\begin{pmatrix} x_1 & \dots & x_n \\ y_1 & \dots & y_n \end{pmatrix} = \det_{i,j=1}^{n}\left(\mathbf{A}(x_i, y_j)\right).$$

We make the following definitions (including F_2 again for easy reference):

$$F_2(t) = 1 + \sum_{n=1}^\infty \frac{(-1)^n}{n!}\int_t^\infty \dots \int_t^\infty \det_{\mathbf{A}}\begin{pmatrix} x_1 & \dots & x_n \\ x_1 & \dots & x_n \end{pmatrix} dx_1 \dots dx_n,$$

$$\begin{aligned}
\mathbf{H}(x,y,t) &= \mathbf{A}(x,y) + \sum_{n=1}^{\infty} \frac{(-1)^n}{n!} \int_t^{\infty} \ldots \int_t^{\infty} \det_{\mathbf{A}} \begin{pmatrix} x\ x_1 \ldots x_n \\ y\ x_1 \ldots x_n \end{pmatrix} dx_1 \ldots dx_n \\
&= \mathbf{H}(y,x,t), \\
\mathbf{R}(x,y,t) &= \frac{\mathbf{H}(x,y,t)}{F_2(t)} = \mathbf{R}(y,x,t), \\
Q(x,t) &= \mathrm{Ai}(x) + \int_t^{\infty} \mathbf{R}(x,y,t)\,\mathrm{Ai}(y)\,dy, \\
P(x,t) &= \mathrm{Ai}'(x) + \int_t^{\infty} \mathbf{R}(x,y,t)\,\mathrm{Ai}'(y)\,dy, \\
q(t) &= Q(t,t) = \mathrm{Ai}(t) + \int_t^{\infty} \mathbf{R}(t,y,t)\,\mathrm{Ai}(y)\,dy, \\
p(t) &= P(t,t) = \mathrm{Ai}'(t) + \int_t^{\infty} \mathbf{R}(t,y,t)\,\mathrm{Ai}'(y)\,dy, \\
u(t) &= \int_t^{\infty} Q(x,t)\,\mathrm{Ai}(x)\,dx \\
&= \int_t^{\infty} \mathrm{Ai}(x)^2\,dx + \int_t^{\infty}\int_t^{\infty} \mathbf{R}(x,y,t)\,\mathrm{Ai}(x)\,\mathrm{Ai}(y)\,dx\,dy, \\
v(t) &= \int_t^{\infty} Q(x,t)\,\mathrm{Ai}'(x)\,dx \\
&= \int_t^{\infty} \mathrm{Ai}(x)\,\mathrm{Ai}'(x)\,dx + \int_t^{\infty}\int_t^{\infty} \mathbf{R}(x,y,t)\,\mathrm{Ai}(x)\,\mathrm{Ai}'(y)\,dx\,dy \\
&= \int_t^{\infty} P(x,t)\,\mathrm{Ai}(x)\,dx.
\end{aligned}$$

Fix $t \in \mathbb{R}$. If $\mathbf{K}\colon \mathbb{R} \times \mathbb{R} \to \mathbb{R}$ is a kernel decaying sufficiently fast as its arguments tend to $+\infty$, we can think of it as a linear operator acting on functions $f \in L^2[t,\infty)$ by

$$(\mathbf{K}f)(x) = \int_t^{\infty} \mathbf{K}(x,y) f(y)\,dy.$$

We shall use this interpretation for the Airy kernel $\mathbf{A}(\cdot,\cdot)$, and denote the corresponding operator on $L^2[t,\infty)$ by $\mathbf{A}_t$, to emphasize the dependence on t. As we will see later, for fixed t the functions $\mathbf{H}_t = \mathbf{H}(\cdot,\cdot,t)$ and $\mathbf{R}_t = \mathbf{R}(\cdot,\cdot,t)$ also have a natural interpretation as linear operators on $L^2[t,\infty)$, so it makes sense to think of them as kernels, similarly to $\mathbf{A}(\cdot,\cdot)$, and denote them using boldface letters. Note that operators $\mathbf{K},\mathbf{M}\colon \mathbb{R} \times \mathbb{R} \to \mathbb{R}$ on

$L^2[t,\infty)$ can be multiplied according to the usual formula

$$(\mathbf{KM})(x,y) = \int_t^\infty \mathbf{K}(x,z)\mathbf{M}(z,y)\,dz,$$

and that this multiplication has the usual properties of matrix multiplication.

Lemma 2.34 *The sums and integrals in the preceding definitions all converge absolutely, and uniformly on compacts, as do all of their partial derivatives. Consequently, they are smooth functions where they are defined and partial differentiation operators can be applied termwise and under the integral sign. Furthermore, we have the following bounds which hold asymptotically as $t \to \infty$:*

$$\begin{aligned} F_2(t) &= 1 - O\left(e^{-t}\right), \\ \mathbf{H}(x,y,t) &= \mathbf{A}(x,y) + O\left(e^{-x-y-t}\right) \qquad \textit{(uniformly for } x,y \geq 0\textit{)}, \\ q(t) &= \left(1 + O\left(e^{-t}\right)\right)\mathrm{Ai}(t). \end{aligned}$$

Proof The convergence claims follow using Lemma 2.30 in combination with Hadamard's inequality (2.30) and the bound (2.75), in a similar way to how we showed that the series defining $F_2(t)$ converges (Lemma 2.25). To get the asymptotic bounds one looks also at the dependence of each of the terms on t. For example, for $F_2(t)$ we have using (2.76) that

$$|F_2(t) - 1| \leq \sum_{n=1}^\infty \frac{C^n n^{n/2}}{n!}\left(\int_t^\infty e^{-2x^{3/2}}\,dx\right)^n \leq \sum_{n=1}^\infty \frac{C^n e^n}{n^{n/2}} e^{-nt} = O(e^{-t}).$$

Similarly, for $\mathbf{H}(x,y,t)$ we have that

$$\begin{aligned} |\mathbf{H}(x,y,t) - \mathbf{A}(x,y)| &\leq \sum_{n=1}^\infty \frac{(n+1)^{(n+1)/2}C^{n+1}}{n!} e^{-2x^{3/2}-2y^{3/2}}\left(\int_t^\infty e^{-2u^{3/2}}\,du\right)^n \\ &\leq e^{-x-y}\sum_{n=1}^\infty \frac{(n+1)^{(n+1)/2}C^{n+1}}{n!} e^{-nt}, \end{aligned}$$

and this is $O(e^{-x-y-t})$ uniformly for $x,y \geq 0$. The estimate for $|q(t) - \mathrm{Ai}(t)|$ is proved similarly and is omitted. □

Lemma 2.35

$$\frac{d}{dt}(\log F_2(t)) = \frac{F_2'(t)}{F_2(t)} = \mathbf{R}(t,t,t) = \frac{\mathbf{H}(t,t,t)}{F_2(t)}.$$

Proof For $n \geq 1$ denote

$$\Delta_n(t) = \int_t^\infty \cdots \int_t^\infty \det_{\mathbf{A}} \begin{pmatrix} x_1 \dots x_n \\ x_1 \dots x_n \end{pmatrix} dx_1 \dots dx_n,$$

$$H_n(x,y,t) = \int_t^\infty \cdots \int_t^\infty \det_{\mathbf{A}} \begin{pmatrix} x \; x_1 \dots x_n \\ y \; x_1 \dots x_n \end{pmatrix} dx_1 \dots dx_n,$$

and denote $\Delta_0(t) = 1$, $H_0(x,y,t) = \operatorname{Ai}(x,y)$, so that we have $F_2(t) = \sum_{n=0}^\infty \frac{(-1)^n}{n!}\Delta_n(t)$ and $\mathbf{H}(x,y,t) = \sum_{n=0}^\infty \frac{(-1)^n}{n!} H_n(x,y,t)$. Clearly it will be enough to prove that $\Delta_n'(t) = -nH_{n-1}(t,t,t)$ for $n \geq 1$. This is obvious for $n = 1$. For $n \geq 2$, observe that $\Delta_n'(t)$ is equal to

$$-\sum_{j=1}^n \iint_{[t,\infty)^{n-1}} \det_{\mathbf{A}} \begin{pmatrix} x_1 \dots x_{j-1} \; t \; x_{j+1} \dots x_n \\ x_1 \dots x_{j-1} \; t \; \; x_j \; \dots x_n \end{pmatrix} dx_1 \dots dx_{j-1}\, dx_{j+1} \dots dx_n,$$

and that each of the integrals in this sum is equal to $H_{n-1}(t,t,t)$. □

Lemma 2.36 *$\mathbf{R}(x,y,t)$ satisfies the integral equations*

$$\mathbf{R}(x,y,t) - \mathbf{A}(x,y) = \int_t^\infty \mathbf{A}(x,z)\mathbf{R}(z,y,t)\,dz$$
$$= \int_t^\infty \mathbf{R}(x,z,t)\mathbf{A}(z,y)\,dz. \qquad (2.102)$$

Equivalently, in the language of linear operators we have the operator identities

$$\mathbf{R}_t - \mathbf{A}_t = \mathbf{A}_t\mathbf{R}_t = \mathbf{R}_t\mathbf{A}_t. \qquad (2.103)$$

Proof By the symmetries $\mathbf{A}(y,x) = \mathbf{A}(x,y)$ and $\mathbf{R}(y,x,t) = \mathbf{R}(x,y,t)$, the two equations in (2.102) are equivalent, so we prove the first one, in the equivalent form

$$\mathbf{H}(x,y,t) - F_2(t)\mathbf{A}(x,y) = \int_t^\infty \mathbf{A}(x,z)\mathbf{H}(z,y,t)\,dz.$$

The left-hand side can be written as $\sum_{n=0}^\infty \frac{(-1)^n}{n!}(H_n(x,y,t) - \mathbf{A}(x,y)\Delta_n(t))$, and the right-hand side as $\sum_{m=0}^\infty \frac{(-1)^m}{m!}\int_t^\infty \mathbf{A}(x,z)H_m(z,y,t)\,dz$, in the notation of the previous lemma; so it will be enough to show that

$$H_n(x,y,t) - \mathbf{A}(x,y)\Delta_n(t) = -n\int_t^\infty \mathbf{A}(x,z)H_{n-1}(z,y,t)\,dz \qquad (2.104)$$

holds for any $n \geq 1$. Note that, by expanding the determinant appearing in $H_n(x,y,t)$ on the left-hand side of (2.104) by minors along the first row, one gets a term canceling out $\mathbf{A}(x,y)\Delta_n(t)$ from the first column, plus additional terms, which we can write as

$$
\begin{aligned}
&H_n(x,y,t) - \mathbf{A}(x,y)\Delta_n(t) \\
&= \sum_{j=1}^{n}(-1)^j \int_t^\infty \ldots \int_t^\infty \mathbf{A}(x,x_j)\det{}_{\mathbf{A}}\begin{pmatrix} x_1 & x_2 & \ldots & x_j & x_{j+1} & \ldots & x_n \\ y & x_1 & \ldots & x_{j-1} & x_{j+1} & \ldots & x_n \end{pmatrix} dx_1 \ldots dx_n \\
&= -\sum_{j=1}^{n} \int_t^\infty \ldots \int_t^\infty \mathbf{A}(x,x_j)\det{}_{\mathbf{A}}\begin{pmatrix} x_j & x_1 & \ldots & x_{j-1} & x_{j+1} & \ldots & x_n \\ y & x_1 & \ldots & x_{j-1} & x_{j+1} & \ldots & x_n \end{pmatrix} dx_1 \ldots dx_n \\
&= -\sum_{j=1}^{n} \int_t^\infty \mathbf{A}(x,x_j) H_{n-1}(x_j,y,t)\,dx_j \\
&= -n \int_t^\infty \mathbf{A}(x,x_j) H_{n-1}(z,y,t)\,dz,
\end{aligned}
$$

proving the claim. □

Denote by $\mathbf{I}$ the identity operator on $L^2[t,\infty)$ (omitting the dependence on t). By elementary manipulations, (2.103) can be rewritten in the equivalent form

$$
(\mathbf{I} + \mathbf{R}_t)(\mathbf{I} - \mathbf{A}_t) = (\mathbf{I} - \mathbf{A}_t)(\mathbf{I} + \mathbf{R}_t) = \mathbf{I} \tag{2.105}
$$

which means that $\mathbf{I} + \mathbf{R}_t$ and $\mathbf{I} - \mathbf{A}_t$ are two-sided inverses of each other. Knowing that they are invertible, we can also write

$$
\mathbf{R}_t = \mathbf{A}_t(\mathbf{I} - \mathbf{A}_t)^{-1}, \qquad \mathbf{A} = \mathbf{R}_t(\mathbf{I} + \mathbf{R}_t)^{-1}.
$$

For this reason, in the theory of Fredholm determinants $\mathbf{R}_t$ is called the **resolvent** of $\mathbf{A}_t$. Note also that the proof of Lemma 2.36 only used the way $\mathbf{R}(\cdot,\cdot,\cdot)$ was defined in terms of the kernel $\mathbf{A}(\cdot,\cdot)$ and not any information about any of its specific properties, so an analogous resolvent relation holds in much greater generality; see [74], [120] for more details on the general theory.

Lemma 2.37

$$
\frac{\partial}{\partial t}\mathbf{R}(x,y,t) = -\mathbf{R}(x,t,t)\mathbf{R}(t,y,t).
$$

Proof Denote $\mathbf{r}_t(x,y) = \frac{\partial}{\partial t}\mathbf{R}(x,y,t)$. Starting with the first identity in

(2.102), we have that

$$\begin{aligned}\mathbf{r}_t(x,y) &= \frac{\partial}{\partial t}\left(\int_t^\infty \mathbf{A}(x,z)\mathbf{R}(z,y,t)\,dz\right)\\ &= -\mathbf{A}(x,t)\mathbf{R}(t,y,t) + \int_t^\infty \mathbf{A}(x,z)\mathbf{r}_t(z,y)\,dz,\end{aligned}$$

or, in operator notation,

$$((\mathbf{I}-\mathbf{A}_t)\mathbf{r}_t)(x,y) = \mathbf{B}_t(x,y),$$

where $\mathbf{B}_t(x,y) = -\mathbf{A}(x,t)\mathbf{R}(t,y,t)$. Now using the relation (2.105), by multiplying this equation on the left by the operator $\mathbf{I}+\mathbf{R}_t$ we transform it into

$$\begin{aligned}\mathbf{r}_t(x,y) &= ((\mathbf{I}+\mathbf{R}_t)\mathbf{B}_t)(x,y)\\ &= -\mathbf{A}(x,t)\mathbf{R}(t,y,t) - \int_t^\infty \mathbf{R}(x,z,t)\mathbf{A}(z,t)\mathbf{R}(t,y,t)\,dz\\ &= -\mathbf{R}(t,y,t)\left(\mathbf{A}(x,t) + \int_t^\infty \mathbf{R}(x,z,t)\mathbf{A}(z,t)\,dz\right)\\ &= -\mathbf{R}(x,t,t)\mathbf{R}(t,y,t),\end{aligned}$$

using (2.102) once again in the last step. □

Lemma 2.38

$$\mathbf{R}(x,y,t) = \begin{cases} \dfrac{Q(x,t)P(y,t)-Q(y,t)P(x,t)}{x-y} & \text{if } x\neq y,\\ \left(\frac{\partial}{\partial x}Q(x,t)\right)P(x,t) - Q(x,t)\left(\frac{\partial}{\partial x}P(x,t)\right) & \text{if } x=y.\end{cases}$$

Proof It is enough to prove the claim for the case $x\neq y$, since the case $x=y$ follows from it by taking the limit as $y\to x$ and applying L'Hôpital's rule. Consider the kernels $\tilde{\mathbf{R}}_t(\cdot,\cdot)$, $\tilde{\mathbf{A}}_t(\cdot,\cdot)$ defined by

$$\begin{aligned}\tilde{\mathbf{R}}_t(x,y) &= (x-y)\mathbf{R}(x,y,t),\\ \tilde{\mathbf{A}}_t(x,y) &= (x-y)\mathbf{A}(x,y) = \mathrm{Ai}(x)\,\mathrm{Ai}'(y) - \mathrm{Ai}'(x)\,\mathrm{Ai}(y),\end{aligned}$$

(the subscript t on $\tilde{\mathbf{A}}_t(\cdot,\cdot)$ emphasizes that we consider it as a kernel corresponding to an operator acting on $L^2[t, \infty)$). We have that

$$\begin{aligned}
(\tilde{\mathbf{R}}_t\mathbf{A}_t + \mathbf{R}_t\tilde{\mathbf{A}}_t)(x, y) &= \int_t^\infty \tilde{\mathbf{R}}_t(x, z)\mathbf{A}(z, y)\, dz + \int_t^\infty \mathbf{R}_t(x, z)\tilde{\mathbf{A}}(z, y)\, dz \\
&= \int_t^\infty \mathbf{R}_t(x, z)\mathbf{A}(z, y)(x - z + z - y)\, dz \\
&= (x - y)\int_t^\infty \mathbf{R}(x, z, t)\mathbf{A}(z, y)\, dz \\
&= (x - y)(\mathbf{R}(x, y, t) - \mathbf{A}(x, y)) = (\tilde{\mathbf{R}}_t - \tilde{\mathbf{A}}_t)(x, y).
\end{aligned}$$

In other words, we have the operator identity $\tilde{\mathbf{R}}_t\mathbf{A}_t + \mathbf{R}_t\tilde{\mathbf{A}}_t = \tilde{\mathbf{R}}_t - \tilde{\mathbf{A}}$. Multiplying this identity from the left by $\mathbf{I}+\mathbf{R}$, rearranging terms and simplifying using (2.103) yields the relation

$$\tilde{\mathbf{R}}_t = \tilde{\mathbf{A}}_t + \tilde{\mathbf{A}}_t\mathbf{R} + \mathbf{R}\tilde{\mathbf{A}}_t + \tilde{\mathbf{A}}_t\mathbf{R}\tilde{\mathbf{A}}_t = (\mathbf{I} + \mathbf{R}_t)\tilde{\mathbf{A}}(\mathbf{I} + \mathbf{R}_t).$$

Evaluating this kernel identity for given x, y gives

$$\begin{aligned}
(x - y)\mathbf{R}(x, y, t) = \tilde{\mathbf{R}}_t(x, y) &= \tilde{\mathbf{A}}_t(x, y) + \int_t^\infty \mathbf{R}(x, z, t)\tilde{\mathbf{A}}_t(z, y)\, dz \\
&+ \int_t^\infty \tilde{\mathbf{A}}_t(x, z)\mathbf{R}(z, y, t)\, dz \\
&+ \int_t^\infty \int_t^\infty \mathbf{R}(x, z, t)\tilde{\mathbf{A}}_t(z, w)\mathbf{R}(w, y, t)\, dw\, dz.
\end{aligned}$$

One can now easily verify that this is equal to $Q(x, t)P(y, t) - Q(y, t)P(x, t)$ using the definitions of $Q(x, t)$, $P(x, t)$, and $\tilde{\mathbf{A}}_t(x, y)$. □

Lemma 2.39

$$\frac{\partial}{\partial t}Q(x, t) = -\mathbf{R}(x, t, t)Q(t, t),$$

$$\frac{\partial}{\partial t}P(x, t) = -\mathbf{R}(x, t, t)P(t, t).$$

Proof

$$\begin{aligned}
\frac{\partial}{\partial t}Q(x,t) &= -\mathbf{R}(x,t,t)\,\mathrm{Ai}(t) + \int_t^\infty \frac{\partial}{\partial t}(x,y,t)\,\mathrm{Ai}(y)\,dy \\
&= -\mathbf{R}(x,t,t)\,\mathrm{Ai}(t) - \int_t^\infty \mathbf{R}(x,t,t)\mathbf{R}(t,y,t)\,\mathrm{Ai}(y)\,dy \\
&= -\mathbf{R}(x,t,t)\left(\mathrm{Ai}(t) + \int_t^\infty \mathbf{R}(t,y,t)\,\mathrm{Ai}(y)\,dy\right) \\
&= -\mathbf{R}(x,t,t)Q(t,t).
\end{aligned}$$

The verification for the second equation is similar and is omitted. □

Lemma 2.40

$$\frac{\partial}{\partial x}\mathbf{R}(x,y,t) + \frac{\partial}{\partial y}\mathbf{R}(x,y,t) = \mathbf{R}(x,t,t)\mathbf{R}(t,y,t) - Q(x,t)Q(y,t).$$

Proof Denote by ∂_{x+y} the differential operator $\frac{\partial}{\partial x} + \frac{\partial}{\partial y}$, and let $\rho_t(x,y) = \partial_{x+y}\mathbf{R}(x,y,t)$. An easy computation, which is equivalent to (2.80), gives that $\partial_{x+y}\mathbf{A}(x,y) = -\mathrm{Ai}(x)\,\mathrm{Ai}(y)$. This gives, applying ∂_{x+y} to the first equality in (2.102), that

$$\begin{aligned}
\rho_t(x,y) + \mathrm{Ai}(x)\,\mathrm{Ai}(y) &= \partial_{x+y}\left(\mathbf{R}(x,y,t) - \mathbf{A}(x,y)\right) \\
&= \int_t^\infty \left(\frac{\partial}{\partial x}\mathbf{A}(x,z)\right)\mathbf{R}(z,y,t)\,dz + \int_t^\infty \mathbf{A}(x,z)\left(\frac{\partial}{\partial y}\mathbf{R}(z,y,t)\right)dz \\
&= \int_t^\infty \left(\left(\frac{\partial}{\partial x} + \frac{\partial}{\partial z}\right)\mathbf{A}(x,z)\right)\mathbf{R}(z,y,t)\,dz \\
&\quad + \int_t^\infty \mathbf{A}(x,z)\left(\left(\frac{\partial}{\partial z} + \frac{\partial}{\partial y}\right)\mathbf{R}(z,y,t)\right)dz \\
&\quad - \int_t^\infty \left[\left(\frac{\partial}{\partial z}\mathbf{A}(x,z)\right)\mathbf{R}(z,y,t) + \mathbf{A}(x,z)\left(\frac{\partial}{\partial z}\mathbf{R}(z,y,t)\right)\right]dz \\
&= \int_t^\infty -\mathrm{Ai}(x)\,\mathrm{Ai}(z)\mathbf{R}(z,y,t)\,dz + \int_t^\infty \mathbf{A}(x,z)\rho_t(z,y)\,dz \\
&\quad - \int_t^\infty \frac{\partial}{\partial z}\left(\mathbf{A}(x,z)\mathbf{R}(z,y,t)\right)dz \\
&= -\mathrm{Ai}(x)(Q(y,t) - \mathrm{Ai}(y)) + \int_t^\infty \mathbf{A}(x,z)\rho_t(z,y)\,dz + \mathbf{A}(x,t)\mathbf{R}(t,y,t).
\end{aligned}$$

Cancelling out the common term $\mathrm{Ai}(x)\,\mathrm{Ai}(y)$ and rearranging the remaining terms, we get in operator notation the relation

$$(\mathbf{I} - \mathbf{A}_t)\rho_t = \mathbf{C}_t,$$

where $\mathbf{C}_t(x,y) = -\operatorname{Ai}(x)Q(y,t) + \mathbf{A}(x,t)\mathbf{R}(t,y,t)$. Multiplying this by $\mathbf{I}+\mathbf{R}_t$ and using (2.105), we get that

$$\begin{aligned}
\rho_t(x,y) &= ((\mathbf{I}+\mathbf{R}_t)\mathbf{C}_t)(x,y) \\
&= -\operatorname{Ai}(x)Q(y,t) + \mathbf{A}(x,t)\mathbf{R}(t,y,t) \\
&\quad + \int_t^\infty \mathbf{R}(x,z,t)\Big(-\operatorname{Ai}(z)Q(y,t) + \mathbf{A}(z,t)\mathbf{R}(t,y,t)\Big)\,dz \\
&= \mathbf{R}(t,y,t)\left(\mathbf{A}(x,t) + \int_t^\infty \mathbf{R}(x,z,t)\mathbf{A}(z,t)\,dz\right) \\
&\quad - Q(y,t)\left(\operatorname{Ai}(x) + \int_t^\infty \mathbf{R}(x,z,t)\operatorname{Ai}(z)\,dz\right) \\
&= \mathbf{R}(x,t,t)\mathbf{R}(t,y,t) - Q(x,t)Q(y,t),
\end{aligned}$$

as claimed. □

Lemma 2.41

$$\begin{aligned}
\frac{\partial}{\partial x}Q(x,t) &= P(x,t) + \mathbf{R}(x,t,t)Q(t,t) - Q(x,t)u(t), \\
\frac{\partial}{\partial x}P(x,t) &= xQ(x,t) + \mathbf{R}(x,t,t)P(t,t) + P(x,t)u(t) - 2Q(x,t)v(t).
\end{aligned}$$

Proof We use the statement and notation of the previous lemma, and an integration by parts, and get that

$$\begin{aligned}
\frac{\partial}{\partial x}Q(x,t) &= \operatorname{Ai}'(x) + \int_t^\infty \left(\frac{\partial}{\partial x}\mathbf{R}(x,y,t)\right)\operatorname{Ai}(y)\,dy \\
&= \operatorname{Ai}'(x) + \int_t^\infty \rho_t(x,y)\operatorname{Ai}(y)\,dy - \int_t^\infty \left(\frac{\partial}{\partial y}\mathbf{R}(x,y,t)\right)\operatorname{Ai}(y)\,dy \\
&= \operatorname{Ai}'(x) + \int_t^\infty (\mathbf{R}(x,t,t)\mathbf{R}(t,y,t) - Q(x,t)Q(y,t))\operatorname{Ai}(y)\,dy \\
&\quad + \mathbf{R}(x,t,t)\operatorname{Ai}(t) + \int_t^\infty \mathbf{R}(x,y,t)\operatorname{Ai}'(y)\,dy \\
&= \operatorname{Ai}'(x) + \mathbf{R}(x,t,t)(Q(t,t) - \operatorname{Ai}(t)) - Q(x,t)u(t) \\
&\quad + \mathbf{R}(x,t,t)\operatorname{Ai}(t) + P(x,t) - \operatorname{Ai}'(x) \\
&= P(x,t) + \mathbf{R}(x,t,t)Q(t,t) - Q(x,t)u(t).
\end{aligned}$$

For the second claim, we use similar calculations together with the Airy differential equation $\mathrm{Ai}''(x) = x\,\mathrm{Ai}(x)$, to write

$$\begin{aligned}
\frac{\partial}{\partial x}P(x,t) &= \mathrm{Ai}''(x) + \int_t^\infty \left(\frac{\partial}{\partial x}\mathbf{R}(x,y,t)\right)\mathrm{Ai}'(y)\,dy \\
&= x\,\mathrm{Ai}(x) + \int_t^\infty \rho_t(x,y)\,\mathrm{Ai}'(y)\,dy - \int_t^\infty \left(\frac{\partial}{\partial y}\mathbf{R}(x,y,t)\right)\mathrm{Ai}'(y)\,dy \\
&= x\,\mathrm{Ai}(x) + \int_t^\infty (\mathbf{R}(x,t,t) - \mathbf{R}(t,y,t) - Q(x,t)Q(y,t))\,\mathrm{Ai}'(y)\,dy \\
&\quad + \mathbf{R}(x,t,t)\,\mathrm{Ai}'(t) + \int_t^\infty \mathbf{R}(x,y,t)\,\mathrm{Ai}''(y)\,dy \\
&= x\,\mathrm{Ai}(x) + \mathbf{R}(x,t,t)(P(t,t) - \mathrm{Ai}'(t)) - Q(x,t)v(t) \\
&\quad + \mathbf{R}(x,t,t)\,\mathrm{Ai}'(t) + \int_t^\infty \mathbf{R}(x,y,t)\,y\,\mathrm{Ai}(y)\,dy.
\end{aligned}$$

To complete the computation, note that the last term can be written in the form

$$\begin{aligned}
&\int_t^\infty \mathbf{R}(x,y,t)\,y\,\mathrm{Ai}(y)\,dy \\
&\quad = x\int_t^\infty \mathbf{R}(x,y,t)\,\mathrm{Ai}(y)\,dy - \int_t^\infty \mathbf{R}(x,y,t)\,(x-y)\,\mathrm{Ai}(y)\,dy \\
&\quad = x(Q(x,t) - \mathrm{Ai}(x)) - \int_t^\infty (Q(x,t)P(y,t) - Q(y,t)P(x,t))\,\mathrm{Ai}(y)\,dy \\
&\quad = x(Q(x,t) - \mathrm{Ai}(x)) - Q(x,t)v(t) + P(x,t)u(t).
\end{aligned}$$

Combining these last two computation gives the claim. □

Lemma 2.42

$$\begin{aligned}
q'(t) &= p(t) - q(t)u(t), \\
p'(t) &= tq(t) + p(t)u(t) - 2q(t)v(t).
\end{aligned}$$

Proof Combining Lemmas 2.39 and 2.41, we have that

$$\begin{aligned}
\left(\frac{\partial}{\partial x} + \frac{\partial}{\partial t}\right)Q(x,t) &= -Q(t,t)\mathbf{R}(t,x,t) + P(x,t) + \mathbf{R}(x,t,t)Q(t,t) \\
&\quad - Q(x,t)u(t) = P(x,t) - Q(x,t)u(t).
\end{aligned}$$

It follows that

$$q'(t) = \frac{d}{dt}Q(t,t) = P(t,t) - Q(t,t)u(t) = p(t) - q(t)u(t).$$

Similarly, we have

$$\left(\frac{\partial}{\partial x}+\frac{\partial}{\partial t}\right)P(x,t) = -P(t,t)\mathbf{R}(t,x,t) + xQ(x,t) + \mathbf{R}(x,t,t)P(t,t)$$
$$+ P(x,t)u(t) - 2Q(x,t)v(t)$$
$$= xQ(x,t) + P(x,t)u(t) - 2Q(x,t)v(t),$$

which implies the claimed formula for $p'(t) = \frac{d}{dt}P(t,t)$ on setting $x = t$. □

Lemma 2.43

$$u'(t) = -q(t)^2,$$
$$v'(t) = -p(t)q(t).$$

Proof

$$\begin{aligned} u'(t) &= \frac{d}{dt}\left(\int_t^\infty Q(x,t)\,\mathrm{Ai}(x)\,dx\right) \\ &= -Q(t,t)\,\mathrm{Ai}(t) + \int_t^\infty \left(\frac{\partial}{\partial t}Q(x,t)\right)\mathrm{Ai}(x)\,dx \\ &= -q(t)\,\mathrm{Ai}(t) + \int_t^\infty -Q(t,t)\mathbf{R}(t,x,t)\,\mathrm{Ai}(x)\,dx \\ &= -q(t)\left(\mathrm{Ai}(t)\int_t^\infty \mathbf{R}(t,x,t)\,\mathrm{Ai}(x)\,dx\right) = -q(t)^2. \end{aligned}$$

Similarly, we have

$$\begin{aligned} v'(t) &= \frac{d}{dt}\left(\int_t^\infty Q(x,t)\,\mathrm{Ai}'(x)\,dx\right) \\ &= -q(t)\,\mathrm{Ai}'(t) + \int_t^\infty \left(\frac{\partial}{\partial t}Q(x,t\right)\mathrm{Ai}'(x)\,dx \\ &= -q(t)\left(\mathrm{Ai}'(t) + \int_t^\infty \mathbf{R}(t,x,t)\,\mathrm{Ai}'(x)\,dx\right) = -p(t)q(t). \end{aligned}$$ □

Lemma 2.44 *$q(t)$ satisfies the Painlevé II differential equation* (2.99).

Proof Using the results of the last few lemmas we have

$$\begin{aligned} \frac{d}{dt}\left(u^2 - 2v - q^2\right) &= 2uu' - 2v' - 2qq' \\ &= 2u(-q^2) + 2pq - 2q(p - qu) \equiv 0, \end{aligned}$$

so $u^2 - 2v - q^2$ is a constant function, which must be 0 since $u, v, q \to 0$ as $t \to \infty$. It follows that

$$\begin{aligned} q'' &= (p - qu)' = p' - q'u - qu' = tq + pu - 2qv - (p - qu)u - q(-q^2) \\ &= tq + q^3 + q(u^2 - 2v) = tq + q^3 + q^3 = tq + 2q^3. \end{aligned} \qquad \square$$

Lemma 2.45

$$(\log F_2(t))'' = -q(t)^2.$$

Proof By Lemmas 2.37 and 2.40 we have

$$\left(\frac{\partial}{\partial x} + \frac{\partial}{\partial y} + \frac{\partial}{\partial t}\right)\mathbf{R}(x, y, t) = -Q(x, t)Q(y, t),$$

so we get that

$$(\log F_2(t))'' = \frac{d}{dt}\mathbf{R}(t, t, t) = \left[\left(\frac{\partial}{\partial x} + \frac{\partial}{\partial y} + \frac{\partial}{\partial t}\right)\mathbf{R}(x, y, t)\right]\Bigg|_{x=y=t} = -q(t)^2. \qquad \square$$

Proof of Theorem 2.33 We defined the function $q(t)$ mentioned in the theorem explicitly and proved that it satisfies (2.99) and (2.100), so it remains to prove the representation (2.101) for $F_2(t)$. Letting $\lambda(t) = \log F_2(t)$, since $\lambda(t) = O(e^{-t})$ and $\lambda'(t) = \mathbf{R}(t, t, t) = O(e^{-t})$ as $t \to \infty$ we get that

$$\begin{aligned} \lambda(t) &= -\int_t^\infty \lambda'(y)\, dy = \int_t^\infty \left(\int_y^\infty \lambda''(x)\, dx\right) dy \\ &= \int_t^\infty \lambda''(x)\left(\int_t^x dy\right) dx = -\int_t^\infty (x - t)q(x)^2\, dx. \end{aligned} \qquad \square$$

2.11 Edge statistics of Plancherel measure

Our remaining goal in this chapter is to prove Theorem 2.3. In fact, in our proof of the weaker Theorem 2.2 we already came very close to proving Theorem 2.3, without noticing. A bit like the ending of a good detective story, all that is needed now is a small additional refinement of our senses (delivered perhaps in a dramatic speech by a colorful French-accented detective), which will enable us to look back at the story so far and finally make sense of the mystery.

The main missing piece in the picture is a better understanding of the

Airy ensemble $X_{\text{Airy}} = (\zeta_1, \zeta_2, \ldots)$, claimed in Theorem 2.3 to be the limiting process to which the scaled largest rows of a Plancherel-random partition $\lambda^{(n)}$ converge as $n \to \infty$. We have not proved that it exists, let alone described its properties. Doing so rigorously would actually require a more detailed digression into the general theory of random point processes, which is beyond the scope of this book. We merely quote the relevant results here without proof and refer the interested reader to Sections 1 and 2 of [121] or Section 4.2 (especially Proposition 4.2.30) of [4] for more details.[10]

To state the result, we use the concept of a Fredholm determinant of the form $\mathbf{I} + \mathbf{T}$, where $\mathbf{T}$ is a kernel $\mathbf{T}\colon E \times E \to \mathbb{R}$ for some Borel set $E \subseteq \mathbb{R}$. This is defined by

$$\det(\mathbf{I} + \mathbf{T}) = 1 + \sum_{n=1}^{\infty} \frac{1}{n!} \int_E \cdots \int_E \det_{i,j=1}^{n} \Big(\mathbf{T}(x_i, x_j)\Big)\, dx_1 \ldots dx_n,$$

provided the integrals and series converge absolutely. This is analogous to our previous definition of Fredholm determinants for kernels acting on $\ell^2(\Omega)$ where Ω is a countable set. The general theory of Fredholm determinants provides a unified framework for treating both of these scenarios simultaneously, but we shall not consider these generalities here.

Theorem 2.46 (a) *There exists a unique random point process X_{Airy} on $\mathbb{R}$ whose correlation functions $\rho^{(n)}_{X_{\text{Airy}}} : \mathbb{R}^n \to \mathbb{R}$ (as defined by* (2.5)*) are given by*

$$\rho^{(n)}_{X_{\text{Airy}}}(x_1, \ldots, x_n) = \det_{i,j=1}^{n} \Big(\mathbf{A}(x_i, x_j)\Big), \qquad (x_1, \ldots, x_n \in \mathbb{R}).$$

(b) *For any disjoint Borel sets $E_1, \ldots, E_k \subseteq \mathbb{R}$, the function*

$$\det\left(\mathbf{I} - \sum_{j=1}^{k} z_j \mathbf{P}_{A_j} \cdot \mathbf{A} \cdot \mathbf{P}_{E_1 \cup \ldots \cup E_k}\right),$$

is defined for any complex numbers $z_1, \ldots, z_k$ and is an entire function of these variables.

(c) *If $E_1, \ldots, E_k \subseteq \mathbb{R}$ are disjoint Borel sets, then for any nonnegative* *integers* $n_1, \dots, n_k$ *we have that*

$$\mathbb{P}\left(\bigcap_{j=1}^{k}\left\{|X_{\text{Airy}} \cap E_j| = n_j\right\}\right)$$
$$= \frac{(-1)^N}{n_1! \dots n_k!} \frac{\partial^N}{\partial z_1^{n_1} \dots \partial z_k^{n_k}}\bigg|_{z_1=\dots=z_k=1} \det\left(\mathbf{I} - \sum_{j=1}^{k} z_j \mathbf{P}_{A_j} \cdot \mathbf{A} \cdot \mathbf{P}_{E_1 \cup \dots \cup E_k}\right). \tag{2.106}$$

Note that (2.106) implies in particular that

$$\mathbb{P}\Big(X_{\text{Airy}} \subseteq (-\infty, t)\Big) = \mathbb{P}\Big(\big|X_{\text{Airy}} \cap [t, \infty)\big| = 0\Big)$$
$$= \det\Big(\mathbf{I} - \mathbf{I}_{[t,\infty)}\mathbf{A}\Big) = F_2(t).$$

The event on the left-hand side is almost surely equal to the event that X_{Airy} has a maximal element ζ_1 that satisfies $\zeta_1 < t$ (the two events differ in the zero-probability event that $X_{\text{Airy}} = \emptyset$). So in particular we get that the Airy ensemble almost surely has a maximal element, whose distribution is the Tracy–Widom distribution $F_2(t)$.

Our tools are now sufficiently sharpened to prove Theorem 2.3. The proof will use the same ideas as were used previously for the proof of Theorem 2.2, with slightly more elaborate continuity arguments.

As before, we first prove an analogous version for Poissonized Plancherel measure, and then use the de-Poissonization principle from Lemma 2.31.

Theorem 2.47 *For each $\theta > 0$, let $\lambda^{\langle\theta\rangle}$ denote a random partition chosen according to Poissonized Plancherel measure with parameter θ, let $\lambda_j^{\langle\theta\rangle}$ denote the length of its jth row, and denote $\bar{\lambda}_j^{\langle\theta\rangle} = \theta^{-1/6}(\lambda_j^{\langle\theta\rangle} - 2\sqrt{\theta})$. Then for each $k \geq 1$, we have the convergence in distribution*

$$(\bar{\lambda}_1^{\langle\theta\rangle}, \dots, \bar{\lambda}_k^{\langle\theta\rangle}) \xrightarrow{d} (\zeta_1, \dots, \zeta_k) \text{ as } \theta \to \infty,$$

where $\{\zeta_1 > \zeta_2 > \dots\}$ denote the elements of the Airy ensemble arranged in decreasing order as in (2.7).

Proof Consider the first k of the "positive" Frobenius coordinates of $\lambda^{\langle\theta\rangle}$, namely

$$p_j^{\langle\theta\rangle} = \lambda_j^{\langle\theta\rangle} - j, \qquad j = 1, \dots, k,$$

and define their rescaled counterparts $\bar{p}_j^{\langle\theta\rangle} = \theta^{-1/6}(p_j^{\langle\theta\rangle} - 2\sqrt{\theta})$. Clearly it will be enough to prove that

$$(\bar{p}_1^{\langle\theta\rangle}, \ldots, \bar{p}_k^{\langle\theta\rangle}) \xrightarrow{d} (\zeta_1, \ldots, \zeta_k) \text{ as } \theta \to \infty. \tag{2.107}$$

To do this, we consider events of a specific type associated with the random process $(\bar{p}_1^{\langle\theta\rangle}, \ldots, \bar{p}_k^{\langle\theta\rangle})$. Fix numbers $t_1 > t_2 > \ldots > t_k$ in $\mathbb{R}$, and define

$$E_\theta = \left\{\bar{p}_k^{\langle\theta\rangle} \le t_k < \bar{p}_{k-1}^{\langle\theta\rangle} \le t_{k-1} < \ldots < \bar{p}_2^{\langle\theta\rangle} \le t_2 < \bar{p}_1^{\langle\theta\rangle} \le t_1\right\}.$$

Similarly, let E be the analogous event for the largest elements $(\zeta_1, \ldots, \zeta_k)$ of the Airy ensemble, namely

$$E = \left\{\zeta_k \le t_k < \zeta_{k-1} \le t_{k-1} < \ldots < \zeta_2 \le t_2 < \zeta_1 \le t_1\right\}.$$

It is not difficult to see that in order to prove the convergence in distribution (2.107) it will be enough to show that

$$\mathbb{P}(E_\theta) \to \mathbb{P}(E) \qquad \text{as } \theta \to \infty. \tag{2.108}$$

The advantage of working with such events is that they can be expressed in terms of the underlying determinantal point processes in a way that does not refer to the labelling of specific elements of the process. More precisely, if we denote by Y_θ the process $\mathrm{Fr}(\lambda^{\langle\theta\rangle})$, then we have

$$E = \left\{|X_{\mathrm{Airy}} \cap (t_1, \infty)| = 0\right\} \cap \bigcap_{j=2}^{k} \left\{|X_{\mathrm{Airy}} \cap (t_j, t_{j-1}]| = 1\right\},$$

$$E_\theta = \left\{|Y_\theta \cap (\tilde{t}_1, \infty)| = 0\right\} \cap \bigcap_{j=2}^{k} \left\{|Y_\theta \cap (\tilde{t}_j, \tilde{t}_{j-1}]| = 1\right\},$$

where as before we denote $\tilde{t}_j = 2\sqrt{\theta} + t_j\theta^{1/6}$. But now we see that such a representation fits perfectly with the framework that was presented in Section 2.4 and with Theorem 2.46, since the probabilities of such events can be expressed in terms of Fredholm determinants. Define intervals $T_1, \ldots, T_n$ by $T_1 = (t_1, \infty)$ and $T_j = (t_j, t_{j-1}]$ for $2 \le j \le k$. Similarly, denote $\tilde{T}_1 = (\tilde{t}_1, \infty)$ and $\tilde{T}_j = (\tilde{t}_j, \tilde{t}_{j-1}]$ for $2 \le j \le k$. Denote $T = \cup_{j=1}^k T_j = (t_k, \infty)$ and $\tilde{T} = \cup_{j=1}^k \tilde{T}_j = (\tilde{t}_k, \infty)$. Then, by (2.106) and Proposition 2.17, we have

that

$$\mathbb{P}(E) = (-1)^{k-1} \frac{\partial^{k-1}}{\partial z_2 \ldots \partial z_k}\Big|_{z_1=\ldots=z_k=1} \det\left(\mathbf{I} - \sum_{j=1}^{k} z_j \mathbf{P}_{\tilde{T}_j} \cdot \mathbf{A}_{\tilde{T}} \cdot \mathbf{P}_{\tilde{T}}\right), \quad (2.109)$$

$$\mathbb{P}(E_\theta) = (-1)^{k-1} \frac{\partial^{k-1}}{\partial z_2 \ldots \partial z_k}\Big|_{z_1=\ldots=z_k=1} \det\left(\mathbf{I} - \sum_{j=1}^{k} z_j \mathbf{P}_{T_j} \cdot \mathbf{K}_\theta \cdot \mathbf{P}_{T}\right). \quad (2.110)$$

It will therefore come as no surprise that the same methods and estimates that we used previously in the proof of Theorem 2.29 will apply to prove the convergence in this case. Here's how to make the necessary changes. Observe that the computations in the proof of Theorem 2.29 can be used with little modification to show that

$$\det\left(\mathbf{I} - \sum_{j=1}^{k} z_j \mathbf{P}_{T_j} \mathbf{K}_\theta \mathbf{P}_{T}\right) \xrightarrow[\theta\to\infty]{} \det\left(\mathbf{I} - \sum_{j=1}^{k} z_j \mathbf{P}_{\tilde{T}_j} \mathbf{A} \mathbf{P}_{\tilde{T}}\right), \quad (2.111)$$

for any complex numbers $z_1, \ldots, z_k$, and that in fact the convergence is uniform if one assumes that $|z_j| \le \xi$ for some fixed ξ. Note also that, according to Proposition 2.17 and Theorem 2.46, the convergence is of a family of entire functions to another entire function. Now use a bit of complex analysis: if $f(z_1, \ldots, z_k)$ is an analytic function of several complex variables in a neighborhood of a point $\mathbf{p} = (p_1, \ldots, p_k) \in \mathbb{C}^k$, it follows by an iterative use of the well-known Cauchy integral formula that its partial derivatives at $\mathbf{p}$ can be expressed as multidimensional contour integrals, namely

$$\frac{\partial^N f(\mathbf{p})}{\partial z_1^{n_1} \ldots \partial z_k^{n_k}} = \frac{\prod_{j=1}^k n_j!}{(2\pi i)^k} \int_{|z_1-p_1|=\epsilon} \cdots \int_{|z_k-p_k|=\epsilon} \frac{f(w_1, \ldots, w_k)}{\prod_{j=1}^k (w_j - p_j)^{n_j+1}}\, dw_1 \ldots dw_k,$$

where ϵ is a small enough positive number and $N = \sum_{j=1}^k n_j$. Applying this to our setting gives the desired claim that $\mathbb{P}(E_\theta) \to \mathbb{P}(E)$ as $\theta \to \infty$, since the uniform convergence (2.111) implies convergence of the multidimensional contour integrals. □

Proof of Theorem 2.3 We derive Theorem 2.3 from Theorem 2.47 in a similar way to how Theorem 2.2 was derived from Theorem 2.29 in the previous section. Again, we apply Lemma 2.31, but the quantities $P_n(t)$ and $Q_\theta(t)$ are replaced by probabilities of events depending on the first k rows of the random Young diagrams $\lambda^{(n)}$ and $\lambda^{\langle\theta\rangle}$. Let $x_1, \ldots, x_k \in \mathbb{R}$, and for $n \geq k$ denote

$$P_n = P_n(x_1, \dots, x_k) = \mathbb{P}\left(\lambda_1^{(n)} \leq x_1,\ \lambda_2^{(n)} \leq x_2,\ \dots,\ \lambda_k^{(n)} \leq x_k\right). \qquad (2.112)$$

Define P_n to be 1 for $n \leq k$. Next, denote

$$Q_\theta = Q_\theta(x_1, \dots, x_k) := e^{-\theta} \sum_{m=0}^{\infty} \frac{\theta^m}{m!} P_m$$
$$= \mathbb{P}\left(\lambda_1^{\langle\theta\rangle} \leq x_1,\ \lambda_2^{\langle\theta\rangle} \leq x_2,\ \dots,\ \lambda_k^{\langle\theta\rangle} \leq x_k\right).$$

Then, similarly to (2.97), we have that $Q_\theta = e^{-\theta} \sum_{n=0}^{\infty} \frac{\theta^n}{n!} P_n$, and, similarly to (2.96), the sequence $P_1, P_2, \dots$ is easily seen to be nonincreasing (Exercise 2.20). Therefore Lemma 2.31 applies and we deduce that

$$Q_{\phi_n}(\tilde{x}_1, \dots, \tilde{x}_k) - f(\phi_n) \leq P_n(\tilde{x}_1, \dots, \tilde{x}_k) \leq Q_{\theta_n}(\tilde{x}_1, \dots, \tilde{x}_k) + f(\theta_n),$$

where we denote $\tilde{x}_i = 2\sqrt{n} + x_i n^{1/6}$. Now continue as in the proof of Theorem 2.2. $\square$

2.12 Epilogue: convergence of moments

We are close to the end of our journey into the study of the permutation statistic $L(\sigma)$, and are almost ready to declare victory on the Ulam–Hammersley problem, which has turned out to be quite formidable and fascinating, in a way probably unsuspected by the story's original protagonists. To close a circle with the discussion at the beginning of Chapter 1, note, however, that we started out thinking about the *average* size of $L(\sigma_n)$, where σ_n is a uniformly random permutation, and only later switched to thinking about its probability distribution. Looking back, we claimed in (1.1) that $\ell_n = \mathbb{E}L(\sigma_n)$ has the asymptotic behavior

$$\ell_n = 2\sqrt{n} + cn^{1/6} + o(n^{1/6}) \quad \text{as } n \to \infty.$$

In light of our current understanding, it seems clear that the constant c is none other than the expected value of the Tracy–Widom distribution F_2, that is,

$$c = \int_{-\infty}^{\infty} x\, dF_2(x),$$

which has been numerically evaluated [18] to equal $-1.7710868074\dots$. Still, from a rigorous point of view we have not actually proved this claim, since (as any probabilist worth her salt knows) convergence in distribution of a sequence of random of variables does not in general imply convergence of expectations.

The question is resolved by the following more general result proved by Baik, Deift, and Johansson [11], which we quote without proof.

Theorem 2.48 (Convergence of moments for $L(\sigma_n)$) *For any integer $k \geq 1$ we have*

$$\mathbb{E}\left(\frac{L(\sigma_n) - 2\sqrt{n}}{n^{1/6}}\right)^k \to \int_{-\infty}^{\infty} x^k \, dF_2(x) \quad \textit{as } n \to \infty.$$

From general facts of probability theory it follows that one can deduce convergence of moments from convergence in distribution if one has sufficiently uniform control of the decay rate of the tail of the distributions of the converging sequence. In our case, it would not be too difficult to use the bound (2.78) to obtain this type of control for the positive part of the tail. However, it is not clear whether the approach we pursued in this chapter could be used to obtain corresponding bounds for the *negative* part of the tail. I leave this as an open problem to stimulate readers' thinking. It should be noted that Baik, Deift, and Johansson's approach to proving Theorems 2.2 and 2.48 was based on different ideas, starting from another remarkable identity relating the distribution of the "Poissonization" of $L(\sigma_n)$ to the Bessel functions (see Exercise 2.21). Using advanced asymptotic analysis techniques they were able to prove the following tail bounds for the distributions of the scaled random variables $X_n = n^{-1/6}(L(\sigma_n) - 2\sqrt{n})$:

$$\mathbb{P}(X_n \geq t) \geq 1 - Ce^{-ct^{3/5}} \qquad (T < t < n^{5/6} - 2n^{1/3}),$$
$$\mathbb{P}(X_n \leq t) \leq Ce^{c|t|^3} \qquad (-2n^{1/3} < t < -T),$$

for some constants $T, C, c > 0$. Combining this with the convergence in distribution of X_n to F_2, standard techniques can be used to deduce convergence of moments.

Exercises

2.1 (☕) If $\lambda^{\langle\theta\rangle}$ denotes a random Young diagram chosen according to Poissonized Plancherel measure P_θ, show that the length $\lambda_1^{\langle\theta\rangle}$ of its first row is equal in distribution to the random variable $Y_{0,\sqrt{\theta}}$ where $Y_{s,t}$ is defined in (1.5) (Section 1.4).

2.2 (☕☕) Prove the Cauchy determinant identity (2.14). There are many proofs; one relatively easy one involves performing a simple elementary operation on columns 2 through n of the matrix, extracting factors common to rows and columns and then performing another simple elementary operation on rows 2 through n.

2.3 Let X be a determinantal point process in a finite set Ω, and let $\mathbf{K}\colon \Omega\times\Omega \to \mathbb{R}$ denote its correlation kernel.

(a) (☕) Show that the complementary point process $X^c = \Omega \setminus X$ is determinantal and identify its correlation kernel.

(b) (☕☕) Show that X has a configuration kernel if and only if $\mathbf{I} - \mathbf{K}$ is invertible. In the case when X has a configuration kernel $\mathbf{L}$, identify the configuration kernel of the complementary process X^c.

2.4 (☕) Let Ω be a finite or countable set. Let $p\colon \Omega \to [0,1]$ be a function. Define a point process X on Ω by specifying that each $x \in \Omega$ is in X with probability $p(x)$, or not in X with the complementary probability $1 - p(x)$, independently for all x (that is, in measure-theoretic language, the distribution of X is a product measure on $\{0,1\}^\Omega$). Show that X is determinantal and find its correlation kernel. Identify when X has a configuration kernel and find its configuration kernel when it has one.

2.5 (a) (☕☕☕) Let $M = (m_{i,j})_{i,j=1}^n$ denote a square matrix, and recall that its **adjugate matrix** $\operatorname{adj}(M)$ is the matrix whose (i,j)th entry is $(-1)^{i+j}$ times the (j,i)-minor of M. Let $A \subseteq \{1,\dots,n\}$, and denote $A^c = \{1,\dots,n\} \setminus A$. Prove the following identity due to Jacobi:

$$\det((\operatorname{adj}(M))_A) = \det(M)^{|A|-1} \det(M_{A^c}) \tag{2.113}$$

(where the meaning of the notation M_E is the same as in Section 2.4). In the case when M is invertible, because of the well-known relation $M^{-1} = (\det(M))^{-1} \operatorname{adj}(M)$, this can be written in the equivalent form

$$\det(M_A) = \det(M)\det((M^{-1})_{A^c}).$$

(b) (☕☕☕) Deduce from Jacobi's identity and Proposition 2.10 a new proof of Proposition 2.9.

2.6 (☕) Prove Proposition 2.12.

2.7 (☕☕☕) The Euler gamma function is defined by

$$\Gamma(t) = \int_0^\infty e^{-x} x^{t-1}\,dx \qquad (t > 0).$$

Prove or look up (e.g., in [5]) the proofs of the following properties it satisfies:

(a) $\Gamma(n+1) = n!$ for integer $n \geq 0$.

(b) $\Gamma(t+1) = t\,\Gamma(t)$ for $t > 0$.

(c) $\Gamma(1/2) = \sqrt{\pi}$.

(d) $\Gamma(t)$ has the infinite product representation

$$\Gamma(t)^{-1} = te^{\gamma t} \prod_{n=1}^{\infty} \left(1 + \frac{t}{n}\right) e^{-t/n},$$

where $\gamma = \lim_{n\to\infty} \left(1 + \frac{1}{2} + \frac{1}{3} + \ldots + \frac{1}{n} - \log n\right)$ is the Euler–Mascheroni constant.

(e) $\psi(t) := \dfrac{\Gamma'(t)}{\Gamma(t)} = -\gamma + \displaystyle\sum_{n=0}^{\infty} \left(\frac{1}{n+1} - \frac{1}{n+t}\right)$ for $t > 0$.

(f) $\Gamma(t) = (1 + o(1)) \sqrt{\dfrac{2\pi}{t}} \left(\dfrac{t}{e}\right)^t$ as $t \to \infty$.

(g) $\psi(t) = \log t + O(1)$ as $t \to \infty$. (**Hint:** Find a formula for $\psi(t)$ when t is an integer).

(h) $\Gamma(t)$ can be analytically continued to a meromorphic function on $\mathbb{C}$ with simple poles at $0, -1, -2, \ldots$ and no zeroes.

(i) $\dfrac{\pi}{\Gamma(t)\Gamma(1-t)} = \sin(\pi t)$ for all $t \in \mathbb{C}$.

2.8 (☕☕) Prove (2.51), by first using the fact that $J_\alpha(z), J_{-\alpha}(z)$ are solutions of the Bessel differential equation (2.41) to show that

$$\frac{\partial}{\partial z}\left[z\left(J_\alpha(2z)J'_{-\alpha}(2z) - J'_{-\alpha}(2z)J_{\alpha-1}(2z)\right)\right] = 0.$$

This implies that the left-hand side of (2.51) is of the form $g(\alpha)/z$ for some function $g(\alpha)$. Then examine the asymptotic behavior of the left-hand side of (2.51) as $z \to 0$, to infer that for noninteger values of α (and hence by continuity also for integer values) $g(\alpha) = -(\Gamma(\alpha)\Gamma(1-\alpha))^{-1}$, which is equal to $-\frac{1}{\pi}\sin(\pi\alpha)$ by a classical identity for the gamma function (see Exercise 2.7).

2.9 (☕☕) Prove (2.52). It is helpful to first prove and then use the following "binomial inversion principle": If $(a_n)_{n\geq 0}$ and $(b_n)_{n\geq 0}$ are two sequences such that $a_n = \sum_{k=0}^{n}(-1)^k\binom{n}{k}b_k$ for all n, then $(b_n)_n$ can be obtained from $(a_n)_n$ using the symmetric expression $b_n = \sum_{k=0}^{n}(-1)^k\binom{n}{k}a_k$.

2.10 (☕☕) Prove (2.53).

2.11 (☕) Show that

$$J_{1/2}(z) = \sqrt{\frac{2}{\pi z}} \sin z,$$
$$J_{-1/2}(z) = \sqrt{\frac{2}{\pi z}} \cos z.$$

2.12 (☕☕) Prove the following identities involving Bessel functions $J_n(z)$ of integer order:

$$\exp\left[\frac{z}{2}\left(t - \frac{1}{t}\right)\right] = \sum_{n=-\infty}^{\infty} J_n(z) t^n,$$
$$\cos(z \sin\theta) = J_0(z) + 2\sum_{n=1}^{\infty} J_{2n}(z)\cos(2n\theta),$$
$$\sin(z \sin\theta) = 2\sum_{n=0}^{\infty} J_{2n+1}(z)\sin((2n+1)\theta),$$
$$\cos(z \cos\theta) = J_0(z) + 2\sum_{n=1}^{\infty} (-1)^n J_{2n}(z)\cos(2n\theta),$$
$$\sin(z \cos\theta) = 2\sum_{n=0}^{\infty} (-1)^n J_{2n+1}(z)\sin((2n+1)\theta),$$
$$J_n(z) = \frac{1}{\pi}\int_0^{\pi} \cos(z\sin\theta - n\theta)\, d\theta \qquad (n \in \mathbb{Z}).$$

2.13 Let $\lambda = (\lambda_1, \ldots, \lambda_k)$ be an integer partition, with its Frobenius coordinates being denoted by $\mathrm{Fr}(\lambda) = \left\{p_1 + \frac{1}{2}, \ldots, p_d + \frac{1}{2}, -q_1 - \frac{1}{2}, \ldots, -q_d - \frac{1}{2}\right\}$ as in (2.9). Denote

$$D(\lambda) = \{\lambda_j - j \mid j = 1, 2, \ldots\},$$

with the convention that $\lambda_j = 0$ if $j > k$.

(a) (☕☕) Show that $\mathrm{Fr}(\lambda) = (D(\lambda) + \frac{1}{2}) \triangle \{-\frac{1}{2}, -\frac{3}{2}, \ldots\}$, where $A \triangle B = (A \setminus B) \cup (B \setminus A)$ denotes the symmetric difference of sets.

(b) (☕☕☕☕) Show that if $\lambda^{\langle\theta\rangle}$ denotes a Young diagram chosen according to Poissonized Plancherel measure P_θ, then $D(\lambda^{\langle\theta\rangle})$ is a determinantal point process and its correlation kernel is $\mathbf{J}_\theta(\cdot, \cdot)$.

2.14 (☕) Verify that the two relations in (2.54) are equivalent. Use the fact that $\mathbf{L}$ and $\mathbf{K}$ have the symmetries $\mathbf{L}(x, y) = -\mathbf{L}(y, x) = -\mathbf{L}(-x, -y)$ and $\mathbf{K}(x, y) = -\mathbf{K}(y, x) = -\mathbf{K}(-x, -y)$.

2.15 (a) (☕) Use (2.67) to show that the Airy function $\mathrm{Ai}(x)$ satisfies the Airy differential equation (2.8).

(b) (☕) Express Ai(0) and Ai′(0) in terms of the Euler gamma function, and use these expressions and the Airy differential equation (2.8) to obtain the power series expansion of Ai(x) around $x = 0$.

2.16 (☕) Prove (2.73).

2.17 (☕) Show that the Airy function has the following representation in terms of Bessel functions.

$$\mathrm{Ai}(x) = \begin{cases} \frac{1}{3}\sqrt{x}\left(I_{-1/3}\left(\frac{2}{3}x^{3/2}\right) - I_{1/3}\left(\frac{2}{3}x^{3/2}\right)\right) & \text{if } x \geq 0, \\ \frac{1}{3}\sqrt{-x}\left(J_{-1/3}\left(\frac{2}{3}(-x)^{3/2}\right) + J_{1/3}\left(\frac{2}{3}(-x)^{3/2}\right)\right) & \text{if } x \leq 0. \end{cases}$$

Here, $I_\alpha(z)$ denotes a modified Bessel function of order α (see Exercise 2.21 for the definition).

2.18 (☕) Use (2.28) to show that the Airy kernel $\mathbf{A}(\cdot,\cdot)$ is a positive-semidefinite kernel, in the sense that for any $x_1, \ldots, x_k \in \mathbb{R}$, the matrix $(\mathbf{A}(x_i, x_j))_{i,j=1}^n$ is a positive-semidefinite matrix.

2.19 (☕☕☕) Show that $f(\theta)$ defined in (2.94) satisfies

$$f(\theta) \leq C\,\theta^{-\alpha}$$

for some constants $C, \alpha > 0$.

2.20 (☕) Prove that for fixed $x_1, \ldots, x_n$, P_n defined in (2.112) is nonincreasing as a function of n.

2.21 (☕☕☕) The goal of this exercise is to present another remarkable identity relating the distribution of the maximal increasing subsequence length in a random permutation and the Bessel functions. This identity, due to Ira M. Gessel [47], was the starting point of Baik, Deift, and Johansson's approach to determining the limiting distribution of $L(\sigma)$ in the seminal paper [11].

To formulate the identity, let σ_n denote as before a uniformly random permutation of order n. Let

$$F_n(t) = \mathbb{P}(L(\sigma_n) \leq t)$$

denote the cumulative distribution function of $L(\sigma_n)$, and for $\theta > 0$ let

$$G_\theta(t) = e^{-\theta} \sum_{n=0}^{\infty} \frac{\theta^n}{n!} F_n(t)$$

denote the "Poissonized average" of the distribution functions $F_n(t)$. As we have seen before, $G_\theta(t)$ can itself be thought of as the distribution function of $\lambda_1^{\langle\theta\rangle}$, the length of the first row of a Young diagram chosen according to Poissonized Plancherel measure P_θ.

One other definition we need is that of the **modified Bessel functions**. For

$\alpha \in \mathbb{R}$, the modified Bessel function of the first kind of order α is defined by

$$I_\alpha(z) = \left(\frac{z}{2}\right)^\alpha \sum_{m=0}^{\infty} \frac{(z/2)^{2m}}{m!\,\Gamma(m+\alpha+1)}.$$

(Trivially, $I_\alpha(\cdot)$ is related to $J_\alpha(\cdot)$ by $I_\alpha(z) = (-i)^\alpha J_\alpha(iz)$.)

Theorem 2.49 (The Gessel–Bessel identity) *For integer $t \geq 1$, the distribution function $G_\theta(t)$ can be expressed as*

$$G_\theta(t) = e^{-\theta} \det_{j,k=0}^{t-1} \left(I_{j-k}(2\sqrt{\theta}) \right). \tag{2.114}$$

Note that the entries of the matrix on the right-hand side are constant along diagonals. A matrix with this property is called a **Toeplitz matrix**.

Below we outline the steps required to prove (2.114). This elegant proof is due to Baik, Deift, and Johansson [11].

(a) Express $F_n(t)$ as a sum over Young diagrams, where, crucially, instead of restricting the length of the first row to be at most t, impose the restriction on the length of the first *column* (or, which is the same thing, on the number of rows).

(b) The sum will involve the ubiquituous function d_λ enumerating Young tableaux of given shape. Express this factor using one of the variants (2.10) we derived for the hook-length formula, and manipulate it to bring it to the form

$$F_n(t) = n! \sum_{r=1}^{t} \frac{1}{r!} \sum_{\substack{s_1,\ldots,s_r \geq 1 \\ s_1+\ldots+s_r = n+r(r-1)/2}} \prod_{1 \leq j < k \leq r} (s_j - s_k)^2 \prod_{j=1}^{r} \frac{1}{(s_j!)^2}.$$

(c) Show that in the Poissonized average the annoying constraint in the r-fold summation disappears, so that we end up with the representation

$$G_\theta(t) = e^{-\theta} \left(1 + \sum_{r=1}^{t} \theta^{-r(r-1)/2} H_r(\theta) \right) \tag{2.115}$$

where

$$H_r(\theta) = \frac{1}{r!} \sum_{s_1,\ldots,s_r=1}^{\infty} \prod_{1 \leq j < k \leq r} (s_j - s_k)^2 \prod_{j=1}^{r} \frac{\theta^{s_j}}{(s_j!)^2}.$$

(d) Show that $H_r(\theta)$ can be rewritten as

$$H_r(\theta) = \det_{j,k=0}^{r-1} \left(\sum_{s=1}^{\infty} s^{j+k} \frac{\theta^s}{(s!)^2} \right)$$

(e) Prove that if $q_0(x), q_1(x), \ldots$ is any sequence of monic polynomials such that $\deg q_j = j$, then

$$H_r(\theta) = \det_{j,k=0}^{r-1}\left(\sum_{s=1}^{\infty} q_j(s)q_k(s)\frac{\theta^s}{(s!)^2}\right)$$

(f) Use the above identity with $q_0(x) = 1$ and $q_j(x) = x(x-1)\ldots(x-j+1)$ to show that

$$H_r(\theta) = \theta^{r(r-1)/2}\left[\det_{j,k=0}^{r-1}\left(I_{j-k}(2\sqrt{\theta})\right) - \det_{j,k=0}^{r-2}\left(I_{j-k}(2\sqrt{\theta})\right)\right],$$

and combine this with (2.115) to get (2.114).

2.22 (☕☕☕☕) Develop a proof of Nicholson's approximation for Bessel functions (Theorem 2.27) that does not rely on the difficult (and seemingly incomplete) details given in Section 8.43 of Watson's book [145]. Here is a suggested approach. First, prove that the asymptotic relations

$$J_\alpha(\alpha) = (1+o(1))\frac{\Gamma(\frac{1}{3})}{2^{2/3}3^{1/6}\pi}\alpha^{-1/3},$$

$$J'_\alpha(\alpha) = (1+o(1))\frac{3^{1/6}\Gamma(\frac{2}{3})}{2^{1/3}\pi}\alpha^{-2/3},$$

hold as $\alpha \to \infty$; this is less difficult – see, for example, Section 8.2 of [145]. Next, observe that in the case $\alpha = 2z + xz^{1/3}$, the recurrence relation (2.45) takes the form

$$J_{\alpha+1}(2z) - 2J_\alpha(2z) + J_{\alpha-1}(2z) = xz^{-2/3}J_\alpha(2z).$$

The left-hand side is a discrete second derivative, so this relation is immediately seen to be a discrete approximation to the Airy differential equation (2.8). Combine these observations with the evaluations of $\mathrm{Ai}(0), \mathrm{Ai}'(0)$ from Exercise 2.15 to obtain the result (or, at least, a reasonably convincing heuristic explanation of why one should expect to see the Airy function emerging in this asymptotic regime).

2.23 (☕☕☕☕☕) Find a guess-free proof of Theorem 2.19, that is, a proof which derives the correct expression for $\mathbf{K}_\theta = \mathbf{L}_\theta(\mathbf{I}+\mathbf{L}_\theta)^{-1}$ without having to know it in advance.

2.24 (☕☕☕☕) Prove Theorem 2.48.

2.25 (a) (☕) If $M = (m_{i,j})_{i,j=1}^n$ is a square matrix, define submatrices $M_{\mathrm{NW}}, M_{\mathrm{NE}}$,

$M_{\text{SW}}, M_{\text{SE}}, M_{\text{NESW}}$ by

$$\begin{aligned} M_{\text{NW}} &= (m_{i,j})_{2\le i,j\le n}, \\ M_{\text{NE}} &= (m_{i,j})_{2\le i\le n,\,1\le j\le n-1}, \\ M_{\text{SW}} &= (m_{i,j})_{1\le i\le n-1,\,2\le j\le n}, \\ M_{\text{SE}} &= (m_{i,j})_{1\le i,j\le n-1}, \\ M_{\text{NESW}} &= (m_{i,j})_{2\le i,j\le n-1}. \end{aligned}$$

Show that a special case of Jacobi's identity (2.113) from Exercise 2.5 gives the relation

$$|M| \cdot |M_{\text{NESW}}| = |M_{\text{NW}}| \cdot |M_{\text{SE}}| - |M_{\text{NE}}| \cdot |M_{\text{SW}}| \tag{2.116}$$

(where $|\cdot|$ denotes the determinant of a matrix).

(b) (☕☕) The above identity gives rise to **Dodgson's condensation method**, a method for computing determinants discovered by Charles L. Dodgson (better known by his nom de plume Lewis Carroll) in 1866. The method works by recursively computing a pyramid of numbers $(d_{i,j}^k)_{1\le k\le n,\,1\le i,j\le n+1-k}$ that has the entries of M at its base $k = 1$ and the determinant of M at the apex $k = n$. The general term $d_{i,j}^k$ of the pyramid is defined by

$$d_{i,j}^k = \det(m_{p,q})_{i\le p\le i+k-1,\; j\le q\le j+k-1}.$$

Prove using (2.116) that the numbers in each successive layer can be computed recursively for $k \ge 2$ by

$$d_{i,j}^k = \frac{d_{i,j}^{k-1} d_{i+1,j+1}^{k-1} - d_{i+1,j}^{k-1} d_{i,j+1}^{k-1}}{d_{i+1,j+1}^{k-2}},$$

assuming that no divisions by zero occur and with the convention that $d_{i,j}^0 = 1$.

Note: Dodgson's condensation method is not a particularly useful method for computing determinants, but underlying it is a remarkable structure involving a family of combinatorial objects known as **alternating sign matrices**, discovered by David P. Robbins and Howard Rumsey in the early 1980s; see [21] for more information.

3

Erdős–Szekeres permutations and square Young tableaux

Chapter summary. We continue our study of longest increasing subsequences in permutations by considering a special class of permutations called **Erdős–Szekeres permutations**, which have the property that their longest monotone subsequence is the shortest possible and are thus extremal cases demonstrating sharpness in the Erdős–Szekeres theorem. These permutations are related via the Robinson–Schensted correspondence to an especially well-behaved class of standard Young tableaux, the **square Young tableaux**. We use the tools developed in Chapter 1 to analyze the behavior of random square Young tableaux, and this leads us to an interesting result on the limiting shape of random Erdős–Szekeres permutations. We also find a mysterious **arctic circle** that appears when we interpret some of the results as describing the asymptotic behavior of a certain **interacting particle system**.

3.1 Erdős–Szekeres permutations

In the previous two chapters we studied the statistical behavior of the permutation statistic $L(\sigma)$ for a typical permutation σ chosen at random from among all permutations of given order. In this chapter we focus our attention instead on those permutations σ whose behavior with regard to longest increasing subsequences, or more precisely longest monotone subsequences, is *atypical* in the most extreme way possible. We refer to these permutations as **Erdős–Szekeres permutations**,[1] because of their role as extremal cases demonstrating the sharpness in the Erdős–Szekeres theorem (Theorem 1.2). For integers $m, n \geq 1$, we call a permutation $\sigma \in S_N$ an (m, n)**-Erdős–Szekeres permutation** if $N = mn$, $L(\sigma) = n$ and $D(\sigma) = m$

(where $D(\sigma)$ denotes the maximal length of a decreasing subsequence as in Section 1.1). Denote the set of (m, n)-Erdős–Szekeres permutations by $\mathrm{ES}_{m,n}$. The most interesting case is when $m = n$. In this case we call a permutation $\sigma \in \mathrm{ES}_{n,n}$ simply an Erdős–Szekeres permutation, and denote the set of such permutations by ES_n. Note that these permutations have a natural max–min property: they minimize the maximal length of a monotone subsequence. The requirement that the order of the permutation be a perfect square may seem somewhat arbitrary but gives the set of Erdős–Szekeres permutations a very nice structure that enables analyzing them in great detail, as we shall see later.

The fact that $\mathrm{ES}_{m,n}$ is nonempty follows from the example given immediately after the Erdős–Szekeres theorem, but we can say much more than that. Let $\sigma \in \mathrm{ES}_{m,n}$, and let (λ, P, Q) be the Young diagram and pair of Young tableaux associated to σ by the Robinson–Schensted correspondence. By the properties of the Robinson–Schensted algorithm proved in Section 1.7, we know that λ_1, the length of the first row of λ, is equal to $L(\sigma) = n$, and that the length λ'_1 of the first column of λ is equal to $D(\sigma) = m$. Furthermore, since $|\lambda| = N = mn = \lambda'_1 \lambda_1$ is the product of the lengths of its first row and first column, λ must be the $m \times n$ rectangular Young diagram, which we denote by $\square_{m,n}$.

Conversely, if P and Q are two Young tableaux whose shape is $\square_{m,n}$, then the permutation associated with $(\square_{m,n}, P, Q)$ via the Robinson–Schensted correspondence is of size mn, and satisfies $L(\sigma) = n$, $D(\sigma) = m$, so it is an (m, n)-Erdős–Szekeres permutation. We have proved the following.

Theorem 3.1 *The Robinson–Schensted algorithm maps the set* $\mathrm{ES}_{m,n}$ *bijectively onto the set of pairs of Young tableaux* (P, Q) *of shape* $\square_{m,n}$. *In particular, we have*

$$|\,\mathrm{ES}_{m,n}\,| = d^2_{\square_{m,n}} = \left(\frac{(mn)!}{\prod_{i=1}^{m} \prod_{j=1}^{n} (i + j - 1)} \right)^2. \tag{3.1}$$

The Erdős–Szekeres permutations in ES_n *are in bijection with the set of pairs of* $n \times n$ *square Young tableaux, and are enumerated by*

$$|\,\mathrm{ES}_n\,| = d^2_{\square_{n,n}} = \left(\frac{(n^2)!}{1 \cdot 2^2 \cdot 3^3 \cdots n^n (n+1)^{n-1} (n+2)^{n-2} \cdots (2n-1)^1} \right)^2. \tag{3.2}$$

A variant of the preceding argument can be used to give an alternative proof of the Erdős–Szekeres theorem. Indeed, if $\sigma \in S_N$ where $N > mn$, then the Robinson–Schensted shape λ associated with σ is of size N, so we must have $L(\sigma) = \lambda_1 > n$ or $D(\sigma) = \lambda'_1 > m$; otherwise λ would be contained in the diagram $\square_{m,n}$, which is of size mn.

Theorem 3.1 can be thought of as a structure theorem for Erdős–Szekeres permutations, in that it gives a computational procedure for generating these permutations from more familiar objects, namely Young tableaux; but the insight that it gives us into their structure is somewhat limited, since this computational procedure involves the Robinson–Schensted algorithm and its inverse, which are in many ways nontrivial and difficult-to-understand procedures. As it turns out, however, for the special case of Erdős–Szekeres permutations the Robinson–Schensted algorithm reduces to a much simpler mapping. This is explained in the next section.

3.2 The tableau sandwich theorem

For a Young diagram λ, denote by $\mathcal{T}(\lambda)$ the set of standard Young tableaux of shape λ.

Theorem 3.2 (The tableau sandwich theorem) *Let $m, n \geq 1$. There is a bijection from $\mathcal{T}(\square_{m,n}) \times \mathcal{T}(\square_{m,n})$ to $\mathrm{ES}_{m,n}$, described as follows. Given the tableaux $P = (p_{i,j})_{\substack{1\le i\le m\\ 1\le j\le n}}, Q = (q_{i,j})_{\substack{1\le i\le m\\ 1\le j\le n}} \in \mathcal{T}(\square_{m,n})$, the permutation $\sigma \in \mathrm{ES}_{m,n}$ corresponding to the pair (P, Q) satisfies*

$$\sigma(q_{i,j}) = p_{m+1-i,j}, \qquad (1 \leq i \leq m,\ 1 \leq j \leq n). \tag{3.3}$$

In the opposite direction, $P, Q \in \mathcal{T}(\square_{m,n})$ can be determined from $\sigma \in \mathrm{ES}_{m,n}$ by the relations

$$\begin{aligned} q_{i,j} &= \text{the unique } 1 \leq k \leq mn \text{ such that } D_k(\sigma) = i \\ &\qquad \text{and } L_k(\sigma) = j, \end{aligned} \tag{3.4}$$

$$\begin{aligned} p_{i,j} &= \text{the unique } 1 \leq k \leq mn \text{ such that } D_k(\sigma^{-1}) = i \\ &\qquad \text{and } L_k(\sigma^{-1}) = j, \end{aligned} \tag{3.5}$$

where $L_k(\sigma)$ and $D_k(\sigma)$ refer (as on p. 7) to the maximal length of an increasing (respectively, decreasing) subsequence of σ that ends with $\sigma(k)$. Moreover, this bijection coincides with the bijection induced by the Robinson–Schensted correspondence as described in Theorem 3.1.

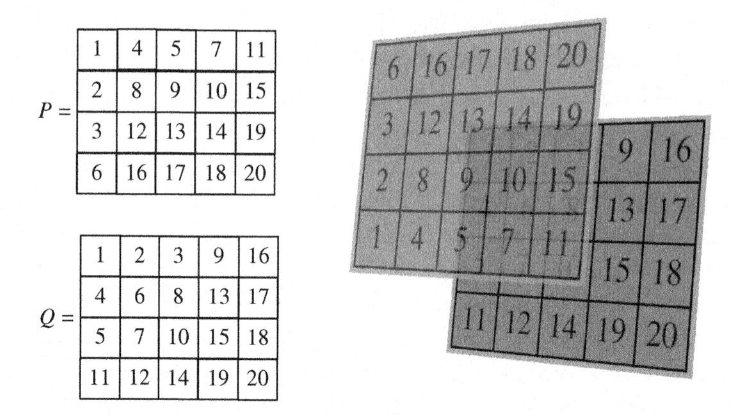

$P =$

1	4	5	7	11
2	8	9	10	15
3	12	13	14	19
6	16	17	18	20

$Q =$

1	2	3	9	16
4	6	8	13	17
5	7	10	15	18
11	12	14	19	20

Figure 3.1 A "tableau sandwich" producing the permutation $\sigma = (6, 16, 17, 3, 2, 12, 8, 13, 18, 9, 1, 4, 14, 5, 10, 20, 19, 15, 7, 11)$.

There is a nice way to graphically visualize the mapping taking the two tableaux P, Q to a permutation $\sigma \in \mathrm{ES}_{m,n}$ as a sort of "tableau sandwich" where the two tableaux are placed one on top of the other, with P being flipped over vertically. The permutation σ is the mapping that matches to each number in Q (which we imagine as the bottom "slice" of the sandwich) the number directly above it in P, the top slice of the sandwich. This is illustrated in Fig. 3.1

What makes the Robinson–Schensted algorithm behave in such a regular fashion? Trying out the algorithm with some numerical examples such as the one above, one is led to observe an interesting phenomenon regarding the behavior of the **bumping sequences**, which turns out to be key to the proof of Theorem 3.2. Recall that the Robinson–Schensted algorithm consists of building the tableaux P and Q in a sequence of insertion steps. At each insertion step one inserts the next value of the permutation σ into P, which results in a cascade of bumping operations. The sequence of positions where the bumping occurs is referred to as the bumping sequence.

Lemma 3.3 *When the Robinson–Schensted algorithm is applied to an (m, n)-Erdős–Szekeres permutation σ, each bumping sequence consists of a vertical column of cells.*

Note that this behavior is in marked contrast to the behavior of the

Robinson–Schensted algorithm for an arbitrary permutation, where the bumping sequences can be very erratic.

Proof of Lemma 3.3 We prove the obviously equivalent claim that when the *inverse* Robinson–Schensted algorithm is applied to two rectangular Young tableaux $P, Q \in \mathcal{T}(\square_{m,n})$, all the bumping sequences are vertical columns. In this case the bumping sequences are "reverse" bumping sequences that arise from a sequence of deletion steps (applied to the insertion tableau P) that are the inverse to the insertion steps we considered before. The claim is proved by induction on k, the number of deletion steps. In the kth deletion step, we choose (based on looking at the recording tableau Q) which corner cell of the tableau P in its current form to start a deletion from, and start bumping numbers up until we reach the first row. In each successive bumping, say from position (i, j), the number being bumped (denote it by x) moves up to row $i-1$ and possibly to the right. However, we claim that it cannot move to the right; for, if it moves to the right that means it is bigger than the number in the adjacent $(i-1, j+1)$ position, which we denote by y. But, by the inductive hypothesis all bumping sequences prior to the kth deletion were vertical columns. This means that in the original tableau P before any deletions started, x and y occupied respective positions of the form (i_0, j) and $(i_1, j+1)$ for some $i_0 \le i_1$; that is, we have $x = p_{i_0,j}$ and $y = p_{i_1,j}$, so by the fact that P is a Young tableau it follows that $x < y$. $\square$

Proof of Theorem 3.2 Lemma 3.3 can be used to prove equation (3.3) as follows. Consider what happens in the kth insertion step for some $1 \le k \le mn$. If the cell that was added to the shape of P and Q in that insertion step was in position (i, j), then we have $q_{i,j} = k$. Because of Lemma 3.3, that means that the number $\sigma(k)$ that was inserted into P at that step settled down in position $(1, j)$, bumping the number in that position one row down to position $(2, j)$, resulting in the number in that position being bumped down to $(3, j)$, and so on, all the way down to (i, j). Now consider where in the final tableau P the number $\sigma(k)$ will end up at the end of the execution of the algorithm. This is illustrated in Fig. 3.2. Again using the lemma, any subsequent bumping will push it, along with the column of numbers below it, directly down. When the algorithm terminates, the number that was in position (i, j) after the kth insertion step will be at the bottom row

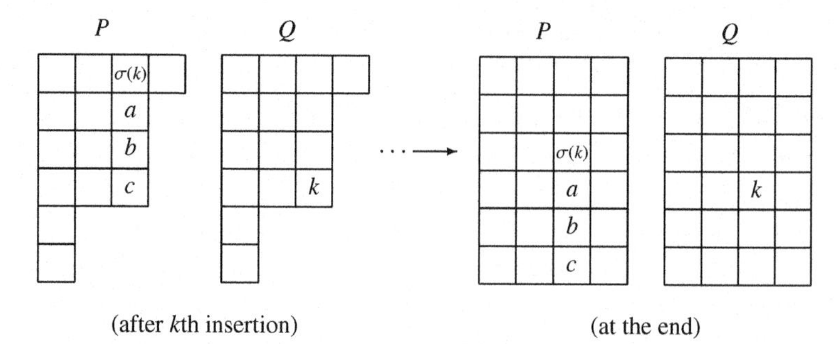

Figure 3.2 Considering the state of the insertion and recording tableaux after the kth insertion and after inserting all elements of σ provides an easy visual explanation of the relation $\sigma(q_{i,j}) = p_{m+1-i,j}$.

in position (m, j), exactly $m - i$ positions below where it was after the kth insertion. Therefore the number $\sigma(k) = \sigma(q_{i,j})$ will also be $m - i$ positions below where it was after the kth insertion, namely in position $(m + 1 - i, j)$. That gives exactly the relation $\sigma(q_{i,j}) = p_{m+1-i,j}$ which is (3.3).

Next, turn to proving (3.4), which would also imply (3.5) by switching the roles of P and Q and using Theorem 1.10(b). Recall from the original proof of the Erdős–Szekeres theorem that the pairs $(D_k(\sigma), L_k(\sigma))$ are all distinct, and for $\sigma \in \mathrm{ES}_{m,n}$ they all lie in the discrete rectangle $[1, m] \times [1, n]$. That proves that for each (i, j) there is indeed a unique k such that $(D_k(\sigma), L_k(\sigma)) = (i, j)$.

Now, if $q_{i,j} = k$, then Lemma 3.3 implies that in the kth insertion step in the Robinson–Schensted algorithm, the number $\sigma(k)$ is inserted into position $(1, j)$ of the insertion tableau, bumping the entries in the jth column one step down to form a column of height i. This has two implications: first, reading this column of the insertion tableau from bottom to top gives a decreasing subsequence of σ of length i that ends in $\sigma(k)$, which shows that $D_k(\sigma) \geq i$; and second, by repeating the argument used in the proof of Lemma 1.7 (modified slightly to account for the fact that we are interested in $L_k(\sigma)$ rather than $L(\sigma)$) we can get an *increasing* subsequence of σ of length j that ends in $\sigma(k)$, so we get that $L_k(\sigma) \geq j$.

We proved that the inequalities $D_{q_{i,j}}(\sigma) \geq i, L_{q_{i,j}}(\sigma) \geq j$ hold for all $1 \leq i \leq m, 1 \leq j \leq n$. But since we noted earlier that the mapping $k \mapsto$

$(D_k(\sigma), L_k(\sigma))$ is injective and takes values in $[1, m] \times [1, n]$, equality must hold for all i, j, which proves (3.4). □

3.3 Random Erdős–Szekeres permutations

Having characterized combinatorially the class of permutations ES_n which have extremal behavior with regard to their longest monotone subsequences, we now ask the related question of how a typical such permutation behaves. By comparing the behavior of a random Erdős–Szekeres permutation to that of a typical *general* permutation (chosen at random from among all permutations in the symmetric group of the same order, namely n^2), we may gain some further insight into the significance of the constraint of having only short monotone subsequences.

Thanks to the connection to square Young tableaux, it is easy to sample a uniformly random Erdős–Szekeres permutation by sampling two random square Young tableaux (using the algorithm described in Section 1.10) and applying the bijection described in the previous section. Fig. 3.3 shows the result for such a random permutation in ES_{50} alongside a standard uniformly random permutation from S_{2500}. In both cases, we represent the permutations graphically as a plot of the set of points $((j, \sigma(j)))_{j=1}^{n^2}$ (i.e., the positions of the 1's in the associated permutation matrix).

The picture of the random S_{2500} permutation in Fig. 3.3(b) has a simple structure that will be obvious to anyone with a background in elementary probability: since each value $\sigma(j)$ is uniformly distributed in the discrete interval $[1, n]$, and pairs of values $(\sigma(j), \sigma(j'))$ for $j \neq j'$ are only very weakly correlated, what we are observing is a kind of random noise that fills the square $[1, n]^2$ with a uniform density (one can ask questions about the local structure of this noise, but those too are answered easily and hold no surprises). The picture in Fig. 3.3(a), however, is much more interesting: it appears that the defining condition of belonging to the set ES_n leads the random permutation to exhibit very specific statistical behavior. The precise statement of what is happening asymptotically as $n \to \infty$ is described in the following limit shape theorem, the proof of which is one of the main goals of this chapter.

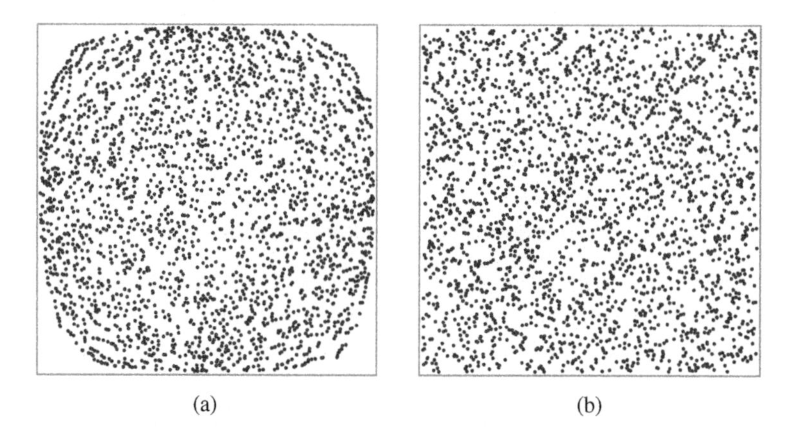

(a) (b)

Figure 3.3 (a) A uniformly random permutation in ES_{50}; (b) a uniformly random permutation in S_{2500}.

Theorem 3.4 (Limit shape theorem for random Erdős–Szekeres permutations) *For each n, let σ_n denote a permutation chosen uniformly at random from* ES_n, *and let A_n denote its graph $\{(j, \sigma_n(j)) : 1 \le j \le n^2\}$. Define a set $A_\infty \subset \mathbb{R}^2$ by*

$$A_\infty = \left\{(x,y) \in \mathbb{R}^2 \,:\, (x^2 - y^2)^2 + 2(x^2 + y^2) \le 3\right\}$$
$$= \left\{(x,y) \in \mathbb{R}^2 \,:\, |x| \le 1,\ |y| \le \sqrt{x^2 - 1 + 2\sqrt{1 - x^2}}\right\}. \tag{3.6}$$

As $n \to \infty$, the scaled random set $\tilde{A}_n = \frac{2}{n^2} A_n - (1,1)$ converges in probability to A_∞, in the sense that the following two statements hold:

(a) *For any $\epsilon > 0$, $\mathbb{P}(\tilde{A}_n \subset (1+\epsilon)A_\infty) \to 1$ as $n \to \infty$.*

(b) *For any open set $U \subset A_\infty$, $\mathbb{P}(\tilde{A}_n \cap U \neq \emptyset) \to 1$ as $n \to \infty$.*

The limit shape A_∞ of random Erdős–Szekeres permutations is illustrated in Fig. 3.4. Note that, curiously, the boundary of A_∞ is a quartic curve, that is, an algebraic curve of degree 4.

In the next section we see how, with the help of Theorem 3.2, Theorem 3.4 can be (easily) deduced from another (difficult) limit shape result that describes the asymptotic behavior of random square Young tableaux.

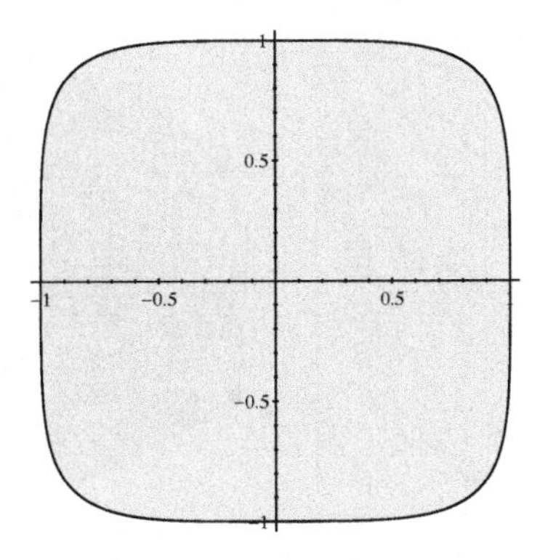

Figure 3.4 The limit shape of random Erdős–Szekeres permutations.

3.4 Random square Young tableaux

In view of the connection between Erdős–Szekeres permutations and square Young tableaux, the problem of understanding the behavior of *random* permutations in ES_n reduces to the problem of getting a good understanding of the behavior of random square Young tableaux of high order, since choosing a random Erdős–Szekeres permutation is equivalent to choosing two uniformly random square Young tableaux and applying the bijection of Theorem 3.2. We will formulate a result that answers the question of how such objects behave in the limit, but once again, before presenting the precise result it makes sense to look at some simulation results. Fig. 3.5 shows some sample results from such simulations. Square Young tableaux are visualized in two different ways. First, since the tableau is an array of numbers $(t_{i,j})_{i,j=1}^n$, we can consider it as the graph of a discrete "stepped surface," where for each i, j the number $t_{i,j}$ represents the height of a stack of unit cubes being placed over the square $[i, i+1] \times [j, j+1]$ in the x–y plane. The second representation shows an alternative way of interpreting a Young tableau. In general, given a Young tableau of shape $\lambda \in \mathcal{P}(n)$, we can interpret the tableau as an increasing sequence of Young diagrams

$$\emptyset = \lambda^{(0)} \nearrow \lambda^{(1)} \nearrow \ldots \nearrow \lambda^{(n)} = \lambda,$$

(or equivalently a path in the Young graph starting from the empty diagram and ending at λ; see Section 1.19). The diagram $\lambda^{(k)}$ in this path consists

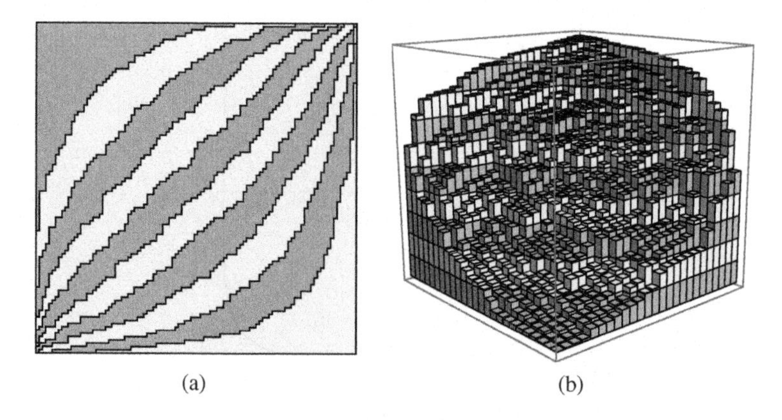

Figure 3.5 (a) A random square Young tableau of order 100, visualized as a growth profile of Young diagrams growing to fill the square. (b) A 3D plot of a random square Young tableau of order 25 shown as a stepped surface.

of all cells of λ where the entry of the tableau is $\leq k$. The association between Young tableaux and such paths is bijective, since we can recover the tableau from the path by filling in the number k in each cell $\mathbf{c}$ of the diagram if $\lambda^k \setminus \lambda^{k-1} = \{\mathbf{c}\}$, that is, if $\mathbf{c}$ was added to $\lambda^{(k-1)}$ to obtain $\lambda^{(k)}$.

With this interpretation, we can visualize the square Young tableau by plotting some of the diagrams in the path encoded by the tableau. This can be thought of as the *growth profile* of the tableau, showing the manner in which one arrives at the final square diagram by successively adding boxes. Fig. 3.5(a) shows such a visualization of a random square Young tableau.

Note that the relationship between the two pictures is that the shapes in the growth profile in Fig. 3.5(a) are sublevel sets of the stepped surface in Fig. 3.5(b).

To formulate the precise results explaining these pictures, we need to define the functions that describe the limiting shape of the random square Young tableau. First, define a one-parameter family of curves $(g_\tau(u))_{0<\tau<1}$

where for each $0 < \tau < 1$, $g_\tau \colon \left[-\sqrt{2\tau(1-\tau)}, \sqrt{2\tau(1-\tau)}\right] \to \mathbb{R}$ and

$$g_\tau(u) = \begin{cases} \frac{2}{\pi} u \tan^{-1}\left(\frac{(1-2\tau)u}{\sqrt{2\tau(1-\tau)-u^2}}\right) + \frac{\sqrt{2}}{\pi}\tan^{-1}\left(\frac{\sqrt{2(2\tau(1-\tau)-u^2)}}{1-2\tau}\right) & \text{if } 0 < \tau < \frac{1}{2}, \\ \sqrt{2} & \text{if } \tau = \frac{1}{2}, \\ \sqrt{2} - g_{1-\tau}(u) & \text{if } \frac{1}{2} < \tau < 1. \end{cases} \tag{3.7}$$

The idea is that the family of curves $v = g_\tau(u)$ for $0 < \tau < 1$ represents the limiting growth profile of the random square Young tableau, in the rotated (a.k.a. Russian) u–v coordinate system defined in (1.23). To talk about convergence we will want to consider each g_τ as belonging to a suitable function space, so we extend it by defining a function $\tilde{g}_\tau \colon \left[-\sqrt{2}/2, \sqrt{2}/2\right] \to [0, \infty)$ given by

$$\tilde{g}_\tau(u) = \begin{cases} g_\tau(u) & \text{if } |u| \le \sqrt{2\tau(1-\tau)}, \\ |u| & \text{if } \sqrt{2\tau(1-\tau)} < |u| \le \sqrt{2}/2,\ \tau < \frac{1}{2}, \\ \sqrt{2} - |u| & \text{if } \sqrt{2\tau(1-\tau)} < |u| \le \sqrt{2}/2,\ \tau > \frac{1}{2}. \end{cases} \tag{3.8}$$

Next, if λ is a Young diagram contained in the square diagram $\square_{n,n}$, we encode it as a continual Young diagram ϕ_λ by defining

$$\phi_{n,\lambda}(x) = n^{-1}\lambda'_{\lfloor nx \rfloor + 1}, \qquad (x \ge 0) \tag{3.9}$$

(compare with (1.17), and note the minor differences in scaling, which make it more convenient to discuss all subdiagrams of $\square_{n,n}$ using the same scale). We can associate with each $\phi_{n,\lambda}$ another function $\psi_{n,\lambda}$ that describes the same continual Young diagram in the rotated coordinate system (that is, $\phi_{n,\lambda}$ and $\phi_{n,\lambda}$ are related as f and g in (1.24)). Note that the condition $\lambda \subseteq \square_{n,n}$ means that instead of considering $\psi_{n,\lambda}$ as a function on $\mathbb{R}$, we can restrict it to the interval $\left[-\sqrt{2}/2, \sqrt{2}/2\right]$ with no loss of information.

Theorem 3.5 (Limiting growth profile of random square Young tableaux[2]) *For each $n \ge 1$, let $T_n = (t^n_{i,j})^n_{i,j=1}$ be a uniformly random Young tableau of shape $\square_{n,n}$, and let*

$$\emptyset = \Lambda^{(n,0)} \nearrow \Lambda^{(n,1)} \nearrow \ldots \nearrow \Lambda^{(n,k)} \nearrow \ldots \nearrow \Lambda^{(n,n^2)} = \square_{n,n} \tag{3.10}$$

denote the growing sequence of random Young diagrams encoded by the tableau T_n. For any $1 \le k \le n^2$ let $\phi_{n,k} = \phi_{n,\Lambda^{(n,k)}}$ denote the continual Young diagram associated with $\Lambda^{(n,k)}$ as defined in (1.17)*, and let $\psi_{n,k}$ be*

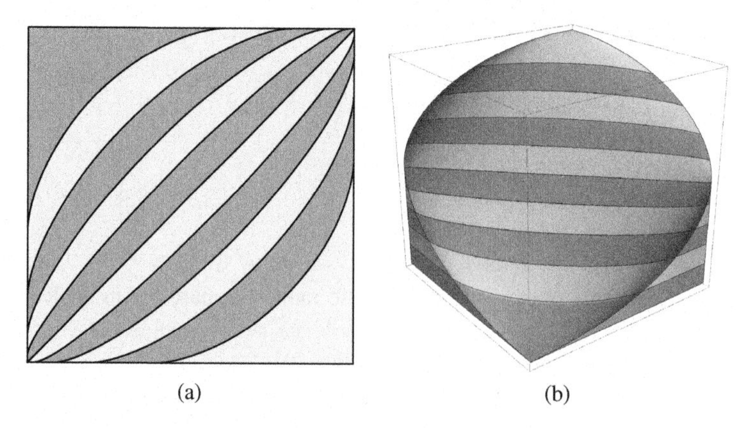

Figure 3.6 (a) The limiting growth profile. (b) Limit shape of random square Young tableaux.

associated the function encoding $\Lambda^{(n,k)}$ in the rotated coordinate system. Then for any $\epsilon > 0$ we have

$$\mathbb{P}\left[\max_{1\le k\le n^2} \sup_{u\in\left[-\sqrt{2}/2,\sqrt{2}/2\right]} |\tilde{g}_{k/n^2}(u) - \psi_{n,k}(u)| > \epsilon\right] \to 0 \text{ as } n \to \infty. \tag{3.11}$$

Next, we consider the representation of the Young tableau as a discrete surface. The limiting object in this case will be a surface $S : [0,1]^2 \to [0,1]$, which we define as the unique continuous function satisfying for any $0 < \tau < 1$

$$\left\{(x,y) : S(x,y) = \tau\right\} = \left\{(u, g_\tau(u)) : |u| \le \sqrt{2\tau(1-\tau)}\right\}, \tag{3.12}$$

where we identify each point (u, v) in the u–v plane with the corresponding point $\frac{1}{\sqrt{2}}(x-y, x+y)$ in the x–y plane. Thus $S(\cdot,\cdot)$ is defined in terms of an implicit equation, but the verification that it is well-defined is straightforward and left to the readers. It is also easy to verify the following explicit formulas for the boundary values of S:

$$S(x,0) = S(0,x) = \frac{1-\sqrt{1-x^2}}{2}, \tag{3.13}$$

$$S(x,1) = S(1,x) = \frac{1+\sqrt{2x-x^2}}{2}. \tag{3.14}$$

Theorem 3.6 (Limit shape theorem for random square Young tableaux) *For each* $n \geq 1$*, let* $T_n = (t^n_{i,j})^n_{i,j=1}$ *be a uniformly random Young tableau of shape* $\square_{n,n}$*. For any* $\epsilon > 0$ *we have*

$$\mathbb{P}\left[\max_{1 \leq i,j \leq n} \left| n^{-2} t^n_{i,j} - S(i/n, j/n) \right| > \epsilon \right] \to 0 \text{ as } n \to \infty.$$

The proofs of Theorems 3.5 and 3.6 will take up most of our efforts in the rest of this chapter. Before we begin, let us see how we can use these results to derive Theorem 3.4.

Proof of Theorem 3.4 Let $P_n = (p^n_{i,j})^n_{i,j=1}$, $Q_n = (q^n_{i,j})^n_{i,j=1}$ be the two Young tableaux associated with the uniformly random Erdős–Szekeres permutation σ_n via the bijection of Theorem 3.2. By (3.3), the set A_n from Theorem 3.4 can also be written as

$$A_n = \{(q^n_{i,j}, p^n_{n+1-i,j}) \; : \; 1 \leq i, j \leq n\}.$$

By Theorem 3.6, with high probability as $n \to \infty$ each point $n^{-2}(q^n_{i,j}, p^n_{n+1-i,j})$ is uniformly close to the point $(S(x,y), S(1-x,y))$, where $x = i/n$ and $y = j/n$. It follows that if we define the set

$$A'_\infty = \{(2S(x,y) - 1, 2S(1-x,y) - 1) \; : \; 0 \leq x, y \leq 1\},$$

then as $n \to \infty$ the values of the scaled set $2n^{-2}A_n - (1,1)$ will intersect each open subset of A'_∞, and will be contained in any open set containing A'_∞, with asymptotically high probability; that is exactly the statement of Theorem 3.4, except for the fact that $A_\infty = A'_\infty$, which we now verify. The set A'_∞ is the image of the square $[0,1] \times [0,1]$ under the mapping

$$\Phi \colon (x,y) \mapsto (2S(x,y) - 1, 2S(1-x,y) - 1).$$

By (3.13) and (3.14), Φ maps the boundary of the square into the four curves described parametrically by

$$\left(-\sqrt{1-t^2}, -\sqrt{2t-t^2}\right)_{0\leq t\leq 1}, \qquad \left(-\sqrt{1-t^2}, \sqrt{2t-t^2}\right)_{0\leq t\leq 1},$$
$$\left(\sqrt{1-t^2}, -\sqrt{2t-t^2}\right)_{0\leq t\leq 1}, \qquad \left(\sqrt{1-t^2}, \sqrt{2t-t^2}\right)_{0\leq t\leq 1}.$$

Setting $x = \pm\sqrt{1-t^2}$, $y = \pm\sqrt{2t-t^2}$, it is easy to verify that

$$(x^2 - y^2)^2 + 2(x^2 + y^2) = 3,$$

so these curves parametrize the boundary of the set A_∞. Furthermore, since the function $(x, y) \mapsto 2S(x, y) - 1$ is strictly increasing in x and y and $(x, y) \mapsto 2S(1 - x, y) - 1$ is strictly increasing in y and decreasing in x, it is easy to see that Φ is one-to-one. Thus the interior of the square $[0, 1]^2$ must be mapped under Φ to the interior of A_∞. This proves that $A'_\infty = A_\infty$ and completes the proof. □

3.5 An analogue of Plancherel measure

We start our attack on Theorem 3.5 by noting in the following lemma a formula for the probability distribution of the Young diagram $\Lambda^{(n,k)}$ in (3.10). This formula highlights a remarkable similarity between the problem of the limiting growth profile of a random square Young tableau and the problem of finding the limit shape of a Young diagram chosen according to Plancherel measure, which we already solved in Chapter 1. Indeed, the tools we developed to solve that problem will be directly applicable to the problem currently under discussion.

Lemma 3.7 *Let $0 \le k \le n^2$. We have*

$$\mathbb{P}\left(\Lambda^{(n,k)} = \lambda\right) = \begin{cases} \dfrac{d_\lambda d_{\square_{n,n} \setminus \lambda}}{d_{\square_{n,n}}} & \text{if } \lambda \in \mathcal{P}(k), \lambda \subset \square_{n,n}, \\ 0 & \text{otherwise.} \end{cases} \tag{3.15}$$

where $d_{\square_{n,n} \setminus \lambda}$ denotes the number of standard Young tableau whose shape is the Young diagram obtained by taking the difference of the two Young digrams $\square_{n,n}$ and λ and rotating it by 180 degrees.

Proof Assume that $\lambda \in \mathcal{P}(k)$ satisfies $\lambda \subset \square_{n,n}$. The condition $\Lambda^{(n,k)} = \lambda$ means that all the entries in the tableau T_n inside λ are $\le k$, and all the entries outside λ are $> k$. The number of square Young tableau satisfying this property is exactly the product $d_\lambda d_{\square_{n,n} \setminus \lambda}$, since we have to fill the cells of λ with the numbers $1, \ldots, k$ in a way that is increasing along both rows and columns, which has d_λ possibilities, and independently fill the cells of $\square_{n,n} \setminus \lambda$ with the numbers $k+1, \ldots, n^2$ in a similarly increasing fashion; this is once again a Young tableau-counting problem in disguise, with $d_{\square_{n,n} \setminus \lambda}$ possibilities. Finally, since T_n is uniformly random, to get the probability one has to divide the expression $d_\lambda d_{\square_{n,n} \setminus \lambda}$ by the total number of Young tableau of shape $\square_{n,n}$, which is $d_{\square_{n,n}}$. □

3.6 An asymptotic hook-length formula for square Young tableaux

Since (3.15) represents the probability distribution of the random Young diagram $\Lambda^{(n,k)}$ in terms of the Young tableau counting function $\lambda \mapsto d_\lambda$, the asymptotic form of the hook-length formula (Theorem 1.14) we proved in Chapter 1 will also be useful for the purpose of analyzing this probability distribution. We now formulate a variant of that result. For a Young diagram $\lambda \in \mathcal{P}(k)$ satisfying $\lambda \subset \square_{n,n}$, let ϕ_λ be defined as in (1.17). As before, the correspondence $\lambda \mapsto \phi_\lambda$ embeds the discrete shape λ into a bigger function space, which in this case we define as

$$\mathcal{F}_\square = \{f\colon [0,1] \to [0,1] \,:\, f \text{ is monotone nonincreasing and left-continuous}\}.$$

If $f \in \mathcal{F}_\square$ and $x, y \in [0,1]$, the hook-length of f at (x,y) is as before denoted by $h_f(x,y)$ and defined by $h_f(x,y) = f(x) - y + f^{-1}(y) - x$. Note that this is defined also for (x,y) lying above the graph of f (or "outside" f when considered as a continuous analogue of a Young diagram). In this case the hook-length will be a negative number, but we will only be interested in its magnitude.

Theorem 3.8 *Define the quantities*

$$\kappa_0 = \tfrac{3}{2} - 2\log 2, \tag{3.16}$$

$$H(\tau) = -\tau \log \tau - (1-\tau)\log(1-\tau), \tag{3.17}$$

$$I^\square_{hook}(f) = \int_0^1 \int_0^1 \log |h_f(x,y)|\, dy\, dx, \qquad (f \in \mathcal{F}_\square). \tag{3.18}$$

Let $0 < \tau < 1$, and let $k = k(n)$ be a sequence of integers such that $k/n^2 \to \tau$ as $n \to \infty$. Then asymptotically as $n \to \infty$ we have that

$$\mathbb{P}\left(\Lambda^{(n,k)} = \lambda\right) = \exp\left[-n^2\left(1 + O\left(\frac{\log n}{n}\right)\right)(I^\square_{hook}(\phi_\lambda) + H(\tau) + \kappa_0)\right], \tag{3.19}$$

uniformly over all diagrams $\lambda \in \mathcal{P}(k)$ satisfying $\lambda \subset \square_{n,n}$.

Proof This can be proved directly by a computation analogous to that used in the proof of Theorem 1.14. However, since we already proved that result, we can deduce the present claim as an easy consequence. The idea is that the product of hook-lengths in the expression d_λ on the right-hand side of (3.15) contributes an exponential factor approximating the hook integral

$$\int_0^1 \int_0^{\phi_\lambda(x)} \log h_f(x,y)\, dy\, dx.$$

Similarly, the product of hook-lengths in $d_{\square_{n,n}\setminus\lambda}$ contributes the complementary hook-type integral

$$\int_0^1 \int_{\phi_\lambda(x)}^1 \log(-h_f(x,y))\, dy\, dx,$$

so that the two integrals combine to give exactly the expression $I^{\square}_{\text{hook}}(\phi_\lambda)$ inside the exponent (note that the factor 2 in front of the hook integral in (1.19) disappears, since on the left-hand side of that equation we have the *square* of d_λ so we need to take a square root). The other factors $H(t)$ and κ_0 appear by taking into account the various numerical factors involving factorials, and the denominator $d_{\square_{n,n}}$ of (3.15). The details are straightforward and left to the reader (Exercise 3.3). □

3.7 A family of variational problems

We proceed using the same methods and ideas we developed in Chapter 1. Viewing the right-hand side of (3.19) as a first-order asymptotic expansion for the probabilities on the left-hand side, we see that it is reasonable to expect (and, as we have seen already in the case of Plancherel measure, not too hard to justify rigorously) that the typical behavior of the scaled diagram $\phi_{\Lambda^{(n,k)}}$ for $k \approx \tau n^2$ will be close to the *most likely* behavior, which is related to the shape $f \in \mathcal{F}_\square$ that minimizes the functional $I^{\square}_{\text{hook}}(\cdot)$ from among all candidate shapes for any given τ. In other words, we have arrived at a new variational problem – or rather, in this case, a family of variational problems indexed by the parameter τ.

The variational problem for square Young tableaux. *For* $0 < \tau < 1$ *denote*

$$\mathcal{F}^{\tau}_{\square} = \left\{ f \in \mathcal{F}_\square \ :\ \int_0^1 f(x)\, dx = \tau \right\}.$$

Find the function $f_\tau \in \mathcal{F}^{\tau}_{\square}$ *that minimizes the functional* $I^{\square}_{hook}(\cdot)$.

As in the case of Plancherel measure, it turns out that the form of the functional simplifies, and further analysis becomes possible, once we transform the coordinate system – first to the Russian coordinates, and then to a variant of the hook coordinates introduced in Section 1.14.

As before, if $f \in \mathcal{F}_\square$, let $v = g(u)$ denote the representation of the curve $y = f(x)$ in the rotated coordinate system. The association $f \mapsto g$ maps the function space $\mathcal{F}_\square$ to a new space, which we denote by $\mathcal{G}_\diamond$, consisting of functions $g\colon \left[-\frac{\sqrt{2}}{2}, \frac{\sqrt{2}}{2}\right] \to \left[0, \sqrt{2}\right]$ satisfying the following conditions:

1. g is 1-Lipschitz;
2. $g\left(\pm\frac{\sqrt{2}}{2}\right) = \frac{\sqrt{2}}{2}$.

If $f \in \mathcal{F}_\square^\tau$ then g satisfies the further property

3. $$\int_{-\frac{\sqrt{2}}{2}}^{\frac{\sqrt{2}}{2}} (g(u) - |u|)\, du = \tau, \tag{3.20}$$

and we denote $g \in \mathcal{G}_\diamond^\tau$.

To derive the new form of the functional $I_{\text{hook}}^\square(f)$ in terms of the rotated function g, we separate it into two parts, namely

$$\begin{aligned} I_{\text{hook}}^\square(f) &= I_1(f) + I_2(f) \\ &:= \int_0^1 \int_0^{f(x)} \log h_f(x,y)\, dy\, dx + \int_0^1 \int_{f(x)}^1 \log\left(-h_f(x,y)\right) dy\, dx, \end{aligned}$$

and set

$$J_1(g) = I_1(f), \quad J_2(g) = I_2(f), \quad J(g) = J_1(g) + J_2(g) = I_{\text{hook}}^\square(f). \tag{3.21}$$

Each of $J_1(g)$, $J_2(g)$, can be represented in terms of a separate system of hook coordinates – the "lower" hook coordinates (coinciding with the hook coordinates defined previously in (1.25)) for $J_1(g)$ and the "upper" hook coordinates for $J_2(g)$. The meaning of these terms is explained in Fig. 3.7.

Consider first $J_1(g)$. By elementary geometry this can be written as

$$J_1(g) = \int_{-\frac{\sqrt{2}}{2}}^{\frac{\sqrt{2}}{2}} \int_{g(u)}^{\sqrt{2}-|u|} \log h_f(x,y)\, du\, dv.$$

Transforming to the lower hook coordinates (t, s) we get

$$J_1(g) = \iint_\Delta \log\left(\sqrt{2}(s-t)\right) \left|\frac{D(u,v)}{D(s,t)}\right| ds\, dt,$$

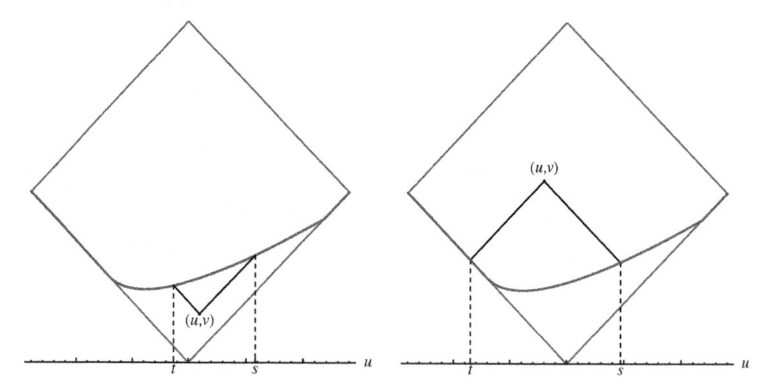

Figure 3.7 Lower hook coordinates and upper hook coordinates.

where $\Delta = \left\{\frac{-\sqrt{2}}{2} \le t \le s \le \frac{\sqrt{2}}{2}\right\}$ and $D(u,v)/D(s,t)$ is the Jacobian of the change of variables, whose magnitude we previously computed to be equal to $\frac{1}{2}(1+g'(t))(1-g'(s))$ (see (1.26)). Therefore we get that

$$J_1(g) = \frac{1}{2}\iint_\Delta \log\left(\sqrt{2}(s-t)\right)(1+g'(t))(1-g'(s))\,ds\,dt.$$

Now perform a similar computation for $J_2(g)$. Here we use the upper hook coordinates (which we still denote by (t,s)), for which the absolute value of the Jacobian can be computed in a similar manner to be $\frac{1}{2}(1+g'(s))(1-g'(t))$, and find that

$$J_2(g) = \frac{1}{2}\iint_\Delta \log\left(\sqrt{2}(s-t)\right)(1+g'(s))(1-g'(t))\,ds\,dt.$$

Adding the two expressions and symmetrizing the integration region, we find after a short computation that

$$\begin{aligned} J(g) &= \frac{1}{2}\iint_\Delta \log\left(\sqrt{2}(s-t)\right)(2-2g'(t)g'(s))\,ds\,dt \\ &= -\frac{1}{2}\int_{-\frac{\sqrt{2}}{2}}^{\frac{\sqrt{2}}{2}}\int_{-\frac{\sqrt{2}}{2}}^{\frac{\sqrt{2}}{2}} \log|s-t|\cdot g'(t)g'(s)\,ds\,dt + \log 2 - \frac{3}{2} \\ &= Q(g) + \log 2 - \frac{3}{2}, \end{aligned} \tag{3.22}$$

where $Q(\cdot)$ is the quadratic functional from Section 1.14.

Thus, we get the following reformulation of the original family of variational problems, which is directly analogous to the variational problem from Chapter 1: for each $0 < \tau < 1$, we need to find the function that minimizes the functional $Q(\cdot)$ on the space $\mathcal{G}_\diamond^\tau$.

3.8 Minimizing the functional

The solution of the family of variational problems presented above is given by the following result, whose proof is the focus of the next few sections.

Theorem 3.9 (Solution of the variational problem) *For each* $0 < \tau < 1$, *the function* $\tilde{g}_\tau$ *defined in* (3.8) *is the unique minimizer of the functional* $Q(\cdot)$ *on* $\mathcal{G}_\diamond^\tau$, *and the value of the minimum is*

$$Q(\tilde{g}_\tau) = -H(\tau) + \log 2, \tag{3.23}$$

where $H(\tau)$ *is defined in* (3.17).

Note that for $\tau = 1/2$, since $\tilde{g}_\tau' \equiv 0$ we have $Q(g_\tau) = 0$, so it is a global minimizer of $Q(\cdot)$ on all of $\mathcal{G}_\diamond$, and in particular on $\mathcal{G}_\diamond^{1/2}$. For $\tau > 1/2$, since $Q(h) = Q(\sqrt{2} - h)$ and $\sqrt{2} - h \in \mathcal{G}_\diamond^{1-\tau}$, the minimizer h_τ over $\mathcal{G}_\diamond^\tau$ is related to the minimizer over $\mathcal{G}_\diamond^{1-\tau}$ via $h_\tau = \sqrt{2} - h_{1-\tau}$. Since the functions $\tilde{g}_\tau$ also satisfy $g_\tau = \sqrt{2} - g_{1-\tau}$, we see that it will be enough to prove the theorem in the case $\tau < 1/2$, and for the rest of the section we assume that τ is in this range.

We start by formulating a sufficient condition, analogous to (1.36), for determining when a given function is the desired minimizer.

Theorem 3.10 *Let* $h_\tau \in \mathcal{G}_\diamond^\tau$. *Assume that for some constant* $\lambda \in \mathbb{R}$, *the function* $p\colon \mathbb{R} \to \mathbb{R}$ *defined by*

$$p(u) = -\int_{-\frac{\sqrt{2}}{2}}^{\frac{\sqrt{2}}{2}} h_\tau'(s) \log|s-u|\,ds + \lambda u \quad \textit{is} \quad \begin{cases} = 0 & \textit{if } -1 < h_\tau'(u) < 1, \\ \geq 0 & \textit{if } h_\tau'(u) = -1, \\ \leq 0 & \textit{if } h_\tau'(u) = 1. \end{cases} \tag{3.24}$$

Then for any $h \in \mathcal{G}_\diamond^\tau$ *we have that*

$$Q(h) \geq Q(h_\tau) + Q(h - h_\tau). \tag{3.25}$$

In particular, since $Q(h - h_\tau) \geq 0$ *by Proposition 1.15, with equality holding iff* $h - h_\tau \equiv 0$, *it follows that* h_τ *is the unique minimizer for* $Q(\cdot)$ *on* $\mathcal{G}_\diamond^\tau$.

Proof As in the proof of Theorem 1.16, from the assumption on $p(u)$ we get that

$$(h'(t) - h'_\tau(t))p(t) \geq 0$$

for all t for which $h'(t), h'_\tau(t)$ exist. By integration we get, in the notation of Section 1.14, that

$$2B(h, h_\tau) - \lambda \int_{-\frac{\sqrt{2}}{2}}^{\frac{\sqrt{2}}{2}} th'(t)\,dt \geq 2B(h_\tau, h_\tau) - \lambda \int_{-\frac{\sqrt{2}}{2}}^{\frac{\sqrt{2}}{2}} th'_\tau(t)\,dt.$$

Since

$$\int t\,h'(t)\,dt = \int t\,h'_\tau(t)\,dt = \tfrac{1}{2} - \tau \tag{3.26}$$

(an equivalent form of the condition (3.20)), this implies that $B(h-h_\tau, h_\tau) \geq 0$, whence we arrive at (3.25) via the relation

$$Q(h) = Q(h_\tau + (h - h_\tau)) = Q(h_\tau) + Q(h - h_\tau) + 2B(h - h_\tau, h_\tau). \qquad \square$$

Our goal is now to find a function h_τ for which (3.24) holds. We shall do this in a constructive way that does not assume prior knowledge of the solution. Instead, we make only the following reasonable assumption (which can be guessed just by looking at the simulated growth profile of a random square Young tableau): for some $\beta = \beta(\tau) \in \left(0, \sqrt{2}/2\right)$, the function h_τ satisfies the condition

$$h'_\tau(u) \text{ is } \begin{cases} = -1 & \text{if } -\frac{\sqrt{2}}{2} < u < -\beta, \\ \in (-1, 1) & \text{if } -\beta < u < \beta, \\ = 1 & \text{if } \beta < u < \frac{\sqrt{2}}{2}. \end{cases} \tag{3.27}$$

Substituting this into the equality in the first case of (3.24) gives that for $-\beta < u < \beta$, we must have

$$\begin{aligned} -\int_{-\beta}^{\beta} h'_\tau(s) \log|s - u|\,ds &= -\lambda u - \int_{-\frac{\sqrt{2}}{2}}^{-\beta} \log(u - s)\,ds + \int_{\beta}^{\frac{\sqrt{2}}{2}} \log(s - u)\,ds \\ &= -\lambda u - \left(\frac{\sqrt{2}}{2} - u\right)\log\left(\frac{\sqrt{2}}{2} - u\right) + \left(\frac{\sqrt{2}}{2} + u\right)\log\left(\frac{\sqrt{2}}{2} + u\right) \\ &\quad + (\beta + u)\log(\beta + u) - (\beta - u)\log(\beta - u). \end{aligned} \tag{3.28}$$

Differentiating, we get

$$-\int_{-\beta}^{\beta} \frac{h_\tau'(s)}{s-u}\,ds = -\lambda + \log \frac{\beta^2 - u^2}{\frac{1}{2} - u^2}. \tag{3.29}$$

The problem of finding h_τ' (from which we can recover h_τ) therefore reduces, as in the analysis of Section 1.15, to the problem of finding a function whose Hilbert transform coincides with the function on the right-hand side of (3.29) on the interval $(-\beta,\beta)$ (compare with (1.39)). In other words, we need to invert a Hilbert transform, except that it is a Hilbert transform on an interval instead of on the entire real line where we would have been able to use the usual inversion formula (1.15).

One possibility would be to look for a result analogous to Lemma 1.17 that would identify the solution. Finding such a result seems like a difficult problem in complex analysis, which in particular requires highly nontrivial intuition that would allow us to "guess" the required analytic function without knowing it in advance. Another approach, which is the one we will use, is to invoke another result from the theory of Hilbert transforms, that solves the general problem of inverting a Hilbert transform on an interval. The result, whose proof can be found in [38], Section 3.2, is as follows.

Theorem 3.11 (Inversion formula for Hilbert transforms on a finite interval) *Given a function* $g\colon [-1,1] \to \mathbb{R}$*, the general solution of the equation*

$$\frac{1}{\pi}\int_{-1}^{1} \frac{f(y)}{y-x} = g(x),$$

where the integral is taken in the sense of the Cauchy principal value, is of the form

$$f(x) = \frac{1}{\pi\sqrt{1-x^2}}\int_{-1}^{1} \frac{\sqrt{1-y^2}\,g(y)}{x-y}\,dy + \frac{C}{\sqrt{1-x^2}}, \qquad (-1 \le x \le 1)$$

(again in the sense of the Cauchy principal value), where $C \in \mathbb{R}$ *is an arbitrary constant.*

Applying Theorem 3.11, we see that on the interval $(-\beta,\beta)$, h_τ' must be of the form

$$h_\tau'(u) = \frac{1}{\pi^2(\beta^2-u^2)^{1/2}}\int_{-\beta}^{\beta}\left(\lambda + \log\frac{\beta^2-u^2}{\frac{1}{2}-u^2}\right)\frac{du}{s-u} + \frac{C}{(\beta^2-u^2)^{1/2}},$$

where C is an arbitrary constant. This integral evaluation is rather tedious, but it can be done; we note it as a lemma and postpone its proof to the next section.

Lemma 3.12 *On $(-\beta,\beta)$, h'_τ is given by*

$$h'_\tau(u) = \frac{C}{(\beta^2-u^2)^{1/2}} + \frac{u}{\pi(\beta^2-u^2)^{1/2}}\left(-\lambda - 2\log\frac{1+\sqrt{1-2\beta^2}}{\sqrt{2}\beta}\right) + \frac{2}{\pi}\tan^{-1}\frac{(1-2\beta^2)^{1/2}u}{(\beta^2-u^2)^{1/2}} \qquad (3.30)$$

Now recall that both λ and C were arbitrary; we are merely looking for a function $h_\tau \in \mathcal{G}^\tau_\diamond$ whose derivative has the form above. Setting $C = 0$ and $\lambda = -2\log\frac{1+\sqrt{1-2\beta^2}}{\sqrt{2}\beta}$ therefore leads to

$$h'_\tau(u) = \frac{2}{\pi}\tan^{-1}\left(\frac{(1-2\beta^2)^{1/2}u}{(\beta^2-u^2)^{1/2}}\right) \qquad (-\beta < u < \beta), \qquad (3.31)$$

which, at the very least, is bounded between -1 and 1, as we would expect for the derivative of a function in $\mathcal{G}^\tau_\diamond$.

We are getting close to finding the minimizer. Two additional computational results that we will need, whose proofs are also postponed to the next section, are contained in the following lemma.

Lemma 3.13 *We have*

$$\int_{-\beta}^{\beta} u\,h'_\tau(u)\,du = \frac{1-2\beta^2-\sqrt{1-2\beta^2}}{2}, \qquad (3.32)$$

and, for $-\beta \le u \le \beta$,

$$h_\tau(u) = \beta + \int_{-\beta}^{u} h'_\tau(s)\,ds = \frac{2}{\pi}u\tan^{-1}\left(\frac{(1-2\beta^2)^{1/2}u}{(\beta^2-u^2)^{1/2}}\right) + \frac{\sqrt{2}}{\pi}\tan^{-1}\left(\frac{\sqrt{2}(\beta^2-u^2)^{1/2}}{(1-2\beta^2)^{1/2}}\right). \qquad (3.33)$$

Note that the value of the parameter β is still unknown, but we can now deduce it from the requirement that h_τ must satisfy the condition (3.26), which we rewrite as

$$\int_{-\beta}^{\beta} u\,h'_\tau(u)\,du = \tau - \beta^2.$$

Combining this with (3.32) we get the relation

$$\tau = \frac{1 - \sqrt{1-2\beta^2}}{2} \iff \beta = \sqrt{2\tau(1-\tau)}.$$

Substituting this value of β in (3.33), we get the expression

$$\begin{aligned} h_\tau(u) &= \frac{2}{\pi} u \tan^{-1}\left(\frac{(1-2\tau)u}{\sqrt{2\tau(1-\tau) - u^2}} \right) \\ &\quad + \frac{\sqrt{2}}{\pi} \tan^{-1}\left(\frac{\sqrt{2(2\tau(1-\tau) - u^2)}}{1 - 2\tau} \right) \qquad \left(|u| \le \sqrt{2\tau(1-\tau)}\right), \end{aligned}$$

and, because of the assumption (3.27), for $\sqrt{2\tau(1-\tau)} \le |u| \le \sqrt{2}/2$ we have the values $h_\tau(u) = |u|$ (note that $h_\tau(-\beta) = \beta$ by assumption and $h_\tau(\beta) = \beta + \int_{-\beta}^{\beta} h'_\tau(u)\,du = \beta$ since the right-hand side of (3.31) is an odd function). Thus, we conclude that our candidate minimizer h_τ coincides with the function $\tilde{g}_\tau$ defined in (3.8) (see also (3.7)). We have seen that h_τ is 1-Lipschitz, satisfies $h_\tau(\pm\sqrt{2}/2) = \sqrt{2}/2$, and satisfies (3.26) (which is equivalent to (3.20)), so $h_\tau \in \mathcal{G}_\diamond^\tau$. From the way we obtained h_τ, we also know that $p'(u) \equiv 0$ for $u \in (-\beta, \beta)$, and furthermore $p(0) = 0$ since h'_τ is odd, so also $p(u) \equiv 0$ on $(-\beta, \beta)$. To conclude that h_τ minimizes the functional $Q(\cdot)$ on $\mathcal{G}_\diamond^\tau$, it will be enough to check the second and third cases of (3.24), namely the inequalities $p(u) \ge 0$ and $p(u) \le 0$ when $u < -\beta$ and $u > -\beta$, respectively.

By symmetry, it is enough to prove that the second condition is satisfied. Interestingly, the proof consists of looking how the function $p(u)$ changes as a function of the parameter τ. Relabel it by $p(u, \tau)$ to emphasize its dependence on τ, and denote similarly the constant λ defined above by $\lambda(\tau)$, its value being

$$\lambda(\tau) = -2\log\left(\frac{1 + \sqrt{1-2\beta^2}}{\sqrt{2}\beta} \right) = -\log\left(\frac{1-\tau}{\tau} \right). \tag{3.34}$$

Fix $0 \le u \le \frac{\sqrt{2}}{2}$, and let $\tilde{\tau} = (1 - \sqrt{1-2u^2})/2$, so that $\beta(\tilde{\tau}) = u$. Because of the first condition in (3.24), which we already verified, $p(u, \tilde{\tau}) = 0$. To finish the proof, we will now show that $\partial p(u,\tau)/\partial\tau > 0$ for $0 < \tau < \tilde{\tau}$. By (3.28),

$$\frac{\partial p(u,\tau)}{\partial\beta} = -\int_{-\beta}^{\beta} \frac{\partial h'_\tau(s)}{\partial\beta} \log|u - s|\,ds - u\frac{d\lambda(\tau)}{d\beta}. \tag{3.35}$$

Using (3.31) and simplifying gives

$$\frac{\partial h'_\tau(s)}{\partial \beta} = -\frac{2}{\pi\beta(1-2\beta^2)^{1/2}} \cdot \frac{s}{(\beta^2-s^2)^{1/2}}. \tag{3.36}$$

It is easy to check that β satisfies the equation

$$\beta'(\tau) = (1-2\beta^2)^{1/2}/\beta, \tag{3.37}$$

so (3.35) becomes

$$\frac{\partial p(u,\tau)}{\partial \tau} = \frac{2}{\pi\beta^2}\int_{-\beta}^{\beta} \frac{s\log|u-s|}{(\beta^2-s^2)^{1/2}}\,ds + \frac{u}{\tau(1-\tau)}.$$

In this expression, the integral can be evaluated, using a result (3.38) that will proved in the next section, as

$$\begin{aligned}\int_{-\beta}^{\beta} \frac{s\log|u-s|}{(\beta^2-s^2)^{1/2}}\,ds &= \Big[-(\beta^2-s^2)^{1/2} - \log|u-s|\Big]\Big|_{s=-\beta}^{s=\beta} \\ &\qquad - \int_{-\beta}^{\beta} \frac{(\beta^2-s^2)^{1/2}}{u-s}\,ds = -\pi\Big(u-(u^2-\beta^2)^{1/2}\Big).\end{aligned}$$

Therefore we get that

$$\begin{aligned}\frac{\partial p(u,\tau)}{\partial \tau} &= -\frac{2}{\beta^2}\Big(u-(u^2-\beta^2)^{1/2}\Big) + \frac{u}{\tau(1-\tau)} \\ &= u\left(\frac{1}{\tau(1-\tau)} - \frac{2}{\beta^2}\right) + \frac{2}{\beta^2}(u^2-\beta^2)^{1/2} \\ &= \frac{2}{\beta^2}(u^2-\beta^2)^{1/2} > 0,\end{aligned}$$

which was our claim.

Summarizing the discussion in this section, we proved the main claim of Theorem 3.9, namely that $\tilde{g}_\tau$ is the unique minimizer of the functional $Q(\cdot)$ on $\mathcal{G}_\diamond^\tau$, except for a few integral evaluations that were deferred until the next section. In Section 3.10 we also prove (3.23) and thereby finish the proof of Theorem 3.9.

3.9 Some integral evaluations

Here, we collect the integral evaluations that were used in the previous section. Readers who has no interest in such computational wizardry will be forgiven for skipping to the next section (or even going directly to Section 3.11, since Section 3.10 contains additional computations of a somewhat similar nature).

Lemma 3.14 *For $x \geq -1$ we have the integral evaluations*

$$\int_{-1}^{1} \frac{\sqrt{1-y^2}}{x-y}\,dy = \begin{cases} \pi x & \text{if } -1 < x < 1, \\ \pi x - \pi\sqrt{x^2-1} & \text{if } x \geq 1, \end{cases} \tag{3.38}$$

where in the case $-1 < x < 1$ the left-hand side is defined as a principal value integral.

Proof Using the substitutions $y = \sin z$ and $w = \tan(z/2)$, we get

$$\begin{aligned} \int_{-1}^{1} \frac{\sqrt{1-y^2}}{x-y}\,dy &= \int_{-\pi/2}^{\pi/2} \frac{\cos^2 z}{x-\sin z}\,dz = \int_{-\pi/2}^{\pi/2} \left(x + \sin z + \frac{1-x^2}{x-\sin z} \right) dz \\ &= \pi x + \int_{-\pi/2}^{\pi/2} \frac{1-x^2}{x-\sin z}\,dz = \pi x + 2(1-x^2) \int_{-1}^{1} \frac{dw}{xw^2-2w+x} \end{aligned}$$

(where all integrals are in the principal value sense if $-1 < x < 1$). The denominator in the last integral $xw^2 - 2w + x$ vanishes when $w = w_\pm = x^{-1}(1 \pm \sqrt{1-x^2})$. In the case $-1 < x < 1$, the root w_- will be in $(-1, 1)$ (leading to a singularity in the integral) and w_+ will not; in this case one can show using a simple computation that the principal value integral $\int_{-1}^{1}[xw^2 - 2w + x]^{-1}\,dw$ is 0 (see Exercise 3.4). In the second case $x \geq 1$, both roots $w_\pm$ lie off the real line, in which case the integrand has no singularities and can be evaluated simply as

$$\begin{aligned} \int_{-1}^{1} \frac{dw}{xw^2-2w+x} &= \int_{-1}^{1} \frac{dw}{x(w-x^{-1})^2 + (x^2-1)} \\ &= \frac{1}{\sqrt{x^2-1}} \left[\tan^{-1} \frac{\sqrt{x-1}}{\sqrt{x+1}} + \tan^{-1} \frac{\sqrt{x+1}}{\sqrt{x-1}} \right] = \frac{\pi}{2\sqrt{x^2-1}} \end{aligned}$$

(since $\tan^{-1} t + \tan^{-1} t^{-1} \equiv \pi/2$). Therefore the left-hand side of (3.38) is equal to $\pi x - \pi\sqrt{1-x^2}$, as claimed. □

Lemma 3.15 *For $x \in [-1, 1]$ and $a \geq 1$ we have*

$$\int_{-1}^{1} \frac{\sqrt{1-y^2}}{x-y} \log \frac{1+y}{a+y}\, dy = \pi \Bigg[1 - a + \sqrt{a^2-1} - x \log\left(a + \sqrt{a^2-1}\right) - 2\sqrt{1-x^2} \tan^{-1} \sqrt{\frac{(a-1)(1-x)}{(a+1)(1+x)}} \Bigg]. \tag{3.39}$$

Proof Denote the left-hand side of (3.39) by $F(x, a)$. Note that $F(x, 1) = 0$, and for $a > 1$, using both evaluations in (3.38), we have that

$$\begin{aligned}
\frac{\partial F(x,a)}{\partial a} &= -\int_{-1}^{1} \frac{\sqrt{1-y^2}}{(x-y)(a+y)}\, dy \\
&= -\frac{1}{a+x} \int_{-1}^{1} \sqrt{1-y^2} \left(\frac{1}{x-y} + \frac{1}{a+y} \right) dy \\
&= -\frac{\pi x}{a+x} - \frac{1}{a+x} \int_{-1}^{1} \frac{\sqrt{1-y^2}}{a+y}\, dy \\
&= -\frac{\pi x}{a+x} - \frac{\pi a - \sqrt{a^2-1}}{a+x} = -\pi + \frac{\pi\sqrt{a^2-1}}{a+x}.
\end{aligned}$$

Integrating over a, we get that

$$F(x, a) = \int_1^a \frac{\partial F(x,r)}{\partial r}\, dr = -\pi(a-1) + \pi \int_1^a \frac{\sqrt{r^2-1}}{r+x}\, dr. \tag{3.40}$$

The integral can be computed using a variant of the substitutions used in the proof of the previous lemma. Setting $r = \cosh z$ and later $w = e^z$, we have

$$\begin{aligned}
\int_1^a \frac{\sqrt{r^2-1}}{r+x}\, dr &= \int_0^{\cosh^{-1}(a)} \frac{\sinh^2 z}{x + \cosh z}\, dz \\
&= \int_0^{\cosh^{-1}(a)} \left(\cosh z - x + \frac{x^2-1}{x+\cosh z} \right) dz \\
&= \sinh z \Big|_0^{\cosh^{-1}(a)} - x \cosh^{-1}(a) + \int_0^{\cosh^{-1}(a)} \frac{x^2-1}{x+\cosh z}\, dz \\
&= \sqrt{a^2-1} - x \log\left(a + \sqrt{a^2-1}\right) + \int_0^{\cosh^{-1}(a)} \frac{x^2-1}{x+\cosh z}\, dz \\
&= \sqrt{a^2-1} - x \log\left(a + \sqrt{a^2-1}\right) \\
&\quad + 2(x^2-1) \int_0^{a+\sqrt{a^2-1}} \frac{dw}{w^2 + 2xw + 1},
\end{aligned} \tag{3.41}$$

and the last integral equals

$$\frac{1}{\sqrt{1-x^2}}\tan^{-1}\frac{w+x}{\sqrt{1-x^2}}\Bigg|_{w=1}^{w=a+\sqrt{a^2-1}}$$
$$= \frac{1}{\sqrt{1-x^2}}\tan^{-1}\frac{(a-1+\sqrt{a^2-1})\sqrt{1-x^2}}{1-x^2+(x+a+\sqrt{a^2-1})(1+x)}$$
$$= \frac{1}{\sqrt{1-x^2}}\tan^{-1}\left[\frac{a-1+\sqrt{a^2-1}}{a+1+\sqrt{a^2-1}}\sqrt{\frac{1+x}{1-x}}\right]$$
$$= \frac{1}{\sqrt{1-x^2}}\tan^{-1}\sqrt{\frac{(a-1)(1-x)}{(a+1)(1+x)}}. \tag{3.42}$$

Combining (3.40), (3.41) and (3.42) gives (3.39). □

Proof of Lemma 3.12 Using (3.38), we have

$$\frac{1}{\pi^2(\beta^2-u^2)^{1/2}}\int_{-\beta}^{\beta}\left(\lambda+\log\frac{\beta^2-u^2}{\frac{1}{2}-u^2}\right)\frac{du}{s-u}+\frac{C}{(\beta^2-u^2)^{1/2}}$$
$$= -\frac{\lambda u}{\pi(\beta^2-u^2)^{1/2}}+\frac{\beta}{\pi(\beta^2-u^2)^{1/2}}\left[F\left(\frac{s}{\beta},\frac{\sqrt{2}}{2\beta}\right)-F\left(-\frac{s}{\beta},\frac{\sqrt{2}}{2\beta}\right)\right]$$
$$+\frac{C}{(\beta^2-u^2)^{1/2}},$$

where as before $F(x,a)$ denotes the left-hand side of (3.39). By (3.39) this evaluates to

$$\frac{u}{\pi(\beta^2-u^2)^{1/2}}\left(-\lambda-2\log\frac{1+\sqrt{1-2\beta^2}}{\sqrt{2}\beta}\right)$$
$$+\frac{2}{\pi}\left(\tan^{-1}\sqrt{\frac{(a-1)(1+x)}{(a+1)(1-x)}}-\tan^{-1}\sqrt{\frac{(a-1)(1-x)}{(a+1)(1+x)}}\right)+\frac{C}{(\beta^2-u^2)^{1/2}} \tag{3.43}$$

where $x=u/\beta, a=\sqrt{2}/(2\beta)$, which then further simplifies to the right-hand side of (3.30) using the arctangent addition formula $\tan^{-1}s+\tan^{-1}t=\tan^{-1}((s+t)/(1-st))$. □

Proof of Lemma 3.13 It is more convenient to work with the representation of h'_τ in which the arctangent addition formula isn't used after (3.43), namely,

$$h'_\tau(u) = \frac{2}{\pi}\left(\tan^{-1}\sqrt{\frac{(a-1)(1+x)}{(a+1)(1-x)}} - \tan^{-1}\sqrt{\frac{(a-1)(1-x)}{(a+1)(1+x)}}\right), \tag{3.44}$$

where x and a are defined after (3.43). The claims of the lemma therefore reduce to performing the indefinite integrations

$$\int \tan^{-1}\sqrt{\frac{(a-1)(1+x)}{(a+1)(1-x)}}\,dx, \qquad \int x\tan^{-1}\sqrt{\frac{(a-1)(1+x)}{(a+1)(1-x)}}\,dx \tag{3.45}$$

(note that the second arctangent term in (3.44) is obtained from the first by substituting $-x$ for x). These integral evaluations aren't as difficult as they seem: for the first, perform an integration by parts to get that

$$\int \tan^{-1}\sqrt{\frac{(a-1)(1+x)}{(a+1)(1-x)}}\,dx = x\tan^{-1}\sqrt{\frac{(a-1)(1+x)}{(a+1)(1-x)}} - \frac{1}{2}\sqrt{a^2-1}\int \frac{x\,dx}{(a-x)\sqrt{1-x^2}} + C.$$

(Here and below, C denotes a generic integration constant.) The integral on the right-hand side can then be evaluated using the substitutions $x = \sin z$ and $w = \tan(z/2)$, which gives

$$\int \frac{x\,dx}{(a-x)\sqrt{1-x^2}} = \int \frac{\sin z}{a-\sin z}\,dz = -z + \int \frac{dz}{a-\sin z} + C = -z + \frac{2}{a}\int \frac{dw}{1+w^2-2w/a} + C,$$

which is an elementary integral. The same technique will work for the second integral in (3.45). We leave it to readers to complete the details. □

3.10 Evaluation of $Q(\tilde{g}_\tau)$

In the last two sections we showed that the minimizer h_τ of $Q(\cdot)$ on $\mathcal{G}^\tau_\diamond$ coincides with the function g_τ defined in (3.8). We now compute the minimum value $Q(\tilde{g}_\tau) = Q(h_\tau)$. For convenience, denote $h = h_\tau$ throughout the computation. Start with the observation that the function $p(t)(h'(t) - \operatorname{sgn} t)$ (where $p(t)$ was defined in (3.24)) vanishes on the interval $\left[-\sqrt{2}/2, \sqrt{2}/2\right]$, since $p(u) = 0$ on $(-\beta, \beta)$ (that condition was the starting point of our derivation of h_τ) and $h'(t) - \operatorname{sgn} t = 0$ when $\beta \le |t| \le \sqrt{2}/2$. Thus we have (making use of (3.26)) that

$$\begin{aligned}
0 &= \int_{-\sqrt{2}/2}^{\sqrt{2}/2} p(t)(h'(t) - \operatorname{sgn} t) \\
&= \int_{-\sqrt{2}/2}^{\sqrt{2}/2} \left(\lambda t - \int_{-\sqrt{2}/2}^{\sqrt{2}/2} h'(s) \log |s - t| \, ds \right) (h'(t) - \operatorname{sgn} t) \\
&= \lambda \left(\int_{-\sqrt{2}/2}^{\sqrt{2}/2} t h'(t) \, dt - \frac{1}{2} \right) + 2Q(h) \\
&\quad + \int_{-\sqrt{2}/2}^{\sqrt{2}/2} h'(s) \left(\int_{-\sqrt{2}/2}^{\sqrt{2}/2} \operatorname{sgn} t \cdot \log |t - s| \, dt \right) ds \\
&= -\lambda\tau + 2Q(h) + \int_{-\sqrt{2}/2}^{\sqrt{2}/2} h'(s) Z(s) \, ds,
\end{aligned} \tag{3.46}$$

where $\lambda = -\log((1 - \tau)/\tau)$ as in (3.34), and where we denote

$$Z(s) = \int_{-\sqrt{2}/2}^{\sqrt{2}/2} \operatorname{sgn} t \cdot \log |t - s| \, dt,$$

an integral that evaluates to

$$Z(s) = 2s \log |s| - (\sqrt{2}/2 + s) \log(\sqrt{2}/2 + s) + (\sqrt{2}/2 - s) \log(\sqrt{2}/2 - s).$$

Denote $W = \int_{-\sqrt{2}/2}^{\sqrt{2}/2} h'(s) Z(s) \, ds$. Remembering that $W = W_\tau$ is a function of τ, we differentiate, involving $\beta = 2\sqrt{\tau(1 - \tau)}$ as an intermediate variable and making use of (3.36) and (3.37). This gives

$$\begin{aligned}
\frac{dW}{d\tau} &= \frac{d\beta}{d\tau} \frac{dW}{d\beta} = \frac{\sqrt{1 - 2\beta^2}}{\beta} \int_{-\sqrt{2}/2}^{\sqrt{2}/2} \frac{\partial h'(s)}{\partial \beta} Z(s) \, ds \\
&= \frac{\sqrt{1 - 2\beta^2}}{\beta} \int_{-\beta}^{\beta} \frac{-2}{\pi\beta \sqrt{1 - 2\beta^2}} \cdot \frac{s}{\sqrt{\beta^2 - s^2}} Z(s) \, ds \\
&= \frac{-2}{\pi\beta^2} \int_{-\beta}^{\beta} \frac{s}{\sqrt{\beta^2 - s^2}} Z(s) \, ds = \frac{-2}{\pi\beta^2} \int_{-\beta}^{\beta} \sqrt{\beta^2 - s^2} Z'(s) \, ds \\
&= \frac{-2}{\pi\beta^2} \int_{-\beta}^{\beta} \sqrt{\beta^2 - s^2} \left(2 \log |s| - \log\left(\sqrt{2}/2 + s\right) - \log\left(\sqrt{2}/2 - s\right) \right) ds.
\end{aligned} \tag{3.47}$$

Now denote

$$E_\beta(u) = \int_{-\beta}^{\beta} \sqrt{\beta^2 - s^2} \log|s-u|\,ds,$$

so that (3.47) can be rewritten as

$$\begin{aligned}
\frac{dW}{d\tau} &= \frac{-2}{\pi\beta^2}\left(2E_\beta(0) - E_\beta\left(\sqrt{2}/2\right) - E_\beta\left(-\sqrt{2}/2\right)\right) \\
&= \frac{2}{\pi\beta^2}\left(\int_0^{\sqrt{2}/2} E'_\beta(x)\,dx - \int_{-\sqrt{2}/2}^{0} E'_\beta(x)\,dx\right) = \frac{4}{\pi\beta^2}\int_0^{\sqrt{2}/2} E'_\beta(x)\,dx,
\end{aligned}$$

and observe that $E'_\beta(x)$ can be evaluated via a simple rescaling of (3.38), which gives

$$E'_\beta(x) = \int_{-\beta}^{\beta} \frac{\sqrt{\beta^2-s^2}}{u-s}\,ds = \begin{cases} \pi x & \text{if } |x| < \beta, \\ \pi x - \pi\sqrt{x^2-\beta^2} & \text{if } x \ge \beta. \end{cases}$$

This leads to

$$\begin{aligned}
\frac{dW}{d\tau} &= \frac{4}{\beta^2}\left(\int_0^{\sqrt{2}/2} x\,dx - \int_\beta^{\sqrt{2}/2} \sqrt{x^2-\beta^2}\,dx\right) \\
&= \frac{4}{\beta^2}\left(\frac{1}{4} - \frac{1}{2}\left[x\sqrt{x^2-\beta^2} - \beta^2 \log\left(x + \sqrt{x^2-\beta^2}\right)\right]_{x=\beta}^{x=\sqrt{2}/2}\right) \\
&= \frac{4}{\beta^2}\left(\frac{1}{4} - \frac{1}{2}\left(\frac{1-2\tau}{2} - \tau(1-\tau)\log\left(\frac{1-\tau}{\tau}\right)\right)\right) \\
&= \frac{2}{2\tau(1-\tau)}\left(\tau + \tau(1-\tau)\log\left(\frac{1-\tau}{\tau}\right)\right) \\
&= \frac{1}{1-\tau} - \log\tau + \log(1-\tau).
\end{aligned}$$

Finally, since $h'_{\tau=1/2} \equiv 0$, we have that $W_{\tau=1/2} = 0$, and therefore we get that

$$\begin{aligned}
W &= \int_{1/2}^{\tau} \left(\frac{1}{1-v} - \log v + \log(1-v)\right) dv \\
&= -2\log 2 - \log(1-\tau) - \tau\log\tau - (1-\tau)\log(1-\tau).
\end{aligned}$$

Combining this with the relation $Q(h) = \frac{1}{2}(\lambda\tau - W)$, which follows from (3.46), gives the result that

$$Q(h) = \log 2 + \tau\log\tau + (1-\tau)\log(1-\tau),$$

which confirms (3.23). The proof of Theorem 3.9 is complete. ◻

3.11 The limiting growth profile

We can now prove Theorem 3.5. We start by proving a slightly weaker result without uniformity in k; later we show how this easily implies the claim of Theorem 3.5 by simple monotonicity considerations.

Proposition 3.16 *Let $0 < \tau < 1$, and let $k = k(n)$ be a sequence such that $1 \le k \le n^2$ and $k/n^2 \to \tau$ as $n \to \infty$. For any $\epsilon > 0$ we have*

$$\mathbb{P}\left[\sup_{u\in[-\sqrt{2}/2,\sqrt{2}/2]} |\tilde{g}_{k/n^2}(u) - \psi_{n,k}(u)| > \epsilon\right] \to 0 \text{ as } n \to \infty. \tag{3.48}$$

Proof Let $\epsilon > 0$ be given. As in the proof of Theorem 1.20, denote by $\mathcal{M}_{n,k}$ the set of Young diagrams $\lambda \in \mathcal{P}(k)$ that are contained in the square diagram $\square_{n,n}$ and for which

$$||\psi_{n,\lambda} - \tilde{g}_{k/n^2}||_Q > \epsilon,$$

where $||h||_Q = Q(h)^{1/2}$ as in (1.46) and $\psi_{n,\lambda}$ is defined in the paragraph after (3.9). For $\lambda \in \mathcal{M}_{n,k}$, by Theorem 3.9 and (3.25) we have that

$$Q(\psi_{n,\lambda}) \ge Q(\tilde{g}_{k/n^2}) + ||\psi_{n,\lambda} - \tilde{g}_{k/n^2}||_Q^2 > -H(k/n^2) + \log 2 + \epsilon^2,$$

so that (using (3.21) and (3.22)) $I_{\text{hook}}^{\square}(\phi_{n,\lambda}) > -H(k/n^2) + 2\log 2 - \frac{3}{2} + \epsilon^2$ (where $\phi_{n,\lambda}$ is defined in (3.9)). Therefore by (3.19),

$$\mathbb{P}\left(\Lambda^{(n,k)} = \lambda\right) < \exp\left(-\epsilon^2 n^2 + O(n\log n)\right),$$

uniformly over all $\lambda \in \mathcal{M}_{n,k}$. Since $|\mathcal{M}_{n,k}| \le |\mathcal{P}(k)| = p(k) \le e^{C\tau n}$ for some constant $C > 0$, we get that

$$\begin{aligned}\mathbb{P}\left[||\psi_{n,k} - \tilde{g}_{k/n^2}||_Q > \epsilon\right] &= \mathbb{P}\left[\Lambda^{(n,k)} \in \mathcal{M}_{n,k}\right] \\ &\le \exp\left(-\epsilon^2 n^2 + C\tau n + O(n\log n)\right) \to 0 \text{ as } n \to \infty.\end{aligned}$$

This implies (3.48) by using Lemma 1.21 to replace the Q-norm bound by a bound on the uniform $||\cdot||_\infty$ norm. □

Proof of Theorem 3.5 The idea is to apply simultaneously for several sequences $k_j = k_j(n)$ and then use the fact that both $\psi_{n,k}$ and $\tilde{g}_{k^2/n}$ are monotone nondecreasing in k to deduce a uniform estimate. Fix an integer $m \ge 1$, and set

$$k_j = k_j(n) = 1 + \left\lfloor \frac{j}{m}(n^2 - 1)\right\rfloor, \qquad j = 0, 1, \ldots, m.$$

Define an event

$$E_n = \left\{ \max_{0 \le j \le m} \|\psi_{n,k} - \tilde{g}_{k^2/n}\|_\infty < \epsilon/2 \right\}.$$

By Theorem 3.16, $\mathbb{P}(E_n) \to 1$ as $n \to \infty$. Assume that E_n occurred. Given some k satisfying $1 \le k \le n^2$, we have $k_j \le k \le k_{j+1}$ for some $0 \le j < m$. By the monotonicity properties mentioned earlier, we have for all $u \in \left[-\sqrt{2}/2, \sqrt{2}/2\right]$ the chain of inequalities

$$\tilde{g}_{k_j/n^2}(u) - \frac{\epsilon}{2} \le \psi_{n,k_j}(u) \le \psi_{n,k}(u) \le \psi_{n,k_{j+1}}(u) \le \tilde{g}_{k_{j+1}/n^2}(u) + \frac{\epsilon}{2},$$

and furthermore, by continuity of $\tilde{g}_\tau(u)$ as a function of τ it is easy to see that if m is fixed to be large enough (as a function of ϵ but not depending on n), we will have for large n the inequalities $\tilde{g}_{k_{j+1}}(u) \le \tilde{g}_{k/n^2} + \epsilon/2$ and $\tilde{g}_{k/n^2} - \epsilon/2 \le \tilde{g}_{k_j}(u)$, so we get that

$$\tilde{g}_{k/n^2}(u) - \epsilon \le \psi_{n,k}(u) \le \tilde{g}_{k/n^2}(u) + \epsilon,$$

or equivalently that $\|\psi_{n,k} - \tilde{g}_{k/n^2}\|_\infty < \epsilon$. This shows that occurrence of E_n (which has asymptotically high probability) implies occurrence of the event

$$\max_{1 \le k \le n^2} \sup_{u \in \left[-\sqrt{2}/2, \sqrt{2}/2\right]} |\tilde{g}_{k/n^2}(u) - \psi_{n,k}(u)| \le \epsilon,$$

(the event complementary to the one that appears in (3.11)), and therefore finishes the proof. □

3.12 Growth of the first row of $\Lambda^{(n,k)}$

Our next major goal is to prove Theorem 3.6. When trying to deduce it from Theorem 3.5, we run into a difficulty that is similar to the one we had in Chapter 1 when we tried to derive the result $\Lambda = 2$ from the limit shape theorem for Plancherel-random partitions. The problem is that because the limit shape result proves proximity to the limiting shape in the Russian coordinate system, it does not guarantee that the length $\Lambda_1^{(n,k)}$ of the first row of the Young diagram $\Lambda^{(n,k)}$ is asymptotically close to the value predicted by the limiting shape. To proceed we need to separately prove the following result.

Theorem 3.17 *For any $\epsilon > 0$ we have*

$$\mathbb{P}\left[\max_{1\le k\le n^2} \left|n^{-1}\Lambda_1^{(n,k)} - q(k/n^2)\right| > \epsilon\right] \to 0 \ \ as\ n \to \infty, \tag{3.49}$$

where $q(\cdot)$ is defined by

$$q(\tau) = \begin{cases} 2\sqrt{\tau(1-\tau)} & if\ 0 \le \tau \le \frac{1}{2}, \\ 1 & if\ \frac{1}{2} \le \tau \le 1. \end{cases}$$

Note that since $\Lambda_1^{(n,k)}$ is increasing in k, it is enough to consider the values $1 \le k \le n^2/2$ in (3.49). Theorem 3.5 implies half of this result, namely a lower bound for $\Lambda_1^{(n,k)}$.

Lemma 3.18 *For any $\epsilon > 0$,*

$$\mathbb{P}\left[n^{-1}\Lambda_1^{(n,k)} > q(k/n^2) - \epsilon\ for\ all\ 1 \le k \le n^2/2\right] \to 1 \ \ as\ n \to \infty. \tag{3.50}$$

Proof Fix $\epsilon > 0$. We claim that there is a $\delta > 0$ such that if the event

$$\left\{\max_{1\le k\le n^2} \sup_{u\in\mathbb{R}} |\tilde{g}_{k/n^2}(u) - \psi_{n,k}(u)| < \delta\right\}$$

occurs then for any $1 \le k \le n^2/2$, the inequality

$$n^{-1}\Lambda_1^{(n,k)} > q(k/n^2) - \epsilon \tag{3.51}$$

must hold; if true, this implies the result by (3.11). The fact that the claim is true, and the relationship between δ and ϵ, are illustrated in Fig. 3.8. One can see immediately that the fact that the graph of $\psi_{n,k}$ is constrained to lie above the graph of $\max(\tilde{g}_\tau(u)-\delta, |u|)$ (where we denote $\tau = k/n^2$) forces the inequality $n^{-1}\Lambda_1^{(n,k)} > q(\tau) - \epsilon$ to hold. We leave it to readers to formalize the dependence between δ and ϵ. (In particular, it should be noted that for a given ϵ, a single δ would work for any $1 \le k \le n^2$.) □

Corollary 3.19 *Denote $A_n(k) = \mathbb{E}\Lambda_1^{(n,k)}$. For any $\epsilon > 0$, there is an integer $N \ge 1$ such that for any $n \ge N$, we have*

$$n^{-1}A_n(k) > q(k/n^2) - \epsilon\ \ for\ all\ 1 \le k \le n^2/2. \tag{3.52}$$

Proof Denote by $E_{n,\epsilon}$ the event

$$E_{n,\epsilon} = \left\{n^{-1}\Lambda_1^{(n,k)} > q(k/n^2) - \epsilon/2 \text{ for all } 1 \le k \le n^2/2\right\}, \tag{3.53}$$

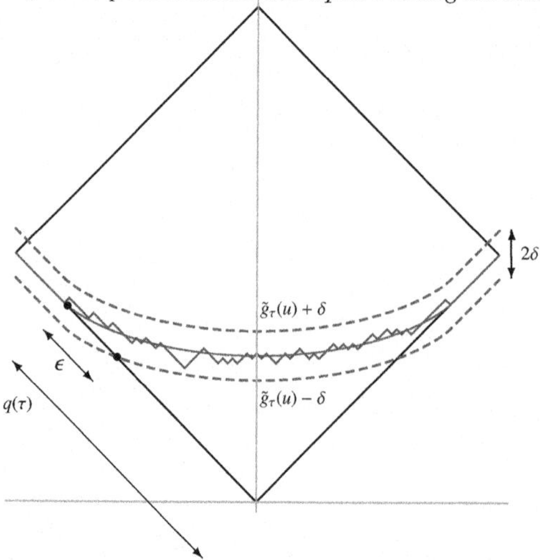

Figure 3.8 Illustration of the proof of Lemma 3.18: proximity of $\psi_{n,k}$ to the limit shape $\tilde{g}_{k/n^2}$ forces a lower bound – but not an upper bound – on the length of the first row of $\Lambda^{(n,k)}$.

and note that for any $1 \le k \le n^2/2$,

$$n^{-1}A_n(k) \ge n^{-1}\mathbb{E}\left[\Lambda_1^{(n,k)}\mathbf{1}_{E_{n,\epsilon}}\right] \ge (q(k/n^2) - \epsilon/2)\mathbb{P}(E_{n,\epsilon}).$$

By (3.50), the right-hand side is $\ge q(k/n^2) - \epsilon$ if n is large enough. □

We now turn to the task of establishing the less obvious matching upper bound to (3.51). First, we note an interesting combinatorial identity involving a product of hook-lengths.

Lemma 3.20 *Let $1 \le k \le n^2$. Let $\lambda \nearrow \nu$ be a pair of Young diagrams contained in the square diagram $\square_{n,n}$ such that λ is of size $k-1$, ν is of size k, and ν is obtained from λ by adding a new cell to the first row of λ. Then we have*

$$\frac{d_\lambda d_{\square_{n,n}\setminus\nu}}{d_{\square_{n,n}\setminus\lambda} d_\nu} = \frac{n^2 - \lambda_1^2}{k(n^2 - k + 1)}. \tag{3.54}$$

Proof For the duration of the proof denote $\bar{\lambda} = \square_{n,n} \setminus \lambda$ and $\bar{\nu} = \square_{n,n} \setminus \nu$.

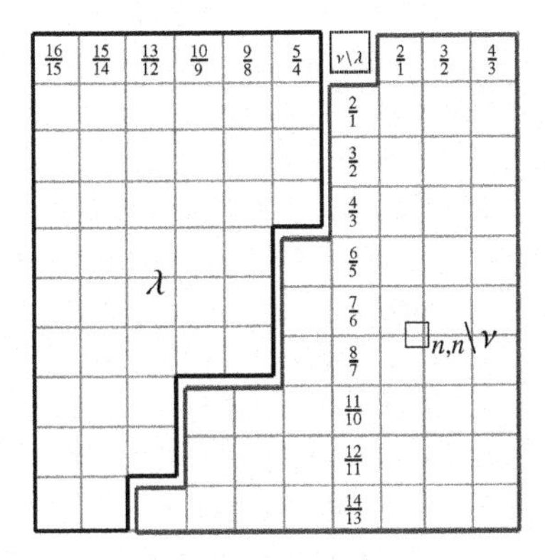

Figure 3.9 An illustration of the four hook-length products participating in the identity (3.54).

By the hook-length formula we have

$$\frac{d_\lambda d_{\bar{\nu}}}{d_{\bar{\lambda}} d_\nu} = \frac{(k-1)!(n^2-k)!}{(n^2-k+1)!k!} \cdot \frac{\prod_{(i,j)\in\bar{\lambda}} h_{\bar{\lambda}}(i,j) \prod_{(i,j)\in\nu} h_\nu(i,j)}{\prod_{(i,j)\in\lambda} h_\lambda(i,j) \prod_{(i,j)\in\bar{\nu}} h_{\bar{\nu}}(i,j)}$$

$$= \frac{1}{k(n^2-k+1)} \cdot \frac{\prod_{(i,j)\in\bar{\lambda}} h_{\bar{\lambda}}(i,j) \prod_{(i,j)\in\nu} h_\nu(i,j)}{\prod_{(i,j)\in\lambda} h_\lambda(i,j) \prod_{(i,j)\in\bar{\nu}} h_{\bar{\nu}}(i,j)},$$

so it remains to explain how the factor $n^2 - \lambda_1^2$ arises out of the four hook-length products, with all other factors cancelling out. This phenomenon is illustrated in Fig. 3.9, which shows the "T"-shaped set of cells contributing to the four hook-length products, with all other hook-lengths cancelling out in a trivial manner because the hook-length does not change in the process of adding the new cell to get from λ to ν (or, symmetrically, removing a cell to get from $\bar{\nu}$ to $\bar{\lambda}$). For each box (i, j) from that set of cells, Fig. 3.9 shows the ratio of hook-lengths $h_\nu(i, j)/h_\lambda(i, j)$ or $h_{\bar{\lambda}}(i, j)/h_{\nu(\bar{i},\bar{j})}$, so that the product above is obtained by multiplying all the fractions. By studying the figure, one can see that now we get further cancellation that arises as follows. In the "right arm" of the set of cells we simply have the fraction of factorials

$$\frac{(n-\lambda_1)!}{(n-\lambda_1-1)!} = n - \lambda_1.$$

In each of the left arm and the leg of the "T" we have a partial cancellation along contiguous stretches where the hook-lengths increase by 1, where the product of fractions is telescoping and leaves only the denominator of the first cell and the numerator of the last cell in each such stretch (counting in the direction leading away from the cell $\nu \setminus \lambda$); for example, in the leg in the example of Fig. 3.9 we have the product $\frac{6}{5} \cdot \frac{7}{6} \cdot \frac{8}{7} = \frac{8}{5}$. Call the remaining hook-lengths after this cancellation the "exceptional" hook-lengths. However, as can be seen in the example in the figure, the same numerators and denominators that remain in the product of hook ratios $h_\nu(i,j)/h_\lambda(i,j)$ in the left arm appear with inverse powers in the dual product of ratios $h_{\bar{\lambda}}(i,j)/h_{\bar{\nu}}(i,j)$ in the leg, and vice versa, with only two hook-lengths remaining, namely the smallest hook-length 1 (in the denominator) and the largest hook-length $n+\lambda_1$ (in the numerator). The reason for this is that, except for these smallest and largest hook-lengths, the exceptional hook-lengths in λ are in one-to-one correspondence with identical exceptional hook-lengths in $\bar{\nu}$; and similarly the exceptional hook-lengths in ν can be matched with identical ones in $\bar{\lambda}$. Note that in the above example, the largest hook-length came from the hook product of hook-lengths of ν, but there is also a separate case in which that hook-length will appear in the product of hook-lengths of $\bar{\nu}$. This happens when λ has less than n parts, but it is easy to verify that the end result is the same in both cases.

We conclude that the product of hook ratios indeed leaves out exactly two factors, which combine to give the expression $(n-\lambda_1)(n+\lambda_1) = n^2 - \lambda_1^2$. This was exactly what we needed to finish the proof. □

Using the identity (3.54), we can show that the numbers $A_n(k) = \mathbb{E}\Lambda_1^{(n,k)}$ satisfy an interesting discrete difference inequality.

Lemma 3.21 *Denote $a_n(k) = A_n(k) - A_n(k-1)$ (where $A_n(0)$ is defined to be 0). We have*

$$a_n(k)^2 \leq \frac{n^2 - A_n(k)^2}{k(n^2 - k + 1)}. \tag{3.55}$$

Proof The "discrete derivative" $a_n(k)$ can be interpreted as the probability that $\Lambda^{(n,k)}$ was obtained from $\Lambda^{(n,k-1)}$ by adding a cell to the first row. Let $\mathcal{L}_{n,k}$ be the set of pairs (λ,ν) of Young diagrams contained in $\square_{n,n}$, where λ is of size $k-1$, ν is of size k, and ν is obtained from λ by adding a cell to the first row of λ. We can write

$$a_n(k) = \sum_{(\lambda,\nu)\in\mathcal{L}_{n,k}} \frac{d_\lambda d_{\square_{n,n}\setminus\nu}}{d_{\square_{n,n}}} = \sum_{(\lambda,\nu)\in\mathcal{L}_{n,k}} \frac{d_\nu d_{\square_{n,n}\setminus\nu}}{d_{\square_{n,n}}} \cdot \frac{d_\lambda}{d_\nu}.$$

This is an average of the quantity d_λ/d_ν (thought of as a function of ν) with respect to the probability measure $\nu \mapsto \frac{d_\nu d_{\square_{n,n}\setminus\nu}}{d_{\square_{n,n}}}$, that is, the measure given in (3.15); except that the summation is not over all possible ν, since not all $\nu \subset \square_{n,n}$ of size k have a corresponding λ such that $(\lambda,\nu) \in \mathcal{L}_{n,k}$. Applying the Cauchy–Schwarz inequality and then using Lemma 3.20, we therefore get that

$$\begin{aligned} a_n(k)^2 &\leq \sum_{(\lambda,\nu)\in\mathcal{L}_{n,k}} \frac{d_\nu d_{\square_{n,n}\setminus\nu}}{d_{\square_{n,n}}} \cdot \frac{d_\lambda^2}{d_\nu^2} = \sum_{(\lambda,\nu)\in\mathcal{L}_{n,k}} \frac{d_\lambda d_{\square_{n,n}\setminus\lambda}}{d_{\square_{n,n}}} \cdot \frac{d_\lambda d_{\square_{n,n}\setminus\nu}}{d_{\square_{n,n}\setminus\lambda} d_\nu} \\ &= \sum_{(\lambda,\nu)\in\mathcal{L}_{n,k}} \frac{d_\lambda d_{\square_{n,n}\setminus\lambda}}{d_{\square_{n,n}}} \cdot \frac{n^2 - \lambda_1^2}{k(n^2-k+1)} \\ &= (k(n^2-k+1))^{-1}\mathbb{E}\left[n^2 - \left(\Lambda_1^{(n,k)}\right)^2\right] \leq \frac{n^2 - A_n(k)^2}{k(n^2-k+1)}. \qquad \square \end{aligned}$$

The difference inequality (3.55) is the key to deriving an upper bound for $A_n(k)$ that matches the lower bound in (3.52).

Lemma 3.22 *For any $\epsilon > 0$, there is an integer $N \geq 1$ such that for any $n \geq N$, we have*

$$n^{-1}A_n(k) < q(k/n^2) + \epsilon \text{ for all } 1 \leq k \leq n^2/2. \tag{3.56}$$

Proof Note that if (3.55) were an equality rather than an inequality, it would be a discrete analogue of the differential equation

$$q'(\tau)^2 = \frac{1 - q(\tau)^2}{\tau(1-\tau)} \tag{3.57}$$

which (check!) is satisfied by $q(\cdot)$ for $0 < \tau < \frac{1}{2}$. Since what we have is an inequality, it will give an upper bound on the growth of $A_n(k)$ (which is what we want), by a self-referential quantity that can nonetheless be made explicit by the fact that we already derived a *lower* bound for $A_n(k)$. To make this argument precise, fix $\epsilon > 0$, and let $\delta > 0$ depend on ϵ in a manner that will be determined shortly. Using (3.52) and (3.55), for

$1 \le k \le n^2/2$ and n large enough we can write

$$A_n(k) = \sum_{j=1}^{k} a_n(j) \le \sum_{j=1}^{k} \left(\frac{n^2 - A_n(j)^2}{j(n^2 - j + 1)} \right)^{1/2}$$

$$\le \sum_{j=1}^{k} \left(\frac{n^2 - n^2 \left(q\left(\frac{j}{n^2}\right) - \delta \right)^2}{j(n^2 - j + 1)} \right)^{1/2} = n \cdot \sum_{j=1}^{k} \left(\frac{1 - \left(q\left(\frac{j}{n^2}\right) - \delta \right)^2}{\frac{j}{n^2}\left(1 - \frac{j-1}{n^2}\right)} \right)^{1/2} \cdot \frac{1}{n^2}$$

$$\le n \left(\int_0^{\tau} \left(\frac{1 - (q(t) - \delta)^2}{t(1-t)} \right)^{1/2} dt + o(1) \right)$$

where we denote $\tau = k/n^2$ as before, and where the $o(1)$ is uniform in k as $n \to \infty$. Taking δ to be sufficiently small as a function of ϵ, we can further upper-bound this for large n (making use of the differential equation (3.57)) by

$$n \left(\int_0^{\tau} \left(\frac{1 - q(t)^2}{t(1-t)} \right)^{1/2} dt + \tfrac{1}{2}\epsilon + o(1) \right) = n \left(q(\tau) + \tfrac{1}{2}\epsilon + o(1) \right)$$
$$= n \left(q(k/n^2) + \tfrac{1}{2}\epsilon + o(1) \right) \le n \left(q(k/n^2) + \epsilon \right),$$

as claimed. □

Proof of Theorem 3.17 Fix $0 < \tau \le 1/2$, and let $k = k(n)$ be a sequence such that $k(n) \le n^2/2$ and $k/n \to \tau$ as $n \to \infty$.

$$\mathbb{P}\left(\left| n^{-1}\Lambda^{(n,k)} - q(k/n^2) \right| > \epsilon \right) \le \mathbb{P}\left(n^{-1}\Lambda^{(n,k)} - q(k/n^2) < -\epsilon \right) + \mathbb{P}\left(n^{-1}\Lambda^{(n,k)} - q(k/n^2) > \epsilon \right) \tag{3.58}$$

The first probability on the right goes to 0 as $n \to \infty$, by Lemma 3.18. The second probability is bounded from above (using Markov's inequality) by $\epsilon^{-1}\left(n^{-1}\mathbb{E}\Lambda_1^{(n,k)} - q(k/n^2) \right)$, and this goes to 0 as $n \to \infty$ by Lemma 3.22. It follows that the right-hand side of (3.58) goes to 0. Consequently, a similar statement is true when considering several sequences $k_1(n), \dots, k_m(n)$ simultaneously, namely that

$$\mathbb{P}\left(\max_{1 \le j \le m} \left| n^{-1}\Lambda^{(n,k_j)} - q(k_j/n^2) \right| > \epsilon \right) \to 0 \text{ as } n \to \infty. \tag{3.59}$$

(with m fixed). Apply this with $k_j = \lfloor (j/m)n^2/2 \rfloor$, $(j = 1, \dots, m)$, and use the facts that $\Lambda_1^{(n,k)}$ is monotone nondecreasing in k and $q(\cdot)$ is continuous, to deduce that if the event $\max_{1\le k\le n^2/2} \left|n^{-1}\Lambda_1^{(n,k)} - q(k/n^2)\right| > \epsilon$ occurs, then we will also have that

$$\max_{1\le j\le m} \left|n^{-1}\Lambda^{(n,k_j)} - q(k_j/n^2)\right| > \epsilon - \delta,$$

where

$$\delta = \delta_m = \sup_{\substack{x,y\in[0,1/2] \\ |x-y|\le 2/m}} |q(x) - q(y)|.$$

Since δ can be made arbitrarily small by taking m large enough, it follows that (3.59) implies (3.49). □

3.13 Proof of the limit shape theorem

We now deduce Theorem 3.6 from Theorems 3.5 and 3.17. First, we prove a claim about convergence of individual entries of the tableau. Let $\epsilon > 0$ and $(x, y) \in [0, 1]^2$, and denote $i = i(n, x) = 1 + \lfloor (n-1)x \rfloor$, $j = j(n, x) = 1 + \lfloor (n-1)y \rfloor$.

Proposition 3.23 *We have*

$$\mathbb{P}\left[\left|n^{-2}t^n_{i,j} - S(i/n, j/n)\right| > \epsilon\right] \to 0 \text{ as } n \to \infty. \tag{3.60}$$

Proof Denote

$$(x', y') = \left(\frac{i-1/2}{n}, \frac{j-1/2}{n}\right) = (x, y) + O(n^{-1}),$$
$$(u, v) = \left(\frac{x-y}{\sqrt{2}}, \frac{x+y}{\sqrt{2}}\right),$$
$$(u', v') = \left(\frac{x'-y'}{\sqrt{2}}, \frac{x'+y'}{\sqrt{2}}\right) = (u, v) + O(n^{-1}).$$

First, note that it is enough to prove the claim when (x, y) is in the triangle $\{x, y \ge 0, x + y \le 1\}$, since the distribution of the uniformly random square Young tableau T_n is invariant under the reflection transformation that replaces the entry $t^n_{i,j}$ in position (i, j) by $n^2+1-t^n_{n+1-i,n+1-j}$, and the limit shape $S(\cdot,\cdot)$ has a corresponding symmetry property $S(x, y) = 1 - S(1-x, 1-y)$.

We now divide into two cases according as whether one of x and y is 0 or both are positive. If $x, y > 0$, we use the following characterization of

$t^n_{i,j}$: recall that if $t^n_{i,j} = k$ that means that the cell (i, j) was added to $\Lambda^{(n,k-1)}$ to obtain $\Lambda^{(n,k)}$. Thus, we can write

$$t^n_{i,j} = \min\left\{1 \le k \le n^2 : \psi_{n,k}(u') > v'\right\}. \tag{3.61}$$

Denote $S(x, y) = s$. By the definition of the limit surface $S(\cdot, \cdot)$, this means that $v = \tilde{g}_s(u)$, and therefore also $v' = \tilde{g}_s(u') + O(n^{-1})$. Since (x, y) lies in the interior of the square, we also have that $|u| \le \sqrt{2s(1-s)}$, that is, (u, v) lies on the "curved" part of the curve $v = \tilde{g}_s(u)$ rather than the "flat" part. In this regime of the parameters, we have the property

$$\frac{\partial \tilde{g}_s(u)}{\partial s} > 0, \tag{3.62}$$

which implies that in a neighborhood of (s, u), $\tilde{g}_s(u)$ is increasing in s (and note that in general $\tilde{g}_s(u)$ is nondecreasing in s). Denote $s_- = (s - \epsilon) \vee 0$, $s_+ = (s + \epsilon) \wedge 1$, and

$$\delta = \min\left(\frac{\tilde{g}_s(u) - \tilde{g}_{s_-}(u)}{2}, \frac{\tilde{g}_{s_+}(u) - \tilde{g}_s(u)}{2}\right).$$

Then we have

$$\tilde{g}_{s_-}(u) + \delta < \tilde{g}_s(u) < \tilde{g}_{s_+}(u) - \delta$$

and $\delta > 0$ (in fact, by (3.62) above, δ behaves approximately like a constant times ϵ when ϵ is small, but this is not important for our purposes), and therefore also

$$\tilde{g}_{s_-}(u') + \delta/2 < v' < \tilde{g}_{s_+}(u') - \delta/2$$

when n is sufficiently large. Now denote $k_- = \lceil s_- n^2 \rceil$ and $k_+ = \lfloor s_+ n^2 \rfloor$, and apply Theorem 3.5 with the parameter ϵ in that theorem being set to $\delta/2$, to see that the event

$$\psi_{n,k_-}(u') < v' < \psi_{n,k+}(u')$$

holds with probability that tends to 1 as $n \to \infty$. But if this event holds, by (3.61) this implies that

$$k_- < t^n_{i,j} \le k_+,$$

that is, that $\left|n^{-2} t^n_{i,j} - S(x, y)\right| < \epsilon$. Since $S(i/n, j/n) = S(x, y) + o(1)$ as $n \to \infty$ (by continuity of $S(\cdot, \cdot)$), we get (3.60).

Next, consider the case when one of x and y is 0, say $y = 0$. In this case

$j = j(n,y) = 1$, and we use a different characterization of $t^n_{i,j}$ in terms of the length of the first rows of the sequence of diagrams $\Lambda^{(n,k)}$, namely

$$t^n_{i,j} = t^n_{i,1} = \min\left\{1 \le k \le n^2 \,:\, \Lambda^{(n,k)}_1 \ge k\right\}. \tag{3.63}$$

In this case Theorem 3.17 can be invoked and combined with (3.63) using standard ϵ-δ arguments similar to those used above to deduce that the event

$$\left|n^{-2} t^n_{i,1} - r(x)\right| < \epsilon \tag{3.64}$$

occurs with asymptotically high probability as $n \to \infty$, where we define the function $r\colon [0,1] \to [0,1/2]$ by

$$r(x) = \min\{0 \le \tau \le 1/2 \,:\, q(\tau) \ge x\}.$$

The details are left to the reader as an exercise. Finally, it is easy to check that

$$r(x) = q^{-1}(x) = \frac{1 - \sqrt{1-x^2}}{2} = S(x,0),$$

showing that (3.64) is just what is needed to imply (3.60). □

Proof of Theorem 3.6 Fix $\epsilon > 0$. By continuity of $S(\cdot,\cdot)$, there is a $\delta > 0$ such that if we take partitions

$$0 = x_0 < x_1 < \ldots < x_k = 1, \qquad 0 = y_0 < y_1 < \ldots < y_\ell = 1$$

of $[0,1]$ with a mesh smaller than δ then we have

$$\max\left(|S(x_{p+1}, y_q) - S(x_p, y_q)|, |S(x_q, y_{p+1}) - S(x_q, y_p)|\right) < \frac{\epsilon}{4} \tag{3.65}$$

for all $0 \le p < k, 0 \le q \le \ell$. Fix n, fix two such partitions, and assume that the event

$$\left\{\left|n^{-2} t^n_{i(n,x_p),j(n,y_q)} - S\left(\frac{i(n,x_p)}{n}, \frac{j(n,y_q)}{n}\right)\right| < \epsilon /4 \right.$$
$$\left. \text{for all } 0 \le p \le k, 0 \le q \le \ell\right\} \tag{3.66}$$

occurred, where the notation $i(n,x)$ and $j(n,y)$ from Proposition 3.23 is used. Then for any $1 \le i_0, j_0 \le n$, we can find some p,q with $0 \le p < k, 0 \le q < \ell$, and

$$i(n,x_p) \le i_0 \le i(n,x_{p+1}), \quad j_0 \le j(n,y_q) \le j_0 < j(n,y_{q+1}).$$

Therefore by monotonicity we have that

$$t^n_{i(n,x_p),j(n,x_{p+1})} \le t_{i_0,j_0} \le t^n_{i(n,x_{p+1}),j(n,y_{q+1})},$$

and thus, using (3.65), (3.66), and the fact that $S(x,y)$ is increasing in x and y, we deduce that also

$$\left|n^{-2}t^n_{i_0,j_0} - S(i_0/n, j_0/n)\right| < \epsilon .$$

We have shown that occurrence of the event (3.66), which by Proposition 3.23 has probability that tends to 1 as $n \to \infty$, forces the event

$$\left\{ \max_{1\le i_0,j_0\le n} \left|n^{-2}t^n_{i_0,j_0} - S(i_0/n, j_0/n)\right| < \epsilon \right\}$$

to occur, so this event too has asymptotically high probability, and the proof is complete. □

3.14 Random square Young tableaux as an interacting particle system

In this section we consider a new way of thinking about random square Young tableaux as a system of randomly interacting particles. This ties the study of this probabilistic model to a more general notion of **interacting particle systems**, which are random processes consisting of a configuration of particles evolving over time in a random manner according to some set of rules. Usually the particles occupy a space with some discrete geometry (e.g., a lattice or a graph), and the evolution of each particle depends primarily on the state of nearby particles. Such systems appear naturally in connection with many applied and theoretical problems in probability theory, statistical physics, and other fields. We will encounter another interesting interacting particle system in Chapter 4.

We start by describing the combinatorial structure of the particle system, without specifying the nature of the probabilistic behavior. Let $n \ge 1$. We consider n particles arranged on the discrete interval $[1, 2n]$. Initially, the particles occupy the positions $1, \ldots, n$. Subsequently, particles start jumping to the right at discrete times $t = 1, 2, \ldots$; at each time step exactly one particle jumps one step to the right. The jumps are constrained by the rule that a particle may only jump into a vacant position, and all particles remain in the interval $[1, 2n]$. The process continues until all particles have

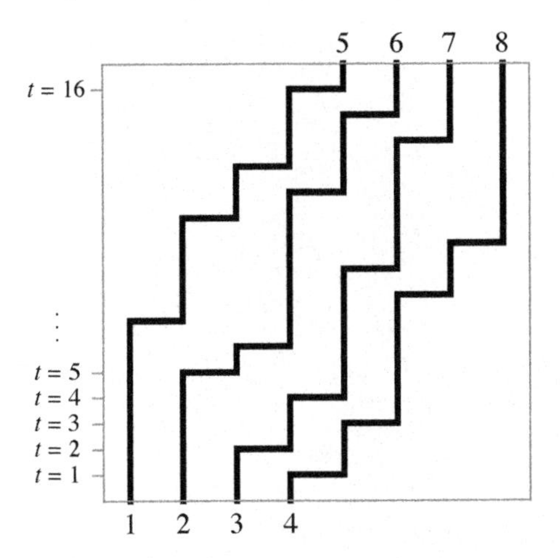

Figure 3.10 Particle trajectories in a square Young tableau jump process with four particles.

moved as far to the right as they can, so no more jumps may occur. At this point, each particle has moved n steps to the right, so the total number of jumps is n^2, and the particles occupy the lattice positions in $[n + 1, 2n]$.

Fig. 3.10 illustrates a way of visualizing an instance of such a system in a space–time diagram that shows the trajectories each of the particles goes through as a function of time. In this example, there are just four particles.

Let Ω_n be the set of possible ways in which the system can move from the initial to the final state. It is not so important how this information is encoded, so we omit a formal definition. Since the set Ω_n is a finite set, we can now add random behavior to the system by considering for each n an element of Ω_n drawn uniformly at random. In other words, we equip Ω_n with the uniform probability measure to get a probability space. Because the system will turn out to be related to square Young tableaux, we call it the **square Young tableau jump process** (or for convenience just **jump process**) with n particles. Fig. 3.11 shows a visualization of such a process of order 100; the trajectories of a few selected particles are highlighted to give a better feeling as to the overall behavior of the process. The picture is rather striking, and suggestive of an interesting asymptotic result waiting to be proved. But in fact, as readers may have guessed, what we are looking at

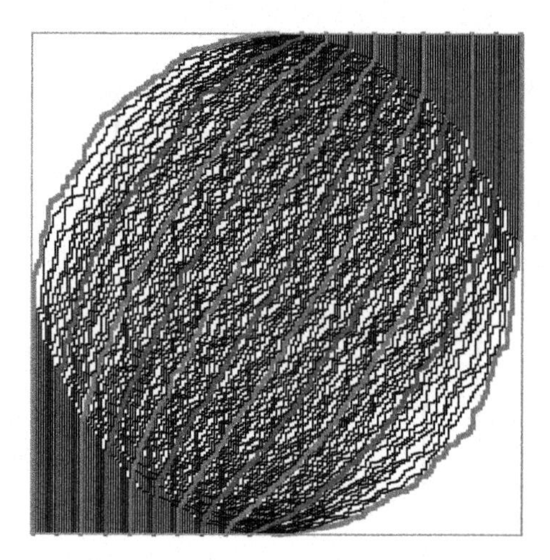

Figure 3.11 A square Young tableau jump process with 100 particles.

is simply a different way of visualizing a uniformly random square Young tableau, and the asymptotic result will be just an alternative way (though quite interesting in its own right) of thinking about the limit shape theorem for those tableaux. To make this precise, we note that the space Ω_n is in bijection with the set of square Young tableaux of order n. The bijection works as follows: for each $1 \leq k \leq n$, the kth row of the tableau lists the times during which the kth leading particle (i.e., the particle that was initially in position $n + 1 - k$) jumps. This works because each particle needs to jump exactly n times (and each row of the tableau has n cells), and the requirement that particles can only jump into vacant spots corresponds precisely to the fact that the Young tableau encodes a growing sequence of Young diagrams. For example, Fig. 3.12 shows the square Young tableau that corresponds to the four-particle jump process from Fig. 3.10. The fact that the first row of the tableau is $(1, 3, 8, 10)$ corresponds to the fact that the leading particle in the example jumped at times 1, 3, 8, and 10, which may be seen in the picture.

We can now translate our knowledge of the asymptotic behavior of random square Young tableaux into an interesting result on the limiting behavior of the jump process as $n \to \infty$.

1	3	8	10
2	4	9	14
5	6	12	15
7	11	13	16

Figure 3.12 The square Young tableau associated with the jump process shown in Fig. 3.10.

Theorem 3.24 (Limiting trajectories in the square Young tableau jump process) *For each $n \geq 1$, let X_n be a random square Young tableau jump process with n particles. For any $1 \leq j \leq n$ and $0 \leq k \leq n^2$, denote by $x_{n,j}(k)$ the position of the particle with index j (meaning the particle that starts at position j and ends in position $n + j$) at time k. Define the one-parameter family of curves $(\psi_x(t))_{0 \leq x \leq 1}$ by*

$$\psi_x(t) = \begin{cases} x & \text{if } 0 \leq t \leq \frac{1-\sqrt{2x-x^2}}{2}, \\ x + \max\{0 \leq y \leq 1 \,:\, S(1-x,y) \leq t\} & \text{if } \frac{1-\sqrt{2x-x^2}}{2} \leq t \leq \frac{1+\sqrt{1-x^2}}{2}, \\ x+1 & \text{if } \frac{1+\sqrt{1-x^2}}{2} \leq t \leq 1. \end{cases} \tag{3.67}$$

Then, as $n \to \infty$, the scaled trajectories of the particles in the processes X_n converge uniformly in probability to the curves $(\psi_x(t))_{0 \leq x \leq 1}$. That is, for any $\epsilon > 0$, we have that

$$\mathbb{P}\left[\max_{1 \leq j \leq n, 1 \leq k \leq n^2} \left| n^{-1} x_{n,j}(k) - \psi_{j/n}(k/n^2) \right| > \epsilon \right] \to 0. \tag{3.68}$$

Before proving Theorem 3.24, let us state another result that will follow from it. One aspect of the theorem that is not immediately apparent from looking at the formulas, but is visually obvious from the simulation picture, is the emergence of a roughly circular curve, or interface, in the scaled space–time square in which the process acts out its life. The area in the interior of this curve has a "chaotic" or random look to it, whereas the behavior of the process outside the curve is deterministic (and is itself of two types: two connected regions where particles are clustered together in a "traffic jam" where there is no movement, and two connected regions which are completely free of particles. In the limit the interface between the two types of behavior approaches the shape of the circle inscribed within the square. We refer to this circle as the **arctic circle**.[3]

We can define the interface curve as follows. Given a square Young tableau jump process of order n, let $\tau_n^-(k)$ and $\tau_n^+(k)$ denote respectively the first and last times at which a particle k jumped from or to position k. Clearly, in the time period $[0, \tau_n^-(k)]$, nothing happens in position k (i.e., for $1 \le k \le n$ it is constantly occupied by a particle, and for $n+1 \le k \le 2n$ it is vacant), and similarly, in the period $[\tau_n^+(k), n^2]$, nothing happens in position k. Therefore the discrete "curves" $\tau_n^\pm(k)$ represent the lower and upper boundaries of the interface. We will prove the following result as an easy corollary of Theorem 3.24.

Theorem 3.25 (The arctic circle theorem for square Young tableau jump processes[4]) *Denote*

$$\varphi_\pm(x) = \tfrac{1}{2} \pm \sqrt{x(1-x)}.$$

Then for any $\epsilon > 0$, *we have*

$$\mathbb{P}\Big[\Big\{ \max_{1\le k\le 2n} \left| n^{-2}\tau_n^-(k) - \varphi_-(k/2n)\right| < \epsilon \Big\} \cap \Big\{ \max_{1\le k\le 2n} \left| n^{-2}\tau_n^+(k) - \varphi_+(k/2n)\right| < \epsilon \Big\}\Big] \to 1 \text{ as } n\to\infty.$$

The arctic circle and limiting particle trajectories in the jump process are shown in Fig. 3.13.

Proof of Theorem 3.24 The proof is based on a representation of the position $x_{n,j}(k)$ of the jth particle at time k in terms of the random Young tableau $T_n = (t_{i,j}^n)_{i,j=1}^n$ corresponding to the jump process. It is easy to see given the description above of the bijection between the two families of objects that in terms of the tableau entries, $x_{n,j}(k)$ is given by

$$x_{n,j}(k) = \begin{cases} j & \text{if } k < t_{1,n+1-j}, \\ j + \max\left\{0 \le i \le n \,:\, t_{i,n+1-j}^n \le k\right\} & \text{if } t_{1,n+1-j} \le k \le t_{n,n+1-j}, \\ j+n & \text{if } t_{n,n+1-j} < k \le n^2 \end{cases} \tag{3.69}$$

(the third case in this equation is superfluous and can be unified with the second, but is added for extra clarity). Note that this is a discrete analogue

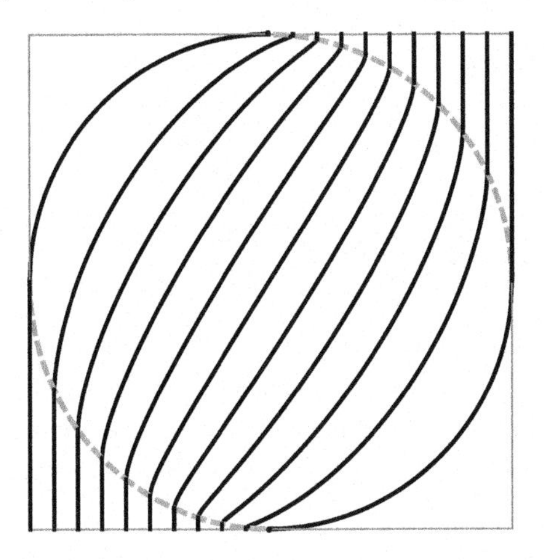

Figure 3.13 Limiting trajectories in the square Young tableau jump process and the arctic circle.

of (3.67). Indeed, when considering the difference

$$\left|n^{-1}x_{n,j}(k) - \psi_{j/n}(k/n^2)\right|,$$

the "x" term in (3.67) cancels out the "j" term in (3.69), and we are left with the difference of two maxima

$$\left|n^{-1}\max\left\{0 \le i \le n \,:\, t^n_{i,n+1-j} \le k\right\} - \max\{0 \le y \le 1 \,:\, S(1-x,y) \le t\}\right|,$$

with the convention that the maximum of an empty set is 0, and with the notation $x = j/n, t = k/n^2$. Now, Theorem 3.6 says that $t^n_{i,n+1-j}$ is (with asymptotically high probability) uniformly close to $S(1-x, i/n)$ – say, closer in distance than some $\delta > 0$. It is easy to see that δ may be chosen small enough as a function of ϵ so that, assuming this event, the absolute value of the difference of the maxima will be smaller than ϵ. We omit the details of this verification, which are straightforward and similar to many of the estimates we encountered in previous sections. □

Proof of Theorem 3.25 It is enough to note the following simple observations that express the times $\tau^-_n(k)$ and $\tau^+_n(k)$ in terms of the random tableau entries $(t_{i,j}^n)_{i,j=1}^n$:

$$\tau_n^-(k) = \begin{cases} t_{n+1-k,1}^n & \text{if } 1 \le k \le n, \\ t_{1,k-n}^n & \text{if } n+1 \le k \le 2n, \end{cases}$$

$$\tau_n^+(k) = \begin{cases} t_{n,k} & \text{if } 1 \le k \le n, \\ t_{2n+1-k,n} & \text{if } n+1 \le k \le 2n. \end{cases}$$

For example, when $1 \le k \le n$, $\tau_n^-(k)$ is simply the first time at which the particle stating at position k (which corresponds to row $n+1-k$ in the tableau) jumps. The three remaining cases are equally simple to verify. The result now follows easily using (3.13), (3.14) and Theorem 3.6. □

Exercises

3.1 (☕) Let $\sigma = (\sigma(1), \dots, \sigma(mn))$ be an (m,n)-Erdős–Szekeres permutation. Denote $\sigma' = (\sigma(mn), \dots, \sigma(1))$ and $\sigma'' = (mn+1-\sigma(1), \dots, mn+1-\sigma(mn))$. Show that $\sigma^{-1} \in \mathrm{ES}_{m,n}$, $\sigma' \in \mathrm{ES}_{n,m}$ and $\sigma'' \in \mathrm{ES}_{n,m}$. What is the effect of each of these operations on the rectangular Young tableaux P, Q associated with σ?

3.2 (☕) Show that the limit shape $S(x,y)$ from (3.12) satisfies

$$S(x,x) = \frac{1}{2}(1 - \cos(\pi x)) \qquad (0 \le x \le 1).$$

3.3 (☕☕) Fill in the missing details of the proof of Theorem 3.8.

3.4 (☕☕) Prove the claim, stated in the proof of Lemma 3.14, that if $-1 < x < 1$ then

$$\text{P.V.} \int_{-1}^{1} \frac{dw}{xw^2 - 2w + x} = 0,$$

where P.V. denotes an integral in the sense of the principal value.

3.5 (☕☕☕) (Stanley [126]) Show that the area of the limit shape A_∞ (defined in (3.6)) of random Erdős–Szekeres permutations is given by 4α, where α, the fraction between the area of the shape and that of the enclosing square $[-1,1]^2$, can be expressed in terms of elliptic integrals as

$$\alpha = 2\int_0^1 \frac{1}{\sqrt{(1-t^2)(1-(t/3)^2}}\,dt - \frac{3}{2}\int_0^1 \sqrt{\frac{1-(t/3)^2}{1-t^2}}\,dt \doteq 0.94544596.$$

3.6 (☕☕☕) (Pittel–Romik [100]) Show that the number $d_{\square_{n,n}}$ of $n \times n$ square

Young tableaux has the asymptotic behavior

$$d_{\square_{n,n}} = n^{11/12}\sqrt{2\pi}\exp\left[n^2\log n + \left(-2\log 2 + \frac{1}{2}\right)n^2 - \frac{1}{6} + \frac{\log 2}{12} - C + o(1)\right] \quad \text{as } n \to \infty,$$

where the constant C is given by

$$C = \int_0^\infty \left[\frac{1}{e^t - 1} - \frac{1}{t} + \frac{1}{2} - \frac{t}{12}\right]\frac{e^{-t}}{t^2}\,dt.$$

3.7 Let λ denote a Young diagram. It is easy to see that a filling of the numbers $1,\ldots,|\lambda|$ in the cells of λ is a Young tableau if and only if each entry is the smallest in its hook. Similarly, define a **balanced tableau of shape** λ to be a filling of the numbers $1,\ldots,|\lambda|$ in the cells of λ that satisfies the condition that the entry in each box (i, j) is the kth largest in its hook, where k is the size of the "leg" part of the hook (which consists of (i, j) and the cells below (i, j) in the same column).

(a) (☕☕) (Knuth [71], Section 5.1.4) Given a random filling of the numbers $1,\ldots,n$ in the cells of λ chosen uniformly at random from among the $|\lambda|!$ possibilities, compute for each box (i, j) the probability that the entry in box (i, j) is the smallest in its hook. Show that if we assume these events are independent, then the hook-length formula (1.12) would follow easily. Does this argument give a new proof of the hook-length formula? Why, or why not?

(b) (☕☕) Find a bijection between Young tableaux of rectangular shape $\square_{m,n}$ and balanced tableaux of rectangular shape $\square_{m,n}$.

(c) (☕☕☕) (Edelman–Greene [34]) Let λ denote the rectangular diagram $\square_{m,n}$ with the bottom-right corner removed. Find a bijection between Young tableaux of rectangular shape λ and balanced tableaux of rectangular shape λ.

(d) (☕☕☕☕) (Edelman–Greene [34]) Prove that for any Young diagram λ, the number b_λ of balanced tableaux of shape λ is equal to d_λ (and therefore is also given by the hook-length formula).

3.8 For $1 \le j \le n-1$, denote by τ_j the permutation in S_n (known as an adjacent transposition or swap) that exchanges j and $j+1$ and leaves all other elements fixed. Denote by rev_n the "reverse permutation," given (in two-line notation) by $\begin{pmatrix} 1 & 2 & \cdots & n \\ n & n-1 & \cdots & 1 \end{pmatrix}$. An n-**element sorting network** is a sequence $(s_1,\ldots,s_N)$ such that $\text{rev}_n = \tau_{s_1}\tau_{s_2}\ldots\tau_{s_N}$ and $N = \binom{n}{2}$. (Sorting networks are also known as **maximal chains in the weak Bruhat order** and as **reduced decompositions of the reverse word**; see Exercise 185 on pp. 400–401 of [127] and

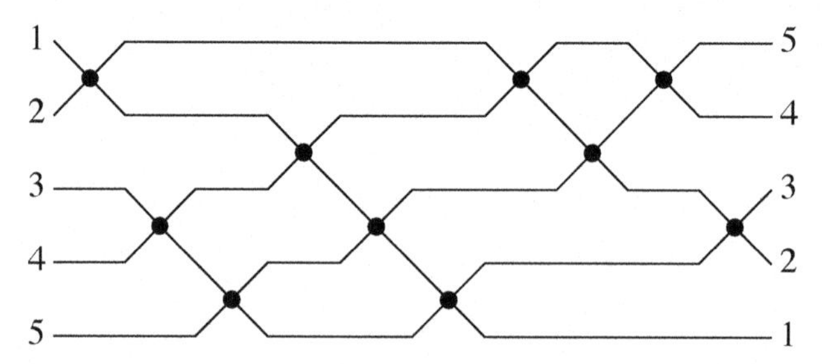

Figure 3.14 The wiring diagram associated with the five-element sorting network $(1, 3, 4, 2, 3, 4, 1, 2, 1, 3)$. The swaps are shown as black discs.

Exercise 7.22 on pp. 454–456 of [125].) Graphically, this means that we start with the list of numbers $(1, 2, \ldots, n)$ in increasing order and successively apply swaps until arriving at the list arranged in decreasing order, a process that can be described schematically in a **wiring diagram** (see Fig. 3.14). Note that $\binom{n}{2}$ is the minimum number of swaps required to change the list from increasing to decreasing order, since one swap is required to exchange any two elements $1 \leq i < j \leq n$.

(a) (☕) For any $n \geq 1$, give an example of an n-element sorting network.

(b) (☕☕) Given a sorting network $(s_1, \ldots, s_N)$ of order n, for any $1 \leq i < j \leq n$ let $\tau(i, j)$ be the unique time when i and j were swapped. More precisely, let $\sigma_k = \tau_{s_1} \ldots \tau_{s_k}$ for $0 \leq k \leq N$ (where σ_0 is the identity permutation), and define $\tau(i, j)$ as the unique number t such that $\sigma_{t-1}^{-1}(i) < \sigma_{t-1}^{-1}(j)$ and $\sigma_t^{-1}(i) > \sigma_t^{-1}(j)$.

Show that $(\tau(i, j))_{1 \leq i < j \leq n}$ is a balanced tableau (defined in Exercise 3.7 above) of shape $⌟_n = (n - 1, n - 2, \ldots, 1)$ (called the **staircase shape**). Conversely, show that any balanced tableau of staircase shape $⌟_n$ arises in such a way from a unique sorting network.

(c) (☕☕☕☕) (Edelman–Greene [34]) Given a Young tableau T of shape λ, denote the coordinates of the cell containing its maximal entry $|\lambda|$ by $(i_{\max}(T), j_{\max}(T))$.

Define the **Schützenberger operator** Φ as a map that takes the Young tableau $T = (t_{i,j})_{(i,j)\in\lambda}$ and returns a new Young tableau $\Phi(T)$ of shape λ, described as follows. Construct inductively a "sliding sequence" of cells $(i_0, j_0), \ldots, (i_d, j_d)$ of λ, where the sequence begins at $(i_0, j_0) = (i_{\max}(T), j_{\max}(T))$ and ends at $(i_d, j_d) = (1, 1)$, and for each k for which

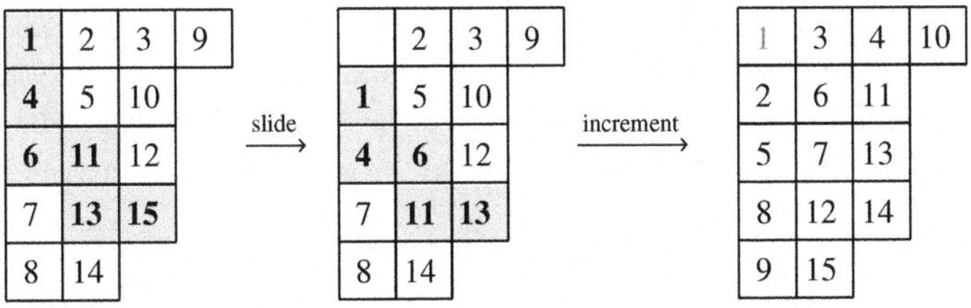

1	2	3	9
4	5	10	
6	**11**	12	
7	**13**	**15**	
8	14		

slide ⟶

	2	3	9
1	5	10	
4	**6**	12	
7	**11**	**13**	
8	14		

increment ⟶

1	3	4	10
2	6	11	
5	7	13	
8	12	14	
9	15		

Figure 3.15 The sliding sequence and the Schützenberger operator.

$(i_k, j_k) \neq (1,1)$ we have

$$(i_{k+1}, j_{k+1}) = \begin{cases} (i_k - 1, j_k) & \text{if } j = 1 \text{ or } t_{i_k-1,j_k} > t_{i_k,j_k-1}, \\ (i_k, j_k - 1) & \text{if } i = 1 \text{ or } t_{i_k 1,j_k} < t_{i_k,j_k-1}. \end{cases}$$

Then the tableau $T' = \Phi(T) = (t'_{i,j})_{(i,j)\in\lambda}$ is defined by setting

$$t'_{i,j} = \begin{cases} 1 & \text{if } (i,j) = (1,1), \\ t_{i,j} + 1 & \text{if } (i,j) \neq (i_k, j_k) \text{ for all } k, \\ t_{i_{k+1},j_{k+1}} + 1 & \text{if } (i,j) = (i_k, j_k) \text{ for } 1 \leq k < d. \end{cases}$$

In words, we start by removing the maximal entry of T; then pull the cells along the sliding sequence, shifting each cell in the sequence to the position vacated by the previous one until the cell $(1,1)$ is left vacant; then increment all entries of the resulting array of numbers by 1, and finally fill the still-vacant cell at position $(1,1)$ with the number 1. It is easy to see that the result is still a Young tableau; Fig. 3.15 shows an example.

Define a map on the set of Young tableaux of staircase shape $\lrcorner_n$ by

$$\mathrm{EG}(T) = \left(j_{\max} \left(\Phi^{\binom{n}{2}-k}(T) \right) \right)_{k=1,\ldots,\binom{n}{2}},$$

where Φ^k denotes the kth functional iterate of Φ. Prove that EG is a bijection from the set of Young tableaux of shape $\lrcorner_n$ to the set of sorting networks of order n. Deduce the result, first proved by Stanley [124], that the number of sorting networks of order n is given by

$$\frac{\binom{n}{2}!}{1^{n-1}3^{n-2}5^{n-3}\ldots(2n-3)^1}.$$

The map EG is known as the **Edelman–Greene bijection**.

(d) (☕☕) (Angel–Holroyd–Romik–Virág [6]) Let $S = (S_1, \ldots, S_N)$ be a uniformly random sorting network of order n. Show that the swap positions $S_1, \ldots, S_N$ are identically distributed random variables, and find a formula for the probability distribution

$$P_{S_1}(m) = \mathbb{P}(S_1 = m) \qquad (1 \le m \le n-1),$$

of the swap positions.

(e) (☕☕) (Reiner [103]) A triple of successive swaps s_k, s_{k+1}, s_{k+2} in a sorting network is called a **Yang–Baxter move** if $(s_k, s_{k+1}, s_{k+2}) = (j, j \pm 1, j)$ for some j. Show that the number X of Yang–Baxter moves in a uniformly random n-element sorting network S defined above satisfies $\mathbb{E}(X) = 1$.

3.9 (☕☕☕☕) (Angel et al. [6]) Let $T = (t_{i,j})_{1 \le i < j \le n}$ be a uniformly random Young tableau of staircase shape ⊿$_n$ (defined in Exercise 3.8 above). State and prove limit shape theorems analogous to Theorems 3.5 and 3.6 for the random tableau T. **Hint:** Exploit a symmetry property of random square Young tableaux and a corresponding symmetry of their limit shapes.

3.10 (a) (☕☕) Let $(x_{i,j})_{i,j=1}^n$ be an array of distinct numbers. Prove that if each row of the array is sorted in increasing order, and then each column of the array is sorted in increasing order, the resulting array has the property that both its rows and columns are arranged in increasing order.

(b) (☕☕☕) Let $(X_{i,j})_{i,j=1}^n$ be the array described above in the case when $X_{i,j}$ are independent and identically distributed random variables with the uniform distribution $U[0,1]$. Let $Y = (Y_{i,j})_{i,j=1}^n$ be the resulting array after sorting the rows and then the columns in increasing order. Define a new array $T = (T_{i,j})_{i,j=1}^n$ by letting $T_{i,j}$ be the ranking of $Y_{i,j}$ in the array Y (i.e., $T_{i,j} = 1$ if $Y_{i,j}$ is the smallest number in the array, 2 if $Y_{i,j}$ is the second smallest, and so on). Note that by the result of part (a) above, T is a square Young tableau.

Investigate the properties of the random square Young tableau T. Can you prove an interesting (or uninteresting, for that matter) limit shape theorem?

3.11 (☕☕☕) (Thrall [134]) Let $\lambda = (\lambda_1, \ldots, \lambda_k)$ be an integer partition with all parts distinct. A **shifted Young diagram** of shape λ is a variant of the ordinary Young diagram of shape λ in which for each j, the jth row of λ is indented $j-1$ positions to the left. For example, the shifted Young diagram associated with the partition $(10, 6, 5, 2, 1)$ is

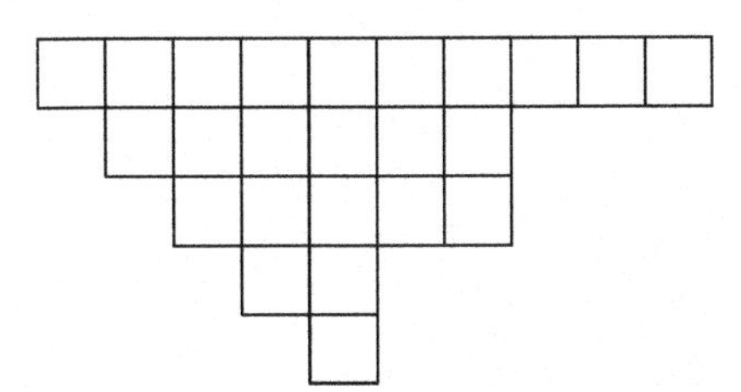

Given a shifted Young diagram (which we denote by λ, identifying it with the underlying partition with distinct parts), a **shifted Young tableau** of shape λ is a filling of the cells of λ with the numbers $1, \ldots, |\lambda|$ that is increasing along rows and columns. The **hook** of a cell (i, j) in the shifted Young diagram is the set

$$H^*_\lambda(i, j) = \{(i, j') \in \lambda : j' \geq j\} \cup \{(i', j) \in \lambda : i' \geq i\} \\ \cup \{(j+1, j') \in \lambda : j' > j\}.$$

(The last part of the hook can be interpreted as a reflection of the actual hook of the cell (i, j) in an augmented Young diagram obtained by glueing λ together with a copy of λ reflected along the northwest–southeast diagonal.) The **hook-length** of (i, j) in λ, denoted $h^*_\lambda(i, j)$, is defined as the number of cells in the hook $H^*_\lambda(i, j)$.

Prove the following version of the hook-length formula for shifted Young tableaux.

Theorem 3.26 (Hook-length formula for shifted Young tableaux) *The number g_λ of shifted Young tableaux of shape λ is given by*

$$g_\lambda = \frac{|\lambda|!}{\prod_{(i,j)\in\lambda} h^*_\lambda(i, j)}.$$

3.12 (a) (☕☕) Let $\sqcup\!\sqcup_n$ denote the shifted Young diagram $(2n-1, 2n-3, \ldots, 3, 1)$. Show that the number of shifted Young tableaux of shape $\sqcup\!\sqcup_n$ is equal to $d_{\square_{n,n}}$, the number of $n \times n$ square Young tableaux.

(b) (☕☕☕☕) (Haiman [53]) Find an explicit bijection between the set of shifted Young tableaux of shape $\sqcup\!\sqcup_n$ and the set of $n \times n$ square Young tableaux.

3.13 (☕☕☕) (Pittel–Romik [101]) Generalize the limit shape results proved in this chapter to the case of uniformly random $m \times n$ rectangular Young tableaux, in the case when $m, n \to \infty$ such that $m/n \to \alpha$ for some fixed $\alpha \in (0, \infty)$.

4

The corner growth process: limit shapes

Chapter summary. In Chapter 1 we discussed the Plancherel growth process, which can be thought of as a type of random walk on the Young graph consisting of all Young diagrams, in which a random Young diagram is "grown" by starting with a diagram with a single cell and successively adding new celles in random places chosen according to a certain probability distribution. In this chapter we consider another natural random walk on the Young graph defined by an even simpler rule: the new cell is always added in a position chosen uniformly at random among the available places. This random walk is called the **corner growth process**. Our analysis of its asymptotic behavior will lead to many interesting ideas such as the use of **Legendre transforms**, a model for **traffic jams** and **hydrodynamic limits** in interacting particle systems.

4.1 A random walk on Young diagrams

In this chapter we study a random process called the **corner growth process**,[1] a natural random walk on the Young graph, which, as you recall from Chapter 1, is the directed graph consisting of all Young diagrams, with the adjacency relation, denoted $\lambda \nearrow \mu$, being that of "adding a cell." For consistency with the existing literature on the subject, throughout this chapter we will use the "French notation" (see p. 35) for drawing Young diagrams and Young tableaux (Fig. 4.1).

Let us fix some notation. Throughout this chapter, we identify a Young diagram λ with a subset of $\mathbb{N}^2$ corresponding to the cell positions in the diagram, that is, the positions (i, j) that satisfy $\lambda_i \leq j$, a relation that we now denote as $(i, j) \in \lambda$. Denote by $\text{ext}(\lambda)$ the set of external corners of λ,

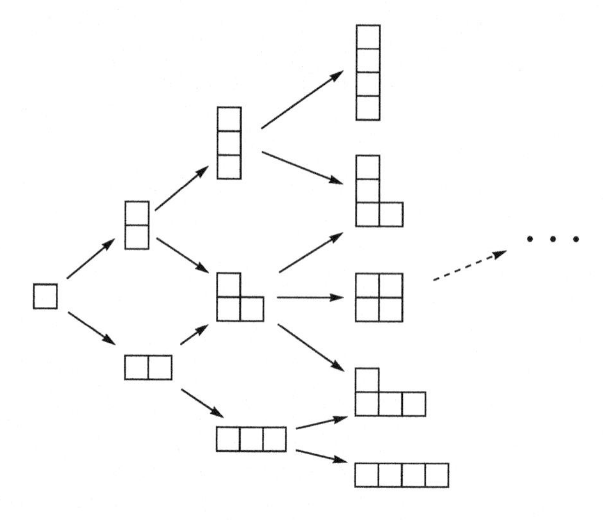

Figure 4.1 The Young graph, French-style.

which are the positions $(i, j) \in \mathbb{N}^2$ where a new cell may be added to λ to form a new Young diagram (which we'll denote $\lambda \cup \{(i, j)\}$); and denote $\text{out}(\lambda) = |\text{ext}(\lambda)|$, the number of external corners of λ (which is the out-degree of λ considered as a vertex in the Young graph).

The definition of the corner growth process is as follows. Let $(\lambda^{(n)})_{n=1}^{\infty}$ be a sequence of random Young diagrams such that $\lambda^{(1)}$ is the unique diagram with one cell, and for each $n \geq 2$, $\lambda^{(n)}$ is generated from $\lambda^{(n-1)}$ by adding a cell in one of the external corners of $\lambda^{(n-1)}$, where the corner is chosen *uniformly at random* from the available ones. Formally, the process can be defined as the Markov chain (see p. 77 for the definition) with transition rule

$$\mathbb{P}\left(\lambda^{(n)} = \mu_n \,\middle|\, \lambda^{(1)} = \mu_1, \ldots, \lambda^{(n-1)} = \mu_{n-1}\right) = \frac{1}{\text{out}(\mu_{n-1})} \tag{4.1}$$

for any Young diagrams $\mu_1 \nearrow \ldots \nearrow \mu_n$ with $|\mu_1| = 1$ (compare with (1.56), and see also Exercise 1.22). Yet another equivalent way of defining the process, in terms that will be familiar to many students of probability theory, is that the corner growth process is the usual (nonweighted) random walk on the Young graph that starts from the one-cell diagram at the "root" of the graph.

The problem we initially consider, which will occupy our attention for a large part of this chapter, is analogous to the one we studied in connection to the Plancherel growth process in Chapter 1, namely that of understanding the shape of the Young diagram $\lambda^{(n)}$ that is typically observed as n grows large. Fig. 4.2(a) shows a simulation of $\lambda^{(n)}$ for $n = 1000$. Once again, we have an interesting limit shape phenomenon, where as $n \to \infty$ the typical shape of the diagram converges to a limiting shape, shown in Fig. 4.2(b). The precise result, which was proved by Hermann Rost [110] in 1981, is as follows.

Theorem 4.1 (Limit shape theorem for the corner growth process) *For a Young diagram λ, let set_λ denote the planar set associated with λ, defined in* (1.57). *Let $(\lambda^{(n)})_{n=1}^{\infty}$ denote the corner growth process, and define a set Δ_{CG} by*

$$\Delta_{\text{CG}} = \left\{(x, y) \in \mathbb{R}^2 \,:\, x, y \geq 0,\ \sqrt{x} + \sqrt{y} \leq 6^{1/4}\right\}.$$

As $n \to \infty$, the scaled planar set $n^{-1/2}\text{set}_{\lambda^{(n)}}$ converges in probability to the set Δ_{CG}, in the same sense as that of Theorem 1.26. That is, for any $0 < \epsilon < 1$ we have that

$$\mathbb{P}\left((1-\epsilon)\Delta_{\text{CG}} \subseteq \frac{1}{\sqrt{n}}\text{set}_{\lambda^{(n)}} \subseteq (1+\epsilon)\Delta_{\text{CG}}\right) \to 1 \textit{ as } n \to \infty.$$

It is easy to verify that the set Δ_{CG} has area 1, which is to be expected, since $\text{set}_{\lambda^{(n)}}$ has area n and we scale both axes by $1/\sqrt{n}$ to get the convergence. The equation $\sqrt{x} + \sqrt{y} = 6^{1/4}$ for the curved part of the boundary is symmetric in x and y, which is also not surprising, since the symmetry is already inherent in the combinatorics of the corner growth process. Rewriting it in terms of the Russian coordinates (1.23) gives, after a short computation, the equation

$$v = \frac{\sqrt{3}}{2} + \frac{1}{2\sqrt{3}}u^2 \qquad \left(|u| \leq \sqrt{3}\right),$$

which shows that the curve is a parabola.

It is interesting to note that Rost's motivation for considering the corner growth process came from his study of a certain interacting particle system (which will be discussed in Section 4.7) and was unrelated to longest increasing subsequences. Indeed, the developments in this chapter will not make any reference to this subject, which has been our focus in earlier

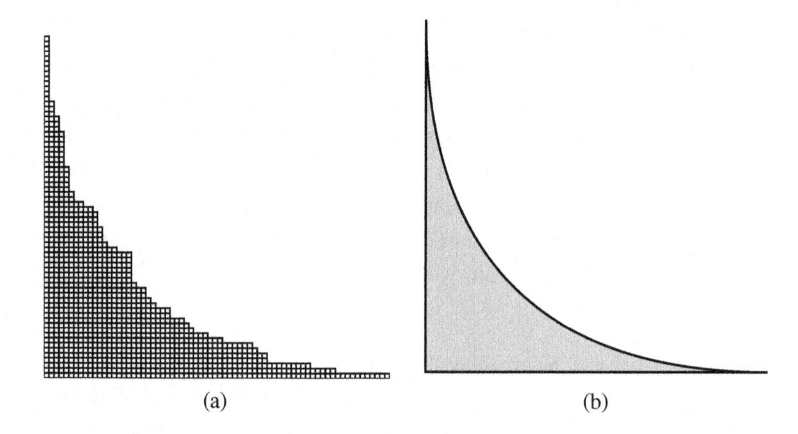

Figure 4.2 (a) A random Young diagram of order 1000 sampled from the corner growth process. (b) The limit shape Δ_{CG} of Young diagrams in the corner growth process. The curved part of the boundary is an arc of a parabola, described by the equation $\sqrt{x} + \sqrt{y} = 6^{1/4}$.

chapters. However, the mathematics that grew out of the analysis of the corner growth process, aside from being quite interesting in its own right, turned out to have direct relevance to the study of longest increasing subsequences, in two separate ways. First, the methods developed to prove Theorem 4.1 inspired a new approach (sometimes referred to as the "hydrodynamic" approach) to the study of longest increasing subsequences, that led to a new proof of Theorem 1.1 on the asymptotics of the maximal increasing subsequence length in a random permutation and to additional new results. We will not cover these developments here; see [2], [52], [116], [117], [118] for more information. Second, it turned out that certain quantities associated with the corner growth process – the so-called passage times, defined in Section 4.3 below – actually correspond to maximal increasing subsequence lengths in random words over a finite alphabet (which differ from permutations in that they can have repetitions). This observation led to a new and very fruitful avenue for exploration, the results of which are described in the next chapter.

In the next section we begin the analysis that will ultimately lead us to a proof of Theorem 4.1. A crucial first step will be to redefine the notion of time in the process, replacing the discrete time variable n with a continuous parameter t.

4.2 Corner growth in continuous time

We now define a continuous-time random process $(\Lambda(t))_{t\geq 0}$ that takes values in the set $\mathcal{P}^*$ of Young diagrams and is equivalent to the original process $(\lambda^{(n)})_{n=1}^\infty$, up to the time parametrization. We show two ways of doing this, and prove that they are equivalent. The first is a general construction that is simple to define, but will add little to our understanding. The second is a more elaborate construction that will provide much better insight into the corner growth process and will play a key role in our subsequent analysis of it.

We start with the simpler method. There is a general recipe to turn a random walk on a graph into a continuous-time process such that the sequence of sites it visits is a realization of the original random walk, and such that the new time parametrization is natural in a certain sense (that is, in technical language, the process is a **Markov process**). We do not need the general theory here, so we simply give the construction in our specific case. Let $(\lambda^{(n)})_{n=0}^\infty$ be the original corner growth process defined above, where (for reasons of future convenience) we now consider the process to start at $n = 0$ and set $\lambda^{(0)}$ to be the empty partition $\emptyset$ of order 0. Let $(X_\lambda)_{\lambda\in\mathcal{P}^*}$ be a family of independent random variables, independent of the process $(\lambda^{(n)})_{n=0}^\infty$, such that for each λ, X_λ has the exponential distribution $\text{Exp}(\text{out}(\lambda))$. (For the empty partition we define $\text{out}(\emptyset) = 1$.)

Now construct an increasing sequence of random variables $(S_n)_{n=0}^\infty$ defined inductively by setting $S_0 = 0$, and for each $n \geq 1$, letting S_n be defined conditionally on the value of $\lambda^{(n)}$ by

$$S_n = S_{n-1} + X_\lambda \quad \text{on the event } \{\lambda^{(n)} = \lambda\}, \quad (\lambda \in \mathcal{P}(n)). \tag{4.2}$$

The idea is that X_λ represents the time it will take the $(n + 1)$th cell to be added to the random partition $\lambda^{(n)} = \lambda$, and that the rate of growth (which is the parameter of the exponential random variable), is equal to the number of possible corners where growth can occur, rather than being constant as in the original discrete-time parametrization. The random variable S_n

represents the total time since the beginning of the process until the nth cell is added to the diagram.

The continuous-time random walk is now defined as the family $(\Lambda(t))_{t\geq 0}$ of random Young diagrams given by

$$\Lambda(t) = \lambda^{(n)} \qquad \text{if } S_n \leq t < S_{n+1}.$$

Note that $\Lambda(t)$ is defined only on the event that $t < \sup_n S_n$. Since it is not a priori obvious that the sequence S_n is unbounded, set $\Lambda(t)$ to be some special symbol, say ∞, if $t \geq \sup_n S_n$, but this is unimportant, since we will show shortly that in fact $\sup_n S_n = \infty$ almost surely, which means that $\Lambda(t)$ is a well-defined random Young diagram for all $t \geq 0$.

This definition of the continuous-time random walk follows a standard recipe from probability theory, but the method is so general that it offers little insight into our specific problem. In particular, it does not explain why we should expect the continuous-time version of the process to be any easier to understand or analyze than the original discrete-time process. Our second construction is better tailored to the problem at hand, and will make much clearer the benefits to be gained from the transition to continuous time. Before introducing it, we should note that this is an example of a more general principle at work, which is the notion of replacing a "global" or "centralized" randomness with a "local" or "decentralized" type of random behavior. This comes up frequently in probability theory. For example, Hammersley's idea, discussed in Section 1.4, of replacing a set of n uniformly random points on a rectangle with a Poisson point process (which is related to the "Poissonization" concept we also encountered in Chapter 2), falls broadly into this line of thinking. The problem with a set of n random points on a rectangle is that when we restrict the set to a subrectangle, whose area is some fraction α of the original rectangle, we get a set with a *random* number of points, which is only approximately equal to αn. Thus, the original process does not scale nicely. The move to a Poisson point process made things simpler, since it created a situation in which, in a very definite sense, each infinitesimal rectangle $[x, x+dx]\times[y, y+dy]$, makes an *independent* random choice as to whether or not it will contain a Poisson point. In this sense, the Poisson process has a "local" type of randomness.

We now wish to apply the same philosophy to our current setting, and replace the corner growth process as we defined it by a variant in which the growth of the Young diagram happens in a "local" way. The way to do this is as follows. We equip each cell position having coordinates $(i, j) \in \mathbb{N}^2$ with a random "clock" that will tell it when to add itself to the growing sequence $(\lambda^{(n)})_{n=1}^{\infty}$ of Young diagrams. Of course, the rules of growth must respect the constraints of the process, namely the requirement that a cell can be added only once it becomes an external corner of the existing Young diagram (i.e., once the cell to its left is already part of the Young diagram if $j > 1$, and the cell below it is part of the Young diagram if $i > 1$). So, we will set things up so that each cell's clock only "starts running" once this condition is satisfied. Furthermore, we also want the competition between the different clocks to be "fair" so that the resulting family of growing Young diagrams grows according to the correct dynamics prescribed by the growth rule (4.1), with the continuous time parametrization also corresponding to our previous construction. The correct way to satisfy these requirements turns out to involve the use of i.i.d. exponentially distributed random variables for the clock times.

To make this more precise, let $(\tau_{i,j})_{i,j\in\mathbb{N}}$ be a family of i.i.d. random variables with the exponential distribution Exp(1). We call $\tau_{i,j}$ the **clock time associated with position** (i, j).[2] We define by induction a sequence of random Young diagrams $(\mu^{(n)})_{n=0}^{\infty}$ and a sequence of random variables $(T_n)_{n=0}^{\infty}$, where, analogously to the sequences $(\lambda^{(n)})_{n=0}^{\infty}$ and $(S_n)_{n=0}^{\infty}$ from our previous construction, $\mu^{(n)}$ will represent a randomly growing Young diagram after n growth steps, starting from the empty diagram, and T_n will denote the time at which the nth growth step occurred. We also denote by (i_n, j_n) the position where the growth occurred in the $(n + 1)$th step, that is, such that $\mu^{(n+1)} = \mu^{(n)} \cup \{(i_n, j_n)\}$.

Set $T_0 = 0$, $\mu^{(0)} = \emptyset$ (the empty diagram), and if T_k and $\mu^{(k)}$ are defined for all $k \leq n$, then T_{n+1} and $\mu^{(n+1)}$ are defined as follows. For each cell position $(i, j) \in \mathbb{N}^2$, let $\kappa(i, j)$ be the minimal index $k \geq 0$ for which (i, j) was an external corner of $\mu^{(k)}$. Given the sequence $\mu^{(0)}, \ldots, \mu^{(n)}$, this is defined at least for cells that are in $\mu^{(n)}$ or that are external corners of $\mu^{(n)}$. Now set

$$(i_n, j_n) = \operatorname{argmin}\left\{T_{\kappa(i,j)} + \tau_{i,j} \ :\ (i, j) \in \operatorname{ext}(\mu^{(n)})\right\}, \tag{4.3}$$

$$\begin{aligned} T_{n+1} &= \min\left\{T_{\kappa(i,j)} + \tau_{i,j} \ :\ (i, j) \in \operatorname{ext}(\mu^{(n)})\right\} \\ &= T_{\kappa(i_n,j_n)} + \tau_{i_n,j_n}, \end{aligned} \tag{4.4}$$

$$\mu^{(n+1)} = \mu^{(n)} \cup \{(i_n, j_n)\}. \tag{4.5}$$

Note that the argmin in the definition of (i_n, j_n) is almost surely well defined. For example, it is well-defined on the event that all the sums of the form $\sum_{(i,j)\in A} \tau_{i,j}$, where A ranges over finite subsets of $\mathbb{N}^2$, are distinct, and this event has probability 1. So $(\mu^{(n)})_{n=0}^{\infty}$ and $(T_n)_{n=0}^{\infty}$ are well-defined random processes.

Lemma 4.2 *The sequence T_n is increasing, and almost surely $T_n \to \infty$ as $n \to \infty$.*

Proof If $\kappa(i_n, j_n) = n$ then $T_{n+1} = T_n + \tau_{i_n,j_n} > T_n$. On the other hand, if $\kappa(i_n, j_n) < n$, that is, (i_n, j_n) was already an external corner of $\mu^{(n-1)}$, then (i_n, j_n) already participated in the minimum defining T_n, so we must have $(i_{n-1}, j_{n-1}) \neq (i_n, j_n)$ (otherwise (i_n, j_n) will have been added to $\mu^{(n-1)}$ and would have been a cell, not an external corner, of $\mu^{(n)}$), and therefore (by the definition of (i_{n-1}, j_{n-1}))

$$T_{n+1} = T_{\kappa(i_n,j_n)} + \tau_{i_n,j_n} > T_{\kappa(i_{n-1},j_{n-1})} + \tau_{i_{n-1},j_{n-1}} = T_n.$$

This proves that T_n is increasing. To prove the second claim, observe that we have the lower bound

$$T_n \geq \max\left(\sum_{1\leq j\leq \mu_1^{(n)}} \tau_{1,j}, \sum_{1\leq i\leq (\mu^{(n)})'_1} \tau_{i,1} \right),$$

where $\mu_1^{(n)}$ and $(\mu^{(n)})'_1$ are the lengths of the first row and first column of $\mu^{(n)}$, respectively, which is easy to prove by induction. Since either the first row or the first column of $\mu^{(n)}$ must be of length at least $\lfloor \sqrt{n} \rfloor$, we get that

$$T_n \geq \min\left(\sum_{1\leq j\leq \sqrt{n}} \tau_{1,j}, \sum_{1\leq i\leq \sqrt{n}} \tau_{i,1} \right),$$

and this implies that $T_n \to \infty$ almost surely as $n \to \infty$ by the law of large numbers. □

Now, as with the first construction, define a family $(\Phi(t))_{t\geq 0}$ of random Young diagrams, by setting $\Phi(t) = \mu^{(n)}$ if $T_n \leq t < T_{n+1}$. Because of the lemma, $\Phi(t)$ is defined for all $t \geq 0$. This completes our second construction of a continuous-time version of the corner growth process.

We now want to prove the equivalence of the two constructions. The following lemma will do most of the necessary heavy lifting.

Lemma 4.3 *Let $n \geq 1$, and let $\emptyset = \nu^{(0)} \nearrow \nu^{(1)} \nearrow \ldots \nearrow \nu^{(n)}$ be a deterministic sequence of Young diagrams. Define an event*

$$E = \{\mu^{(j)} = \nu^{(j)},\ j = 0, \ldots, n\}. \tag{4.6}$$

Then, conditioned on the event E, the set

$$\left\{T_{\kappa(i,j)} - T_n + \tau_{i,j} \,:\, (i, j) \in \text{ext}(\nu^{(n)})\right\}$$

is a family of i.i.d. random variables with distribution Exp(1) *which is independent of the family*

$$\{\tau_{p,q} \,:\, (p, q) \in \nu^{(n)}\}.$$

Proof Denote $K = \text{ext}(\nu^{(n)})$. Start by noting that the occurrence of the event E can be decided by looking at the family of random variables $\boldsymbol{\tau} = (\tau_{i,j})_{(i,j)\in\nu^{(n)}\cup K}$ (in measure-theoretic terminology, E is in the σ-algebra generated by $\boldsymbol{\tau}$), which we will think of as being a vector made up of two parts, writing $\boldsymbol{\tau} = (\boldsymbol{\tau}_{\text{in}}, \boldsymbol{\tau}_{\text{out}})$ where $\boldsymbol{\tau}_{\text{in}} = (\tau_{p,q})_{(p,q)\in\nu^{(n)}}$ and $\boldsymbol{\tau}_{\text{out}} = (\tau_{i,j})_{(i,j)\in K}$.

Furthermore, inside the event E, the random variables $\kappa(i, j)$ for any $(i, j) \in \nu^{(n)} \cup K$ are constant, and the random variables $T_0, \ldots, T_n$ are determined by the $\boldsymbol{\tau}_{\text{in}}$ component (it is easy to see that each T_k is a sum of some subset of the coordinates of $\boldsymbol{\tau}_{\text{in}}$ that is determined by the sequence $\nu^{(0)} \nearrow \ldots \nearrow \nu^{(n)}$). So, we can represent certain conditional probabilities given the event E as integrals on the configuration space of $\boldsymbol{\tau}$, which is the space $\Omega = \mathbb{R}_+^{\nu^{(n)}\cup K} = \mathbb{R}_+^{\nu^{(n)}} \times \mathbb{R}_+^K$ (where $\mathbb{R}_+ = [0, \infty)$), relative to the probability measure $M(d\boldsymbol{\tau}) = \prod_{(i,j)\in\nu^{(n)}\cup K} m(d\tau_{i,j})$ where $m(dx) = e^{-x}\mathbf{1}_{\mathbb{R}_+}(x)\,dx$ is the Exp(1) probability distribution on $\mathbb{R}_+$. The event E itself can be thought of (with a minimal abuse of notation) as a Borel subset of Ω.

Now comes a key observation, which is that, although the occurrence of E depends on both the parts $\boldsymbol{\tau}_{\text{in}}$ and $\boldsymbol{\tau}_{\text{out}}$ of $\boldsymbol{\tau}$, the dependence on $\boldsymbol{\tau}_{\text{out}}$ is a fairly weak one, in the following precise sense: if $(\boldsymbol{\tau}_{\text{in}}, \boldsymbol{\tau}_{\text{out}}) \in E$, then we will also have $(\boldsymbol{\tau}_{\text{in}}, \boldsymbol{\tau}'_{\text{out}}) \in E$ precisely for those $\boldsymbol{\tau}'_{\text{out}} = (\tau'_{i,j})_{(i,j)\in K}$ whose coordinates satisfy

$$\tau'_{i,j} > T_n - T_{\kappa(i,j)} \text{ for all } (i, j) \in K$$

(where $T_n, T_{\kappa(i,j)}$ are evaluated as a function of the first component $\boldsymbol{\tau}_{\text{in}}$; see the comment above). The reason for this is that if these inequalities are satisfied then by the definitions (4.3)–(4.5) the value of the argmin (i_k, j_k) will not change for any $1 \le k \le n-1$, so we remain inside the event E.

It follows from this discussion that E has a kind of "skew-product" structure; more precisely, there is a Borel set $A \subset \mathbb{R}_+^{\nu^{(n)}}$ such that

$$E = \Big\{ (\boldsymbol{\tau}_{\text{in}}, \boldsymbol{\tau}_{\text{out}}) = \big((\tau_{p,q})_{(p,q)\in\nu^{(n)}}, (\tau_{i,j})_{(i,j)\in K} \big) \,:\, \boldsymbol{\tau}_{\text{in}} \in A \\ \text{and } \tau_{i,j} > T_n - T_{\kappa(i,j)} \text{ for all } (i,j) \in K \Big\}.$$

This can be immediately applied to our needs. Let B be a Borel subset of $\mathbb{R}_+^{\nu^{(n)}}$ and let $(t_{i,j})_{(i,j)\in K} \in \mathbb{R}_+^K$. For each $(i,j) \in K$, define a random variable $X_{i,j}$ by $X_{i,j} = T_{\kappa(i,j)} - T_n + \tau_{i,j}$. Let X denote a random variable with the $\text{Exp}(1)$ distribution. We have

$$\begin{aligned}
&\mathbb{P}\Bigg[\bigcap_{(i,j)\in K} \{X_{i,j} > t_{i,j}\} \cap \{\boldsymbol{\tau}_{\text{in}} \in B\} \,\Bigg|\, E \Bigg] \\
&= \frac{1}{\mathbb{P}(E)} \mathbb{P}\Bigg[\{\boldsymbol{\tau}_{\text{in}} \in B \cap A\} \cap \bigcap_{(i,j)\in K} \{\tau_{i,j} > t_{i,j} + T_n - T_{\kappa(i,j)}\} \Bigg] \\
&= \frac{1}{\mathbb{P}(E)} \int \!\cdots\! \int_{B\cap A} \prod_{(p,q)\in\nu^{(n)}} m(d\tau_{p,q}) \\
&\qquad \times \int \!\cdots\! \int_{\prod_{(i,j)\in K}[t_{i,j}+T_n-T_{\kappa(i,j)},\infty)} \prod_{(i,j)\in K} m(d\tau_{i,j}) \\
&= \exp\Bigg(- \sum_{(i,j)\in K} t_{i,j} \Bigg) \\
&\qquad \times \frac{1}{\mathbb{P}(E)} \int \!\cdots\! \int_{B\cap A} \exp\Bigg(\sum_{(i,j)\in K} (T_{\kappa(i,j)} - T_n) \Bigg) \prod_{(p,q)\in\nu^{(n)}} m(d\tau_{p,q}) \\
&= \prod_{(i,j)\in K} \mathbb{P}(X > t_{i,j}) \\
&\qquad \times \frac{1}{\mathbb{P}(E)} \int \!\cdots\! \int_{B\cap A} \exp\Bigg(\sum_{(i,j)\in K} (T_{\kappa(i,j)} - T_n) \Bigg) \prod_{(p,q)\in\nu^{(n)}} m(d\tau_{p,q}).
\end{aligned}$$

This is exactly what was needed to prove the claim of the lemma, since we showed that the conditional probability decomposes into the product of two parts, the first of which is the product of the exponential tail probabilities

$\mathbb{P}(X > t_{i,j})$, and the second part depending only on the set B and not on the numbers $t_{i,j}$. □

Theorem 4.4 *The two constructions are equivalent. That is, we have the equality in distribution of random processes*

$$(\Lambda(t))_{t\geq 0} \stackrel{d}{=} (\Phi(t))_{t\geq 0}.$$

Proof The claim is clearly equivalent to the statement that

$$\left(\lambda^{(n)}, S_n\right)_{n=1}^{\infty} \stackrel{d}{=} \left(\mu^{(n)}, T_n\right)_{n=1}^{\infty},$$

and by induction it is easy to see that it will be enough to show that for each $n \geq 0$, the conditional distribution of $(\mu^{(n+1)}, T_{n+1} - T_n)$ given $(\mu^{(k)}, T_k)_{k=0}^n$ coincides with the conditional distribution of $(\lambda^{(n+1)}, S_{n+1} - S_n)$ given $(\lambda^{(k)}, S_k)_{k=0}^n$.

Fix deterministic partitions $\emptyset = \nu^{(0)} \nearrow \ldots \nearrow \nu^{(n)}$. As in Lemma 4.3, let K denote the set of external corners of $\nu^{(n)}$ and let E be the event defined in (4.6). Let D be a Borel subset of the set $\{(t_1, \ldots, t_n) : 0 < t_1 < t_2 < \ldots < t_n\}$ with positive Lebesgue measure, and denote $B = E \cap \{(T_1, \ldots, T_n) \in D\}$. By Lemma 4.3, the conditional distribution of the family of random variables $\{T_{\kappa(i,j)} - T_n + \tau_{i,j} : (i, j) \in K\}$ given the event B is that of a family of i.i.d. Exp(1) random variables. Furthermore, rewriting (4.3) and (4.4) in the form

$$(i_n, j_n) = \operatorname{argmin}\{T_{\kappa(i,j)} - T_n + \tau_{i,j} : (i, j) \in \operatorname{ext}(\mu^{(n)})\},$$
$$T_{n+1} - T_n = \min\{T_{\kappa(i,j)} - T_n + \tau_{i,j} : (i, j) \in \operatorname{ext}(\mu^{(n)})\},$$

we see that the $(n + 1)$th growth corner (i_n, j_n) and the $(n + 1)$th time increment $T_{n+1} - T_n$ are precisely the argmin and the minimum of this family, respectively. By a well-known and easy property of exponential random variables (see Exercise 4.1), the argmin and the minimum are independent, the argmin is a uniformly random element of K, and the minimum has distribution Exp($|K|$). But this is exactly what we needed, since, by the definitions (4.1) and (4.2), the conditional distribution of $(\lambda^{(n)}, S_{n+1} - S_n)$ conditioned on the event $\{\lambda^{(j)} = \nu^{(j)}, j = 0, \ldots, n\} \cap \{(S_1, \ldots, S_n) \in D\}$ analogous to B is also that of the diagram $\lambda^{(n)}$ grown by adding a uniformly random external corner, and an independent time increment that has the exponential distribution Exp($|K|$). This establishes the inductive claim and finishes the proof. □

From now on we use just the notation $(\Lambda(t))_{t\ge0}$ (and the related quantities $\lambda^{(n)}, T_n$) for the corner growth process in continuous time, and think of $\Lambda(t)$ as being defined by the formulas (4.3)–(4.5) of the second construction. Note that $\lambda^{(n)}$ and T_n can be recovered from $(\Lambda(t))_{t\ge0}$ via the obvious relations

$$T_n = \inf\{t \ge 0 \,:\, |\Lambda(t)| = n\}, \tag{4.7}$$

$$\lambda^{(n)} = \Lambda(T_n). \tag{4.8}$$

Having reformulated the process, we can now reformulate the limit shape theorem which will be our main goal.

Theorem 4.5 (Limit shape theorem for the corner growth process in continuous time) *Let $(\Lambda(t))_{t\ge0}$ denote the corner growth process in continuous time, and define*

$$\tilde{\Delta}_{\mathrm{CG}} = \left\{(x,y) \in \mathbb{R}^2 \,:\, x,y \ge 0,\ \sqrt{x} + \sqrt{y} \le 1\right\}. \tag{4.9}$$

As $t \to \infty$, the scaled planar set $t^{-1}\mathrm{set}_{\Lambda(t)}$ converges in probability to the set $\tilde{\Delta}_{\mathrm{CG}}$. That is, for any $0 < \epsilon < 1$ we have that

$$\mathbb{P}\Big((1-\epsilon)\tilde{\Delta}_{\mathrm{CG}} \subseteq t^{-1}\mathrm{set}_{\Lambda(t)} \subseteq (1+\epsilon)\tilde{\Delta}_{\mathrm{CG}}\Big) \to 1 \text{ as } t \to \infty. \tag{4.10}$$

Since $\tilde{\Delta}_{\mathrm{CG}}$ is obtained by scaling the set Δ_{CG}, which has area 1, by a factor $1/\sqrt{6}$, the area of $\tilde{\Delta}_{\mathrm{CG}}$ is $1/6$. So as a trivial corollary to (4.10) we would get that $\frac{1}{t^2}|\Lambda(t)| \to \frac{1}{6}$ in probability as $t \to \infty$. Note that $|\Lambda(T_n)| = |\lambda^{(n)}| = n$, so if we were allowed to set $t = T_n$ in the relation $|\Lambda(t)| \approx \frac{1}{6}t^2$ we would also get that $T_n \approx \sqrt{6n}$. This argument is made precise in the following lemma.

Lemma 4.6 $T_n/\sqrt{6n} \to 1$ *in probability as $n \to \infty$.*

Proof (assuming Theorem 4.5) Fix some $0 < \epsilon < 1$. For any $t > 0$, if the event

$$C_t = \left\{(1-\epsilon)\tilde{\Delta}_{\mathrm{CG}} \subseteq t^{-1}\mathrm{set}_{\Lambda(t)} \subseteq (1+\epsilon)\tilde{\Delta}_{\mathrm{CG}}\right\}$$

occurred, then

$$(1-\epsilon)^2\frac{t^2}{6} \le |\Lambda(t)| \le (1+\epsilon)^2\frac{t^2}{6}. \tag{4.11}$$

Given an integer $n \geq 1$, define numbers $t_- = t_-(n)$, $t_+ = t_+(n)$ by

$$t_- = \frac{1}{1+\epsilon}\sqrt{6(n-1)},$$
$$t_+ = \frac{1}{1-\epsilon}\sqrt{6(n+1)},$$

so that $n - 1 = (1+\epsilon)^2 t_-^2/6$ and $n + 1 = (1-\epsilon)^2 t_+^2/6$. We then have that on the event $D_n = C_{t_-} \cap C_{t_+}$, the inequalities

$$|\Lambda_{t_-}| \geq n - 1, \qquad |\Lambda_{t_+}| \leq n + 1$$

hold. Combining this with (4.7) (and the fact that the T_n are increasing) gives that $t_- < T_n < t_+$. In other words, we have shown that

$$\frac{1}{1+\epsilon}\sqrt{\frac{n-1}{n}} < \frac{T_n}{\sqrt{6n}} < \frac{1}{1-\epsilon}\sqrt{\frac{n+1}{n}}$$

on the event D_n, for which we also know that $\mathbb{P}(D_n) \to 1$ as $n \to \infty$. Since ϵ was an arbitrary number in $(0, 1)$, this clearly implies the claim. □

The next few sections are devoted to the proof of Theorem 4.5. To conclude this section, we prove that the limit shape theorem for the corner growth process in continuous time implies the result for the original discrete-time process.

Proof of Theorem 4.1 assuming Theorem 4.5 Fix an $0 < \epsilon < 1$. Let $0 < \delta < 1$ be a number such that $(1+\delta)^2 \leq 1+\epsilon$ and $(1-\delta)^2 \geq 1-\epsilon$. For $n \geq 1$ let E_n denote the event

$$\begin{aligned} E_n = &\left\{1 - \delta \leq \frac{T_n}{\sqrt{6n}} \leq 1 + \delta\right\} \\ &\cap \left\{(1-\delta)\tilde{\Delta}_{\mathrm{CG}} \subseteq \frac{1}{(1-\delta)\sqrt{6n}}\mathrm{set}_{\Lambda((1-\delta)\sqrt{6n})}\right\} \\ &\cap \left\{\frac{1}{(1+\delta)\sqrt{6n}}\mathrm{set}_{\Lambda((1+\delta)\sqrt{6n})} \subseteq (1+\delta)\tilde{\Delta}_{\mathrm{CG}}\right\}. \end{aligned}$$

By Lemma 4.6 and Theorem 4.5, $\mathbb{P}(E_n) \to 1$ as $n \to \infty$. Furthermore, on the event E_n we have that

$$\begin{aligned}\frac{1}{\sqrt{n}}\mathrm{set}_{\lambda^{(n)}} &= \frac{1}{\sqrt{n}}\mathrm{set}_{\Lambda(T_n)} \subseteq \frac{1}{\sqrt{n}}\mathrm{set}_{\Lambda((1+\delta)\sqrt{6n})} \\ &= (1+\delta)\sqrt{6}\frac{1}{(1+\delta)\sqrt{6n}}\mathrm{set}_{\Lambda((1+\delta)\sqrt{6n})} \\ &\subseteq (1+\delta)^2\sqrt{6}\tilde{\Delta}_{\mathrm{CG}} = (1+\delta)^2\Delta_{\mathrm{CG}} \subseteq (1+\epsilon)\Delta_{\mathrm{CG}}.\end{aligned}$$

By similar reasoning, we also have that $\frac{1}{\sqrt{n}}\mathrm{set}_{\lambda^{(n)}} \supseteq (1-\epsilon)\Delta_{\mathrm{CG}}$ on the event E_n, so the result is proved. □

4.3 Last-passage percolation

The next step in our analysis will involve yet another way of thinking about the continuous-time corner growth process, using an elegant formalism known as **last-passage percolation** (also known as **directed last-passage percolation**; see the box opposite for a discussion of this terminology and its connection to a more general family of stochastic growth models). The idea is that instead of focusing on the shape of the diagram $\Lambda(t)$ at a given time t, we look at a particular position $(i, j) \in \mathbb{N}^2$ and ask at which point in time the cell at that position was added to the growing family of shapes. That is, we define random variables

$$G(i, j) = \inf\{t \geq 0 : (i, j) \in \Lambda(t)\} \qquad (i, j \geq 1),$$

and call $G(i, j)$ the **passage time to** (i, j).

One of the benefits of working with the passage times $G(i, j)$ is that they satisfy a nice recurrence relation.

Lemma 4.7 *The array* $(G(i, j))_{i,j\geq 1}$ *satisfies the recurrence*

$$G(i, j) = \tau_{i,j} + G(i-1, j) \vee G(i, j-1), \tag{4.12}$$

with the convention that $G(i, j) = 0$ *if* $i = 0$ *or* $j = 0$.

Proof The quantity $G(i-1, j) \vee G(i, j-1)$ represents the first time t when both the cell positions $(i-1, j)$ and $(i, j-1)$ belong to the growing shape. Equivalently, this is the first time at which (i, j) becomes an external corner of $\Lambda(t)$. That is, in our earlier notation, we have

$$G(i-1, j) \vee G(i, j-1) = T_{\kappa(i,j)}.$$

Last-passage percolation and stochastic growth models

Last-passage percolation belongs to a family of stochastic growth models that are widely studied in probability theory. One can consider such models on a lattice such as $\mathbb{Z}^2, \mathbb{Z}^3$, and so on, or more generally on any graph. For illustration purposes we focus on $\mathbb{Z}^2$ here. The idea is that an "infection" starts out occupying some of the sites in the lattice, and proceeds to infect other nearby sites over time according to some probabilistic rules. Thus, the set of infected sites at any given time forms a random "shape" that evolves over time.

The models vary according to the specifics of how the infection spreads. In **first-passage percolation**, each site of the lattice has a random clock, independent of all other clocks – formally, a nonnegative random variable drawn from some distribution – telling it when to become infected, and the clock starts ticking as soon as one of the neighboring sites becomes infected (in some versions of the process, the clock times are associated with the edges rather than sites of the lattice). In **directed first-passage percolation**, the infection spreads only in two directions, east and north, so the rule is that a site's clock starts ticking as soon as the site below it or the site to its left is infected. In **last-passage percolation** – the growth model we are considering, which is defined only in the directed setting – the site can become infected only once *both* the site to its left and the site below it have become infected (the directed models are usually considered as taking place in $\mathbb{N}^2$, with the sites in $\mathbb{Z}^2 \setminus \mathbb{N}^2$ being initially infected).

Another degree of freedom involves the choice of probability distribution for the clock times. Although it is most natural to consider exponential or geometric clock times, due to the "lack of memory" property of these distributions, one can consider arbitrary distributions. Many other variations and generalizations (to other graphs, different infection rules etc.) have been considered in the literature. For more details, see [15], [86].

Similarly, by the definition of $G(i, j)$ we can write

$$G(i, j) = T_{\gamma(i,j)},$$

where $\gamma(i, j)$ is defined as the minimal index $n \geq 0$ for which $(i, j) \in \lambda^{(n)}$ when $i, j \geq 1$, or as 0 if $i < 0$ or $j < 0$. So (4.12) reduces to the claim that $T_{\gamma(i,j)} = T_{\kappa(i,j)} + \tau_{i,j}$. That this is true can be seen by conditioning on the value of $\gamma(i, j)$: for any $n \geq 0$, the equality holds on the event

$$\{(i_{n-1}, j_{n-1}) = (i, j)\} = \{\gamma(i, j) = n\},$$

by (4.4). This finishes the proof. □

The recurrence (4.12) will be one of the main tools in our analysis of the corner growth process. There is yet another useful way of looking at the passage times. It turns out that the recurrence can be solved, leading to a more explicit, nonrecursive expression for the passage times. In fact, this expression also suggests defining a larger family of random variables $G(a,b;i,j)$ that generalize the passage times $G(i,j)$. The idea is as follows. Given two lattice positions $(a,b),(i,j) \in \mathbb{N}^2$, denote the partial order $(a,b) \preceq (i,j)$ if $a \le i$ and $b \le j$, as in Section 1.4. If $(a,b) \preceq (i,j)$, let $\mathcal{Z}(a,b;i,j)$ denote the family of lattice paths of the form $(p_0,q_0),(p_1,q_1),\ldots,(p_k,q_k)$ where $(p_0,q_0)=(a,b),(p_k,q_k)=(i,j)$ and $(p_{\ell+1},q_{\ell+1})-(p_\ell,q_\ell) \in \{(1,0),(0,1)\}$ for all ℓ (and necessarily $k = i-a+j-b$). We refer to the elements of $\mathcal{Z}(a,b;i,j)$ as **up-right paths from** (a,b) **to** (i,j) (see Fig. 4.3 for an illustration). Now define

$$G(a,b;i,j) = \max\left\{ \sum_{\ell=0}^{k} \tau_{p_\ell,q_\ell} \,:\, (p_\ell,q_\ell)_{\ell=0}^k \in \mathcal{Z}(a,b;i,j) \right\}. \tag{4.13}$$

We will refer to $G(a,b;i,j)$ as the **passage time from** (a,b) **to** (i,j). We also define $G(a,b,i,j)$ as 0 if $i<a$ or $j<b$. The similarity in notation and terminology to the earlier passage times, and the connection to the recurrence (4.12), are explained in the folowing lemma.

Lemma 4.8 *For all $(i,j) \in \mathbb{N}^2$ we have $G(i,j) = G(1,1;i,j)$.*

Proof It is enough to observe that the family $(G(1,1;i,j))_{i,j\ge 1}$ satisfies the recurrence (4.12) (with a similar convention that $G(1,1;i,j)=0$ if $i=0$ or $j=0$). The reason for this is that the sum $\sum_{\ell=0}^k \tau_{p_\ell,q_\ell}$ in the maximum defining $G(1,1;i,j)$ always contains $\tau_{i,j}$ as its last term, and after removing $\tau_{i,j}$ one is left with a maximum of sums $\sum_{\ell=0}^{k-1} \tau_{p_\ell,q_\ell}$ over all up-right paths from $(1,1)$ to either of the two positions $(i-1,j)$ or $(i,j-1)$. This corresponds exactly to the maximum $G(i-1,j) \vee G(i,j-1)$ in (4.12). □

Note that for any initial and end positions $(a,b) \preceq (i,j)$, $G(a,b;i,j)$ is equal in distribution to the "simple" passage time $G(i-a+1, j-b+1)$, so the generalization may not seem particularly far-reaching; however, we shall see that considering the more general passage times is nonetheless helpful.

Another observation that we will use is the trivial fact that $G(a,b;i,j)$ is monotone nonincreasing in (a,b) and nondecreasing in (i,j), that is, we

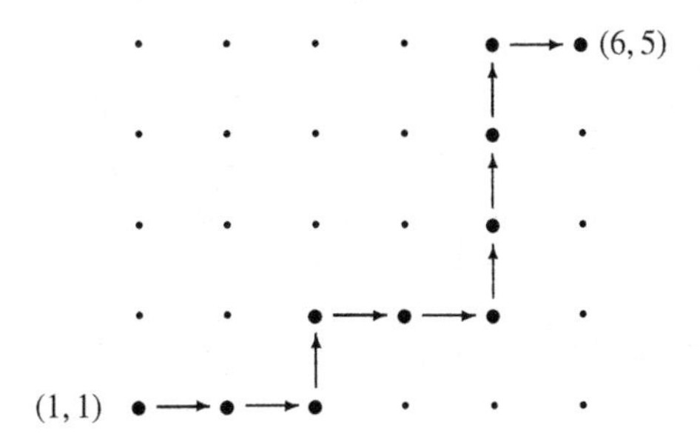

Figure 4.3 An up-right path from $(1, 1)$ to $(6, 5)$.

have

$$G(a', b'; i, j) \geq G(a, b; i, j) \leq G(a, b; i', j') \tag{4.14}$$

whenever $(a', b') \preceq (a, b) \preceq (i, j) \preceq (i', j')$.

The process of passage times $(G(i, j))_{i,j\geq 1}$ has many remarkable properties. One notable result is an explicit formula for the marginal distribution function of $G(i, j)$ for each (i, j), which we discuss in the next chapter. This formula relates $G(i, j)$ to maximal eigenvalue distributions from random matrix theory and has other useful and surprising consequences. In this chapter we use other nice properties of the passage times to derive our limit shape theorem for the corner growth process. A key idea, which we now introduce, is to compare the passage times with those of a modified "slowed-down" version of the growth process.

The idea is as follows. Fix a parameter $0 < \alpha < 1$. We augment the array of clock times $(\tau_{i,j})_{i,j\geq 1}$ with additional clock times in the row and column with index 0, by letting $\tau_{0,0}, (\tau_{i,0})_{i\geq 1}, (\tau_{0,j})_{j\geq 1}$ be a family of independent random variables, independent of the existing $\tau_{i,j}$, whose distributions are given by

$$\tau_{0,0} = 0, \tag{4.15}$$

$$\tau_{i,0} \sim \text{Exp}(\alpha) \qquad (i \geq 1), \tag{4.16}$$

$$\tau_{0,j} \sim \text{Exp}(1 - \alpha) \qquad (j \geq 1). \tag{4.17}$$

With these new clock times, we consider the passage times in (4.13) as being defined for any $a, b, i, j \geq 0$. For $i, j \geq 0$, denote

$$G_\alpha(i, j) = G(0, 0; i, j).$$

We refer to $G_\alpha(i,j)$ as the **slowed down passage time to** (i,j) **with parameter** α. (Note that the value of α will be fixed throughout most of the discussion, until a crucial moment at which we optimize a certain inequality over all values of α. To avoid tedious language and notation, we usually speak of the process of slowed down passage times without explicit mention of the value of α, and similarly omit α from the notation for various quantities where there is no risk of confusion.) The terminology is justified by the fact that

$$G(i,j) \le G_\alpha(i,j) \tag{4.18}$$

(a special case of (4.14)). Another, more precise, connection between the original and slowed down passage times is given in the relation

$$\begin{aligned} G_\alpha(k,\ell) = &\max_{1\le i\le k}\Big\{G_\alpha(i,0) + G(i,1;k,\ell)\Big\} \\ &\vee \max_{1\le j\le \ell}\Big\{G_\alpha(0,j) + G_\alpha(1,j;k,\ell)\Big) \qquad (k,\ell \ge 1), \end{aligned} \tag{4.19}$$

which follows directly from the definition (4.13), encoding the fact that an up-right path from $(0,0)$ to (k,ℓ) must travel a certain number of steps along either the x- or y-axes before visiting a point of the form $(i,1)$ or $(1,i)$ with $i \ge 1$. Note that the slowed down passage times $(G_\alpha(i,j))_{i,j\ge 0}$ also satisfy the recurrence (4.12) for all $i,j \ge 0$, except of course that the values along the x- and y-axes are not zero.

Our next goal is to reach a good understanding of the behavior of the slowed down passage times $(G_\alpha(i,j))_{i,j\ge 0}$. We pursue this in the next section. When combined with the relations (4.18) and (4.19), these insights will later allow us to analyze the original passage times and eventually to prove the limit shape result.

4.4 Analysis of the slowed down passage times

Consider the increments of the slowed down passage times $G_\alpha(i,j)$ in the x- and y-directions, namely

$$\begin{aligned} X(i,j) &= G_\alpha(i,j) - G_\alpha(i-1,j), \\ Y(i,j) &= G_\alpha(i,j) - G_\alpha(i,j-1), \end{aligned} \qquad (i,j \ge 0).$$

The next lemma recasts the recurrence (4.12) into a new form involving the arrays $(X(i,j))_{i,j\ge 0}$, $(Y(i,j))_{i,j\ge 0}$.

Lemma 4.9 *For a real number s denote* $s^+ = s \vee 0$. *The increments* $X(i,j)$ *and* $Y(i,j)$ *satisfy the recurrence relations*

$$X(i,j) = \tau_{i,j} + (X(i,j-1) - Y(i-1,j))^+, \tag{4.20}$$

$$Y(i,j) = \tau_{i,j} + (Y(i-1,j) - X(i,j-1))^+. \tag{4.21}$$

Proof Using (4.12) we can write

$$\begin{aligned} X(i,j) &= \tau_{i,j} + G_\alpha(i-1,j) \vee G_\alpha(i,j-1) - G_\alpha(i-1,j) \\ &= \tau_{i,j} + (G_\alpha(i,j-1) - G_\alpha(i-1,j)) \vee 0 \\ &= \tau_{i,j} + (X(i,j-1) - Y(i-1,j)) \vee 0 \\ &= \tau_{i,j} + (X(i,j-1) - Y(i-1,j))^+. \end{aligned}$$

A symmetric computation for $Y(i,j)$ yields (4.21). □

The next lemma states a curious distributional identity that gives a hint as to why the new recurrences (4.20), (4.21) may be interesting.

Lemma 4.10 *Let* $0 < \alpha < 1$. *Let* X, Y, Z *be independent random variables such that* $X \sim \text{Exp}(\alpha)$, $Y \sim \text{Exp}(1-\alpha)$ *and* $Z \sim \text{Exp}(1)$. *Define three random variables* U, V, W *by*

$$\begin{aligned} U &= Z + (X - Y)^+, \\ V &= Z + (Y - X)^+, \\ W &= X \wedge Y. \end{aligned}$$

Then we have the equality in distribution $(U, V, W) \stackrel{d}{=} (X, Y, Z)$.

Proof The joint density of X, Y, Z is $f_{X,Y,Z}(x,y,z) = \alpha(1-\alpha)e^{-\alpha x-(1-\alpha)y-z}$. For the transformed vector (U, V, W) we can write $(U, V, W) = T(X, Y, Z)$ where $T\colon [0,\infty)^3 \to [0,\infty)^3$ is the transformation

$$(u,v,w) = T(x,y,z) = \begin{cases} (z, z+y-x, x) & \text{if } x \le y, \\ (z+x-y, z, y) & \text{if } x > y. \end{cases} \tag{4.22}$$

Note that $u \le v$ if and only if $x \le y$, that T is a piecewise linear map, and that T is invertible and satisfies $T^{-1} = T$ (check this separately for each of the two linear maps). In particular, the Jacobian of T (which is the determinant of the linear maps) is ± 1. By a standard change of variables formula from probability theory ([111], Theorem 7.26), the joint density of U, V, W is given by

$$f_{U,V,W}(u, v, w) = f_{X,Y,Z}(T^{-1}(u, v, w)) = \begin{cases} f_{X,Y,Z}(w, w + v - u, u) & \text{if } u \le v, \\ f_{X,Y,Z}(w + u - v, w, v) & \text{if } u > v. \end{cases}$$

From this it is easy to check directly that $f_{U,V,W}(u, v, w) = f_{X,Y,Z}(u, v, w)$ for all $u, v, w \ge 0$. □

A doubly infinite path $((i_n, j_n))_{n\in\mathbb{Z}}$ in $(\mathbb{N} \cup \{0\})^2$ is called a **down-right path** if for any $n \in \mathbb{Z}$ we have $(i_n, j_n) - (i_{n-1}, j_{n-1}) \in \{(1, 0), (0, -1)\}$. We associate with every such path a family $(W_n)_{n\in\mathbb{Z}}$ of random variables defined by

$$W_n = \begin{cases} X(i_n, j_n) & \text{if } (i_n, j_n) - (i_{n-1}, j_{n-1}) = (1, 0), \\ Y(i_{n-1}, j_{n-1}) & \text{if } (i_n, j_n) - (i_{n-1}, j_{n-1}) = (0, -1). \end{cases} \tag{4.23}$$

Lemma 4.11 *For any down-right path $((i_n, j_n))_{n\in\mathbb{Z}}$, the family $(W_n)_{n\in\mathbb{Z}}$ is a family of independent random variables, and for each n the distribution of W_n is given by*

$$W_n \sim \begin{cases} \mathrm{Exp}(\alpha) & \textit{if } (i_n, j_n) - (i_{n-1}, j_{n-1}) = (1, 0), \\ \mathrm{Exp}(1 - \alpha) & \textit{if } (i_n, j_n) - (i_{n-1}, j_{n-1}) = (0, -1). \end{cases}$$

Proof We prove this first for paths which straddle the x- and y-axes, that is, have the property that for some $M > 0$, $i_n = 0$ if $n < -M$ and $j_n = 0$ if $n > M$. Such a path can be thought of as tracing out the boundary of a Young diagram ν, as illustrated in Fig. 4.4. The proof will be by induction on the size of ν. For the empty diagram, we have

$$(W_n)_{n\in\mathbb{Z}} = \{X(i, 0), i = 1, 2, \ldots\} \cup \{Y(0, j), j = 1, 2, \ldots\}.$$

Since $X(i, 0) = \tau_{i,0}$ and $Y(0, j) = \tau_{0,j}$, the claim is true by the definitions (4.15)–(4.17).

Next, assume the claim is true for ν, and let μ be a Young diagram such that $\nu \nearrow \mu$. Since μ is obtained from ν by adding a cell which has its top-right corner at some position $(i, j) \in \mathbb{N}^2$, the lattice path associated with μ will be obtained from the path of ν by replacing a triple of successive points of the form $(i - 1, j), (i - 1, j - 1), (i, j - 1)$ with the triple $(i - 1, j), (i, j), (i, j-1)$ (see Fig. 4.4). So, in the family $(W_n)_{n\in\mathbb{Z}}$ associated with ν

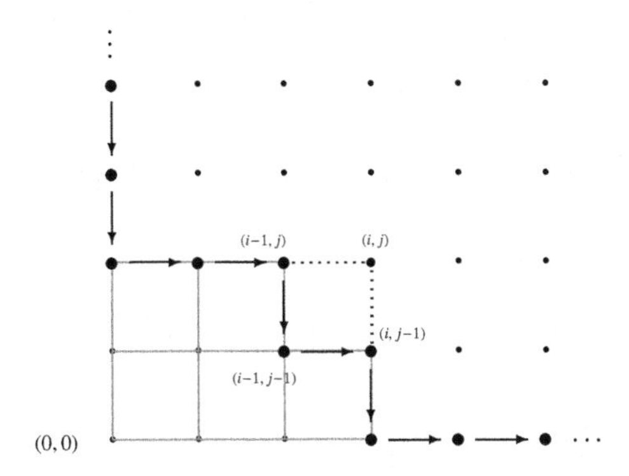

Figure 4.4 A down-right path straddling the x- and y-axes traces out the boundary of a Young diagram. Adding the cell with top-right corner (i, j) will result in a new Young diagram associated with a new down-right path.

(which satisfies the inductive hypothesis), the variables $Y(i-1, j), X(i, j-1)$ will be replaced with $X(i, j), Y(i, j)$ to obtain the new family associated with μ. By the recurrences (4.20), (4.21), the new variables are obtained from the old ones exactly according to the recipe of Lemma 4.10 (where the input for the "Z" variable is $\tau_{i,j}$, which is trivially independent of the family $(W_n)_{n\in\mathbb{Z}}$). We conclude using the lemma that the random variables in the new family are all independent and have the correct distributions.

To conclude, note that for a general down-right path $((i_n, j_n))_{n\in\mathbb{Z}}$, any finite subpath $((i_n, j_n))_{-M\le n\le M}$ is shared with some path from the class discussed above. It follows that the claim of the lemma is true for any finite subset of the variables $(W_n)_{n\in\mathbb{Z}}$, but then it is also true for the entire family since the distribution of a random process is determined by its finite-dimensional marginals. □

From Lemma 4.11 it follows in particular that for any fixed $j \ge 0$, the x-increments along the jth row $(X(i, j))_{i=1}^{\infty}$ form a family of i.i.d. random variables with distribution Exp(α), and similarly, for any fixed $i \ge 0$, the y-increments along the ith column $(Y(i, j))_{j=1}^{\infty}$ form a family of i.i.d. random variables with distribution Exp($1 - \alpha$). This has several important consequences. First, since the slowed down passage time $G_\alpha(i, j)$ can be written as the sum of increments $G_\alpha(i, j) = \sum_{k=1}^{i} X(k, 0) + \sum_{\ell=1}^{j} Y(i, \ell)$, it follows immediately that its expected value is given by

$$\mathbb{E}G_\alpha(i,j) = \frac{i}{\alpha} + \frac{j}{1-\alpha}. \tag{4.24}$$

Second, combining (4.24) with the relation (4.18) and setting $\alpha = 1/2$ gives the rough bound

$$\mathbb{E}G(i,j) \le 2(i+j) \qquad (i,j \ge 1), \tag{4.25}$$

for the rate of growth of the original passage times, which will also be useful later. Third, we have the following important result about the asymptotic behavior of the slowed down passage times.

Theorem 4.12 *Denote* $\Psi_\alpha(x,y) = \frac{x}{\alpha} + \frac{y}{1-\alpha}$. *For any* $x,y > 0$ *we have the almost sure convergence*

$$\frac{1}{n}G_\alpha(\lfloor nx \rfloor, \lfloor ny \rfloor) \to \Psi_\alpha(x,y) \text{ as } n \to \infty. \tag{4.26}$$

Proof Use the representation $G_\alpha(i,j) = \sum_{k=1}^{i} X(k,0) + \sum_{\ell=1}^{j} Y(i,\ell)$ with $i = \lfloor nx \rfloor$, $j = \lfloor ny \rfloor$. For the first sum, the strong law of large numbers gives the almost sure convergence $\frac{1}{n}\sum_{k=1}^{i} X(k,0) \to x/\alpha$ as $n \to \infty$. For the remaining sum $\frac{1}{n}\sum_{\ell=1}^{j} Y(i,\ell)$, the strong law of large numbers in its usual form does not apply, because the variables over which we are summing change with n. Nonetheless, $\frac{1}{n}\sum_{\ell=1}^{j} Y(i,\ell)$ is an empirical average of i.i.d. Exp(1) random variables, so we can go "under the hood" of the machinery of probability theory and use standard estimates from one of the *proofs* of the strong law of large numbers. According to this estimate (see Exercise 4.3), there is a constant $C > 0$ such that for any $\epsilon > 0$ we have

$$\mathbb{P}\left(\left|\frac{1}{n}\sum_{\ell=1}^{j} Y(i,\ell) - \frac{y}{1-\alpha}\right| > \epsilon\right) = \mathbb{P}\left(\left|\frac{1}{n}\sum_{\ell=1}^{j} Y(0,\ell) - \frac{y}{1-\alpha}\right| > \epsilon\right) \le \frac{C}{n^2\epsilon^4}.$$

By the Borel-Cantelli lemma this implies that almost surely $\frac{1}{n}\sum_{\ell=1}^{j} Y(i,\ell) \to y/(1-\alpha)$ as $n \to \infty$, and completes the proof. □

4.5 Existence of the limit shape

In this section we prove the existence of the limit shape in the corner growth process, without yet being able to prove the explicit formula for it. We start by proving a result on the asymptotic behavior of the passage times $G(i,j)$.

A main idea is that the passage times satisfy a superadditivity property, which makes it possible to use Kingman's subadditive ergodic theorem (discussed in the Appendix).

Theorem 4.13 (Asymptotics of the passage times) *There exists a function* $\Psi\colon [0,\infty)^2 \to [0,\infty)$ *such that for all* $x, y \geq 0$, *if* $((i_n, j_n))_{n=1}^{\infty}$ *is a sequence of positions in* $\mathbb{N}^2$ *such that* $\frac{1}{n}(i_n, j_n) \to (x, y)$ *as* $n \to \infty$, *then we have the almost sure convergence*

$$\frac{1}{n}G(i_n, j_n) \to \Psi(x,y) \text{ as } n \to \infty. \tag{4.27}$$

Furthermore, Ψ *has the following properties:*

1. *Continuity:* Ψ *is continuous.*
2. *Monotonicity:* Ψ *is monotone nondecreasing in both arguments.*
3. *Symmetry:* $\Psi(x,y) = \Psi(y,x)$.
4. *Concavity:* Ψ *satisfies*

$$\Psi(\alpha(x_1,y_1) + (1-\alpha)(x_2,y_2)) \geq \alpha\Psi(x_1,y_1) + (1-\alpha)\Psi(x_2,y_2)$$

for all $0 \leq \alpha \leq 1$, $x_1, x_2, y_1, y_2 \geq 0$.

5. *Superadditivity:* Ψ *satisfies*

$$\Psi(x_1 + x_2, y_1 + y_2) \geq \Psi(x_1, y_1) + \Psi(x_2, y_2) \tag{4.28}$$

for all $x_1, x_2, y_1, y_2 \geq 0$.

6. *Homogeneity:* $\Psi(\alpha x, \alpha y) = \alpha\Psi(x,y)$ *for all* $x, y, \alpha \geq 0$.

Proof The proof proceeds in several steps where we initially define Ψ on a limited range of values (x,y) and then extend its definition, proving the claimed properties along the way. In steps 1–3 that follow, x, y are assumed to take positive values, and we prove only a special case of (4.27), namely the fact that almost surely

$$\frac{1}{n}G(\lfloor nx \rfloor, \lfloor ny \rfloor) \to \Psi(x,y) \text{ as } n \to \infty \tag{4.29}$$

(which we then show implies (4.27)).

Step 1: positive integers. Assume that $x, y \in \mathbb{N}$. Define a family $(V_{m,n})_{0 \leq m < n}$ of random variables by

$$V_{m,n} = G(mx+1, my+1; nx, ny).$$

We claim that the array $(-V_{m,n})_{0 \leq m < n}$ satisfies the assumptions of the i.i.d.

case of Kingman's subadditive ergodic theorem (Theorem A.3 in the Appendix). Conditions 2 and 3 of the theorem are obvious. Condition 1 translates to the discrete superadditivity property

$$G(1,1;nx,ny) \geq G(1,1;mx,my) + G(mx+1,my+1;nx,ny), \qquad (4.30)$$

which is immediate from (4.13), since the right-hand side represents the maximum of sums of the form $\sum_j \tau_{k_j,\ell_j}$ over a family of increasing lattice paths that start at $(1,1)$, move north and east until getting to (mx,my), then jumps to $(mx+1,my+1)$ and then again moves north and east until reaching (nx,ny). (Compare to the verification of superadditivity in Exercise 1.2.) Condition 4 follows from (4.25).

The conclusion is that there is a function $\Psi\colon \mathbb{N}^2 \to [0,\infty)$ such that almost surely, for any $x,y \in \mathbb{N}$ we have

$$\frac{1}{n}G(nx,ny) = \frac{1}{n}G(1,1;nx,ny) = V_{0,n} \to \Psi(x,y) \text{ as } n \to \infty.$$

The fact that $G(i,j)$ is symmetric in i,j and monotone nondecreasing in each implies the analogous symmetry and monotonicity properties for Ψ. The homogeneity property follows for $\alpha \in \mathbb{N}$ by setting $x' = \alpha x, y' = \alpha y$ and taking the limit as $n \to \infty$ of the equation $\frac{1}{n}G(nx',ny') = \alpha\frac{1}{\alpha n}G(\alpha nx,\alpha ny)$.

The superadditivity property (4.28) also follows (for $x_1,y_1,x_2,y_2 \in \mathbb{N}$) by writing its discrete analogue

$$\begin{aligned} &\frac{1}{n}G(n(x_1+x_2),n(y_1+y_2)) \\ &\quad\geq \frac{1}{n}G(nx_1,ny_1) + \frac{1}{n}G(nx_1+1,ny_1+1;n(x_1+x_2),n(y_1+y_2)), \end{aligned} \qquad (4.31)$$

which holds for a similar reason as that explained for (4.30), and taking the limit as $n \to \infty$. The left-hand side converges almost surely to the left-hand side of (4.28). The first summand on the right-hand side of (4.31) converges almost surely to $\Psi(x_1,y_1)$, the first summand of (4.28). The second summand on the right-hand side of (4.31) *does not necessarily* converge almost surely to $\Psi(x_2,y_2)$; however, it is equal in distribution to $\frac{1}{n}G(1,1;nx_2,ny_2)$, which does converge almost surely to $\Psi(x_2,y_2)$. We can exploit this convergence by arguing that as a consequence, the random variable $\frac{1}{n}G(nx_1+1,ny_1+1;n(x_1+x_2),n(y_1+y_2))$ converges to $\Psi(x_2,y_2)$ in the weaker sense of convergence in probability (this is true since the limiting random variable is a constant). By a standard fact from probability theory (Exercise 4.2), we can also infer that there is almost sure convergence along a subsequence. In combination with the observations about the other terms, this is enough to imply (4.28).

Step 2: positive rationals. Assume that $x, y \in \mathbb{Q} \cap (0, \infty)$. In this case, write x and y as $x = s/q, y = t/q$ where $q, s, t \in \mathbb{N}$. We extend the definition of Ψ to a function $\Psi\colon (\mathbb{Q} \cap (0, \infty))^2 \to [0, \infty)$ by setting

$$\Psi(x, y) = \frac{1}{q}\Psi(qx, qy).$$

It is easy to check that this definition is independent of the representation of x, y as fractions, by the homogeneity we showed for integer α, x and y, and that the extended function Ψ for rational points also inherits the monotonicity, symmetry, superadditivity, and homogeneity (with rational $\alpha > 0$). The fact that $\frac{1}{n}G(\lfloor nx \rfloor, \lfloor ny \rfloor) \to \Psi(x, y)$ almost surely as $n \to \infty$ follows by taking the limit as $n \to \infty$ in the chain of inequalities

$$\begin{aligned}\frac{1}{n}G(\lfloor n/q \rfloor qx, \lfloor n/q \rfloor qy) &\le \frac{1}{n}G(\lfloor nx \rfloor, \lfloor ny \rfloor)\\ &\le \frac{1}{n}G((\lfloor n/q \rfloor + 1)qx, (\lfloor n/q \rfloor + 1)qy).\end{aligned}$$

Step 3: positive reals. Assume that $x, y \in (0, \infty)$. Extend Ψ to a function $\Psi\colon (0, \infty)^2 \to [0, \infty)$ by setting for arbitrary real $x, y > 0$

$$\Psi(x, y) = \sup\{\Psi(u, v) \,:\, 0 < u < x, 0 < v < y \text{ and } u, v \in \mathbb{Q}\}.$$

First, note that this definition is consistent with the existing definition, since, if $x, y > 0$ are rational, then by the monotonicity property the right-hand side is bounded from above by $\Psi(x, y)$ and (by monotonicity and homogeneity) bounded from below by $\Psi((1 - \epsilon)(x, y)) = (1 - \epsilon)\Psi(x, y)$ for any (rational!) $0 < \epsilon < 1$.

It is now straightforward to show that the monotonicity, symmetry, superadditivity, and homogeneity properties hold also for the extended function. The details are left to the reader.

To conclude the list of claimed properties, note that the superaddivity and homogeneity properties imply that Ψ is concave, since we can write

$$\begin{aligned}\Psi(\alpha(x_1, y_1) + (1 - \alpha)(x_2, y_2)) &\ge \Psi(\alpha(x_1, y_1)) + \Psi((1 - \alpha)(x_2, y_2))\\ &= \alpha\Psi(x_1, y_1) + (1 - \alpha)\Psi(x_2, y_2).\end{aligned}$$

Continuity follows by a general well-known fact that a concave function on an open region of $\mathbb{R}^d$ is continuous (see [105], Theorem 10.1).

We now prove the convergence (4.29) for arbitrary real $x, y > 0$. Fix some rational $\epsilon > 0$. For any rational $u, v > 0$ satisfying $u < x < (1+\epsilon)u$, $v < y < (1+\epsilon)v$, we have the inequalities

$$\frac{1}{n}G(\lfloor nu \rfloor, \lfloor nv \rfloor) \le \frac{1}{n}G(\lfloor nx \rfloor, \lfloor ny \rfloor) \le \frac{1}{n}G(\lfloor n(1+\epsilon)u \rfloor, \lfloor n(1+\epsilon)v \rfloor).$$

Taking the limit as $n \to \infty$ and using the convergence we already know for rational points, and the homogeneity property, shows that almost surely,

$$\begin{aligned}\Psi(u, v) &\le \liminf_{n\to\infty} \frac{1}{n}G(\lfloor nx \rfloor, \lfloor ny \rfloor) \le \limsup_{n\to\infty} \frac{1}{n}G(\lfloor nx \rfloor, \lfloor ny \rfloor) \\ &\le \Psi((1+\epsilon)(u, v)) = (1+\epsilon)\Psi(u, v).\end{aligned}$$

Letting $u \nearrow x$ and $v \nearrow y$ implies that (almost surely)

$$\begin{aligned}\Psi(x, y) &\le \liminf_{n\to\infty} \frac{1}{n}G(\lfloor nx \rfloor, \lfloor ny \rfloor) \\ &\le \limsup_{n\to\infty} \frac{1}{n}G(\lfloor nx \rfloor, \lfloor ny \rfloor) \le (1+\epsilon)\Psi(x, y).\end{aligned}$$

Since ϵ was an arbitrary positive rational number, we get the desired claim that $\frac{1}{n}G(\lfloor nx \rfloor, \lfloor ny \rfloor) \to \Psi(x, y)$ almost surely as $n \to \infty$.

To conclude this part of the proof, we show that (4.29) implies (4.27). If $n^{-1}(i_n, j_n) \to (x, y)$ where $x, y > 0$, then, taking positive real numbers x', x'', y', y'' such that $x' < x < x''$ and $y' < y < y''$, we have that

$$\frac{1}{n}G(\lfloor nx' \rfloor, \lfloor ny' \rfloor) \le \frac{1}{n}G(i_n, j_n) \le G(\lfloor nx'' \rfloor, \lfloor ny'' \rfloor)$$

if n is large enough. Letting $n \to \infty$ shows that almost surely

$$\Psi(x', y') \le \liminf_{n\to\infty} \frac{1}{n}G(i_n, j_n) \le \limsup_{n\to\infty} \frac{1}{n}G(i_n, j_n) \le \Psi(x'', y'').$$

Now using homogeneity as we did in an earlier part of the proof (for example taking $x' = (1-\epsilon)x$, $y' = (1-\epsilon)y$, $x'' = (1+\epsilon)x$, $y'' = (1+\epsilon)y$ for some arbitrary positive ϵ), this implies easily that $\frac{1}{n}G(i_n, j_n) \to \Psi(x, y)$ almost surely.

Step 4: extension to the *x*- and *y*-axes. To prove that Ψ can be extended to a continuous function on $[0, \infty)^2$, we derive upper and lower bounds for $\Psi(x, y)$ that become sharp as the point (x, y) approaches the axes. Let

$x, y > 0$. Use the relation (4.18) with $(i, j) = (\lfloor nx \rfloor, \lfloor ny \rfloor)$. Dividing both sides by n and letting $n \to \infty$ gives (using (4.26)) that

$$\Psi(x, y) \le \Psi_\alpha(x, y) = \frac{x}{\alpha} + \frac{y}{1-\alpha}. \tag{4.32}$$

Now recall that α can take an arbitrary value in $(0, 1)$. Choosing the value $\alpha = \sqrt{x}/\left(\sqrt{x} + \sqrt{y}\right)$ minimizes the right-hand side and gives the bound

$$\Psi(x, y) \le \left(\sqrt{x} + \sqrt{y}\right)^2. \tag{4.33}$$

(Remarkably, it turns out that the right-hand side is the correct formula for $\Psi(x, y)$, but we will prove this only in the next section.) In the other direction, noting the fact that $\frac{1}{n}G(\lfloor nx \rfloor, \lfloor ny \rfloor) \ge \frac{1}{n}\sum_{i=1}^{\lfloor nx \rfloor} \tau_{i,0}$ and letting $n \to \infty$, using the strong law of large numbers we see that $\Psi(x, y) \ge x$. Similar reasoning gives $\Psi(x, y) \ge y$, so together we have the lower bound

$$\Psi(x, y) \ge x \vee y. \tag{4.34}$$

The upper and lower bounds in (4.33) and (4.34) both converge to z as (x, y) converges to a point $(z, 0)$ on the x-axis or to a point $(0, z)$ on the y-axis. It follows that if we set $\Psi(z, 0) = \Psi(0, z) = z$ then Ψ extends to a continuous function on $[0, \infty)^2$. All the properties we claimed for Ψ trivially carry over to the extended function by continuity.

Finally, we prove that (4.27) holds when $n^{-1}(i_n, j_n) \to (x, y)$ where the point (x, y) lies on the x- or y-axes. This involves a monotonicity argument similar to the one used previously. For example, assume that $x = 0$ (the case $y = 0$ is similar). In this case, take arbitrary positive numbers y', y'' satisfying $y' < y < y''$ and a number $x'' > 0$. For large enough n we have that

$$\frac{1}{n}G(1, \lfloor ny' \rfloor) \le \frac{1}{n}G(i_n, j_n) \le \frac{1}{n}G(\lfloor nx'' \rfloor, \lfloor ny'' \rfloor).$$

Letting $n \to \infty$ gives that almost surely,

$$y' = \Psi(0, y) \le \liminf_{n\to\infty} \frac{1}{n}G(i_n, j_n) \le \limsup_{n\to\infty} \frac{1}{n}G(i_n, j_n) \le \Psi(x'', y''),$$

where the limit on the left-hand side follows from the strong law of large numbers (since $G(1, j) = \sum_{k=1}^{j} \tau_{1,j}$). Once again, the limit (4.27) follows by taking the limit as $y' \nearrow y, y'' \searrow y$ and $x'' \searrow x$. □

We will now use Theorem 4.13 to prove an existence result about the

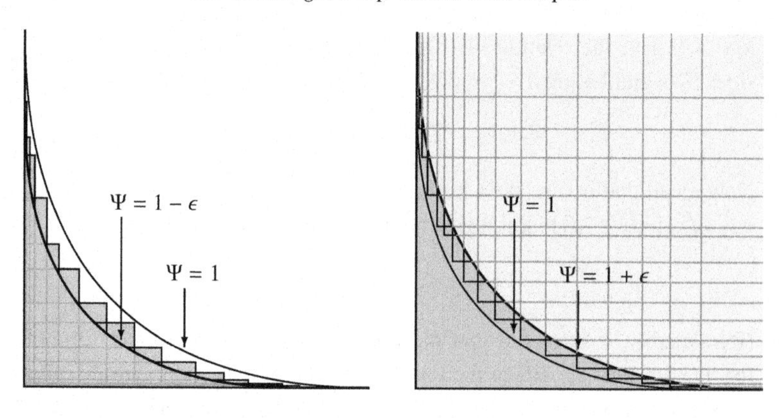

Figure 4.5 Illustration of the claim of Lemma 4.14.

convergence of the corner growth process to a limit shape. The limit shape is defined in terms of the (still unknown) function $\Psi(x, y)$ by

$$D_{\mathrm{CG}} = \left\{(x, y) \in [0, \infty)^2 \ : \ \Psi(x, y) \le 1\right\}. \tag{4.35}$$

Lemma 4.14 *For any* $0 < \epsilon < 1$ *there exist points* $(x_1, y_1), \ldots, (x_k, y_k)$, $(u_1, v_1), \ldots, (u_\ell, v_\ell) \in \mathbb{R}_+^2 = [0, \infty)^2$ *such that* $\Psi(x_i, y_i) < 1$ *for all* $1 \le i \le k$, $\Psi(u_j, v_j) > 1$ *for all* $1 \le j \le \ell$, *and such that the following relations hold:*

$$(1 - \epsilon)D_{\mathrm{CG}} \subseteq \bigcup_{i=1}^{k} \Big([0, x_i] \times [0, y_i]\Big),$$

$$\mathbb{R}_+^2 \setminus (1 + \epsilon)D_{\mathrm{CG}} \subseteq \bigcup_{j=1}^{\ell} \Big((u_j, \infty) \times (v_j, \infty)\Big).$$

The claim of the lemma is illustrated in Fig. 4.5.

Proof By the continuity of Ψ, the set $(1 - \epsilon)D_{\mathrm{CG}}$ is a compact set, and by the homogeneity property it can also be written as $\Psi^{-1}([0, 1 - \epsilon])$. It follows that the family of sets $\{[0, x) \times [0, y) \ : \ \Psi(x, y) < 1\}$ is a covering of $(1 - \epsilon)D_{\mathrm{CG}}$ by sets that are (relatively) open in $[0, \infty)^2$, so the first claim follows from the Heine-Borel theorem.

Similarly, to prove the second claim, let $M > 0$ be a sufficiently large number so that the set $(1 + \epsilon)D_{\mathrm{CG}}$ is contained in $[0, M]^2$. It will be enough (check this!) to show that there are points $(u_1, v_1), \ldots, (u_\ell, v_\ell) \in \mathbb{R}_+^2$ such

that

$$[0, 2M]^2 \setminus (1+\epsilon)D_{\rm CG} \subseteq \bigcup_{j=1}^{\ell} \Big((u_j, 2M] \times (v_j, 2M] \Big).$$

This follows from a compactness argument similar to the one used in the proof of the first claim. □

Theorem 4.15 *As $t \to \infty$, the scaled planar set $t^{-1}\mathrm{set}_{\Lambda(t)}$ converges in probability, in the same sense as before, to the set $D_{\rm CG}$.*

Proof Fix $0 < \epsilon < 1$. Let the points $(x_1, y_1), \ldots, (x_k, y_k), (u_1, v_1), \ldots, (u_\ell, v_\ell)$ be as in Lemma 4.14 above. By Theorem 4.13, the event

$$F_t = \bigcap_{i=1}^{k} \Big\{ G(\lceil tx_i \rceil, \lceil ty_i \rceil) < t \Big\} \cap \bigcap_{j=1}^{\ell} \Big\{ G(\lfloor tu_i \rfloor, \lfloor tv_i \rfloor) > t \Big\}$$

has probability that converges to 1 as $t \to \infty$. But, note that for any $(a, b) \in \mathbb{N}^2$, $G(a, b) < t$ implies that (a, b) is the position of the top-right corner of a cell in $\Lambda(t)$, so the planar set $\mathrm{set}_{\Lambda(t)}$ contains the set $[0, a] \times [0, b]$. It follows that on the event F_t we have

$$\frac{1}{t}\mathrm{set}_{\Lambda(t)} \supseteq \bigcup_{i=1}^{k} \Big(\Big[0, t^{-1}\lceil tx_i \rceil\Big] \times \Big[0, t^{-1}\lceil ty_i \rceil\Big] \Big) \supseteq (1-\epsilon)D_{\rm CG}. \qquad (4.36)$$

Similarly, if $G(a, b) > t$ then $(a-1, b-1)$ is *not* the position of the bottom-left corner of a cell in $\Lambda(t)$, which implies the relation $\mathbb{R}_+^2 \setminus \mathrm{set}_{\Lambda(t)} \supseteq (a-1, \infty) \times (b-1, \infty)$. Thus on the event F_t we have that

$$\mathbb{R}_+^2 \setminus \frac{1}{t}\mathrm{set}_{\Lambda(t)} \supseteq \bigcup_{j=1}^{\ell} \Big([t^{-1}\lfloor tu_j \rfloor, \infty) \times [t^{-1}\lfloor tv_j \rfloor, \infty) \Big) \supseteq \mathbb{R}_+^2 \setminus (1+\epsilon)D_{\rm CG},$$

and therefore that

$$\frac{1}{t}\mathrm{set}_{\Lambda(t)} \subseteq (1+\epsilon)D_{\rm CG}. \qquad (4.37)$$

The fact that for any $\epsilon > 0$, (4.36) and (4.37) hold with asymptotically high probability as $t \to \infty$ is exactly the claim that was to be shown. □

4.6 Recovering the limit shape

Our remaining goal on the path to proving the limit shape theorem for the corner growth process is to show that the limit shape $D_{\rm CG}$ coincides with the set $\tilde{\Delta}_{\text{CG}}$ defined in (4.9). The idea is to take the limit of the relation (4.19) as $n \to \infty$. This will provide us with a relation between the known function $\Psi_\alpha(x, y)$ and the unknown function $\Psi(x, y)$. The result is as follows.

Theorem 4.16 *For any $x, y \geq 0$ and $0 < \alpha < 1$ we have*

$$\Psi_\alpha(x, y) = \max_{0 \leq u \leq x} \left\{ \frac{u}{\alpha} + \Psi(x - u, y) \right\} \vee \max_{0 \leq v \leq x} \left\{ \frac{v}{1 - \alpha} + \Psi(x, y - v) \right\}. \quad (4.38)$$

Proof Let $0 \leq u \leq x$. Using (4.19) we can write that

$$\begin{aligned} G_\alpha(1 + \lfloor nx \rfloor, 1 + \lfloor ny \rfloor) \\ &= \max_{1 \leq i \leq 1 + \lfloor nx \rfloor} \left\{ G_\alpha(i, 0) + G(i, 1; 1 + \lfloor nx \rfloor, 1 + \lfloor ny \rfloor) \right\} \\ &\qquad \vee \max_{1 \leq j \leq \lfloor ny \rfloor} \left\{ G_\alpha(0, j) + G(1, j; \lfloor nx \rfloor, \lfloor ny \rfloor) \right\} \\ &\geq G_\alpha(1 + \lfloor nu \rfloor, 0) + G(1 + \lfloor nu \rfloor, 1; 1 + \lfloor nx \rfloor, 1 + \lfloor ny \rfloor). \end{aligned}$$

Dividing this inequality by n and letting $n \to \infty$, the left-hand side converges almost surely to $\Psi_\alpha(x, y)$. On the right-hand side, the first term $\frac{1}{n} G_\alpha(1 + \lfloor nu \rfloor, 0)$ converges almost surely to $\Psi_\alpha(u, 0) = u/\alpha$. The second term $\frac{1}{n} G(1 + \lfloor nu \rfloor, 1; 1 + \lfloor nx \rfloor, 1 + \lfloor ny \rfloor)$ is handled by observing that it is equal in distribution to $\frac{1}{n} G(1 + \lfloor nx \rfloor - \lfloor nu \rfloor, 1 + \lfloor ny \rfloor - \lfloor nv \rfloor)$, and therefore converges *in probability* to $\Psi(x - u, y)$. As before, this implies (Exercise 4.2) almost sure convergence along a subsequence, so, by looking at the three almost sure limits along that subsequence, we deduce that $\Psi_\alpha(x, y) \geq \frac{u}{\alpha} + \Psi(x - u, y)$. Since this is true for any $0 \leq u \leq x$ we get that

$$\Psi_\alpha(x, y) \geq \max_{0 \leq u \leq x} \left\{ \frac{u}{\alpha} + \Psi(x - u, y) \right\}.$$

A symmetric argument shows that

$$\Psi_\alpha(x, y) \geq \max_{0 \leq v \leq x} \left\{ \frac{v}{1 - \alpha} + \Psi(x, y - v) \right\},$$

so we conclude that the right-hand side of (4.38) is a lower bound for $\Psi(x, y)$.

To establish the same expression as an *upper* bound for $\Psi(x, y)$, start with the observation that the maxima on the right-hand side of (4.19) can be bounded from above by slightly modified maxima where the indices i and j range over sets whose cardinality does not grow with n. More precisely, for any $k, \ell \geq 1$, given integers $1 \leq i_1 < \ldots < i_s \leq k$ and $1 \leq j_1 < \ldots < j_t \leq \ell$, we have that

$$\max_{1\leq i\leq k} \Big\{ G_\alpha(i,0) + G(i,1;k,\ell) \Big\} \leq \max_{1\leq d\leq s-1} \Big\{ G_\alpha(i_{d+1},0) + G(i_d,1;k,\ell) \Big\},$$
$$\max_{1\leq j\leq \ell} \Big\{ G_\alpha(0,j) + G(1,j;k,\ell) \Big\} \leq \max_{1\leq d\leq t-1} \Big\{ G_\alpha(0,j_{d+1}) + G(1,j_d;k,\ell) \Big\}.$$

(These relations follow easily from the monotonicity property (4.14).) So, we have the upper bound

$$G_\alpha(k,\ell) \leq \max_{1\leq d\leq s-1} \Big\{ G_\alpha(i_{d+1},0) + G(i_d,1;k,\ell) \Big\} \vee \max_{1\leq d\leq t-1} \Big\{ G_\alpha(0,j_{d+1}) + G(1,j_d;k,\ell) \Big\}. \tag{4.39}$$

Apply this with the parameters

$$k = 1 + \lfloor nx \rfloor, \quad \ell = 1 + \lfloor ny \rfloor,$$
$$i_d = 1 + \lfloor nxd/q \rfloor, \quad j_d = 1 + \lfloor nyd/q \rfloor \qquad (1 \leq d \leq q),$$

where $q \geq 1$ is some fixed integer. Dividing both sides of (4.39) by n and letting $n \to \infty$, the same subsequence trick used above implies the inequality

$$\Psi_\alpha(x,y) \leq \max_{1\leq d\leq q} \left\{ \frac{(d+1)x/q}{\alpha} + \Psi(x-(dx/q),y) \right\} \vee \max_{1\leq d\leq q} \left\{ \frac{(d+1)y/q}{1-\alpha} + \Psi(x,y-(dy/q)) \right\}. \tag{4.40}$$

(Note that we can get almost sure convergence along the same subsequence simultaneously for all the values of i_d and j_d by taking nested subsequences, since the number of points i_d and j_d does not grow with n.)

This was true for arbitrary $q \geq 1$. Letting $q \to \infty$, the continuity of Ψ ensures that the right-hand side of (4.40) converges to the right-hand side of (4.38), which completes the proof of the upper bound. □

We have reduced the problem of the limit shape in the corner growth process to the problem of finding (and proving uniqueness of) a function $\Psi\colon [0,\infty)^2 \to [0,\infty)$ satisfying the properties listed in Theorem 4.13 and the relation (4.38). It is helpful to think of (4.38) as an (uncountably infinite) system of equations for $\Psi(x,y)$ in terms of Ψ_α. The information encoded in this system turns out to be equivalent to a statement about

Legendre transforms, so we can use well-known properties of these transforms to find Ψ.

Let us start by simplifying (4.38) to bring it to a more tractable form. First, we will actually need to use it only with the values $x = y = 1$ (but arbitrary $0 < \alpha < 1$). In this special case we can write

$$\begin{aligned}\frac{1}{\alpha} + \frac{1}{1-\alpha} &= \Psi_\alpha(1,1)\\ &= \max_{0\le u\le 1}\left(\frac{u}{\alpha} + \Psi(1-u,1)\right) \vee \max_{0\le u\le 1}\left(\frac{u}{1-\alpha} + \Psi(1,1-u)\right).\end{aligned}$$

Because of the symmetry of Ψ, the two maxima differ only in the explicit terms $u/\alpha, u/(1-\alpha)$, so we see that for $\alpha \le 1/2$ the first maximum is the dominant one and for $\alpha > 1/2$ the second maximum dominates. Furthermore, both sides of the equation are symmetric in α and $1-\alpha$, so it is enough to look at values $\alpha \le 1/2$. Denoting $f(u) = -\Psi(1-u,1)$ and replacing the variable α with $s = 1/\alpha$, we therefore have that

$$\frac{s^2}{s-1} = \max_{0\le u\le 1}(su - f(u)) \qquad (s \ge 2). \tag{4.41}$$

Note that f, which we regard as a function on $[0,1]$, is monotone nondecreasing, convex, continuous, and satisfies $f(1) = -1$. As a final step, extend f to a function $\tilde{f}\colon \mathbb{R} \to \mathbb{R} \cup \{\infty\}$ in a rather trivial way by setting

$$\tilde{f}(u) = \begin{cases} f(u) & \text{if } 0 \le u \le 1,\\ \infty & \text{otherwise,}\end{cases}$$

and noting that (4.41) can be rewritten as the statement that

$$\frac{s^2}{s-1} = \sup_{u\in\mathbb{R}}\left(su - \tilde{f}(u)\right) \qquad (s \ge 2). \tag{4.42}$$

The reason this is useful is that the operation that takes a convex function $g\colon \mathbb{R} \to \mathbb{R} \cup \{-\infty, \infty\}$ and returns the function $g^*(s) = \sup_{u\in\mathbb{R}}(su - g(u))$ is a well-known operation, known as the **Legendre transform**. The function $g^*(s)$ is referred to as the Legendre transform, or **convex dual**, of g. It is easy (and recommended as an exercise) to check that $g^*(s)$ is also convex.

A fundamental property of the Legendre transform is that it is its own inverse. Its precise formulation is as follows.

The Legendre transform

The **Legendre transform** of a convex function $g\colon \mathbb{R} \to \mathbb{R} \cup \{\infty\}$ is defined by (4.43). In the simplest case when g is differentiable and strictly convex, g^* can be computed by solving a maximization problem, leading to an equivalent definition in terms of the implicit pair of equations

$$g^*(p) = xp - g(x), \qquad p = g'(x). \tag{4.45}$$

This gives a recipe for computing g^* by solving the equation $p = g'(x)$ for x to obtain $x = x(p) = (g')^{-1}(p)$, which is then substituted into the first equation in (4.45). The equations (4.45) also have a simple geometric interpretation, shown in the figure below. The elegant fact that the same recipe can be used to recover g from g^* was discovered in 1787 by the French mathematician Adrien-Marie Legendre.

The Legendre transform (and its higher-dimensional generalization of the same name) appears in several places in mathematics and physics. Perhaps most notably, it is of fundamental importance in mechanics, where it is used to show the equivalence of the Lagrangian and Hamiltonian formulations of the laws of mechanics. For more details, see [9], [105].

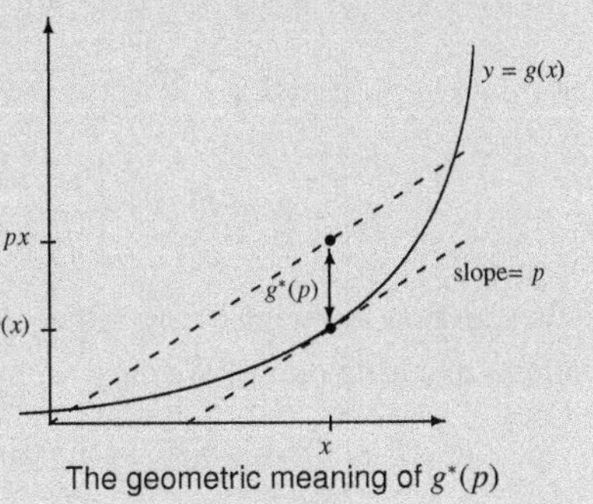

The geometric meaning of $g^*(p)$

Theorem 4.17 *Let $g\colon \mathbb{R} \to \mathbb{R} \cup \{-\infty, \infty\}$ be a convex function that is lower-semicontinuous (that is, $g^{-1}((t,\infty])$ is an open set for any $t \in \mathbb{R}$). Denote*

$$h(s) = g^*(s) = \sup_{u \in \mathbb{R}} \{su - g(u)\} \qquad (s \in \mathbb{R}). \tag{4.43}$$

Then we have that

$$g(u) = h^*(u) = \sup_{s \in \mathbb{R}} \{su - h(s)\} \qquad (u \in \mathbb{R}). \tag{4.44}$$

For the proof of this easy result, see [115, Appendix C] or [105, Section 12]. See the box for some additional background on the Legendre transform, its geometric meaning and its importance in mathematics.

By the properties of f mentioned previously, the extended function $\tilde{f}$ is convex and lower-semicontinuous. Let $h(s) = \tilde{f}^*(s) = \sup\{us - \tilde{f}(u) : u \in$

$\mathbb{R}\}$. We know from (4.42) that $h(s) = s^2/(s-1)$ for $s \geq 2$, but in order to make use of Theorem 4.17 to recover $\tilde{f}$, we need to know $h(s)$ for all real values of s. We claim that $h(s) = h(2) = 4$ for all $s < 2$. To see this, first note that in the particular case $g = \tilde{f}$, the supremum in the definition (4.43) of h can be taken over $u \geq 0$. It follows immediately that h is monotone nondecreasing, and this implies that $h(s) \leq h(2) = 4$ for all $s \leq 2$. On the other hand, since h is convex, it satisfies the inequality

$$\frac{h(y) - h(x)}{y - x} \leq \frac{h(z) - h(y)}{z - y}$$

for any real numbers $x < y < z$. Taking $y = 2$ and letting $z \searrow 2$ gives

$$\frac{4 - h(x)}{2 - x} \leq \lim_{z \searrow 2} \frac{h(z) - 4}{z - 2} = h'(2+) = \frac{d}{ds}\Big|_{s=2} \left(\frac{s^2}{s-1} \right) = 0$$

that is, $h(x) \geq 4$, for all $x < 2$. Thus, we have shown that

$$h(s) = \begin{cases} 4 & \text{if } s < 2, \\ \frac{s^2}{s-1} & \text{if } s \geq 2. \end{cases}$$

We can now compute $\tilde{f}$ (or, rather, f, which is the only part of $\tilde{f}$ that interests us) using (4.44). We have

$$f(u) = \sup_{s \in \mathbb{R}} \{us - h(s)\} = \sup_{s \geq 2} \{us - h(s)\} \qquad (0 \leq u \leq 1),$$

where the second equality holds since $h(s)$ is constant on $(-\infty, 2]$. For $u = 1$ we already know $f(u) = -1$. For $0 \leq u < 1$, the function $g_u(s) = us - h(s)$ has a stationary point (which, it is easy to check, is a global maximum) at the point $s_* = s_*(u)$ satisfying

$$0 = g_u'(s_*) = 1 - u - \frac{1}{(s_* - 1)^2}.$$

This gives $s_* = 1 + 1/\sqrt{1-u}$ and therefore, after a quick computation,

$$f(u) = us_* - h(s_*) = -(1 + \sqrt{1-u})^2.$$

We have found an explicit formula for f. From here, it it a short step to the following important result.

Theorem 4.18 *For $x, y \geq 0$ we have*

$$\Psi(x, y) = (\sqrt{x} + \sqrt{y})^2. \tag{4.46}$$

Proof If $\Phi(x,y) = (\sqrt{x}+\sqrt{y})^2$, then $\Phi(x,1) = (1+\sqrt{x})^2 = -f(1-x) = \Psi(x,1)$ for all $x \in [0,1]$. By the symmetry of Φ and Ψ we also have $\Phi(1,y) = \Psi(1,y)$ for all $y \in [0,1]$. Both Φ and Ψ are homogeneous functions, so, since they coincide on the set $\{(x,0) : 0 \le x \le 1\} \cup \{(0,y) : 0 \le y \le 1\}$, they are equal everywhere. □

Using (4.46) we see that the sets $\tilde{\Delta}_{\mathrm{CG}}$ and D_{CG} defined in (4.9) and (4.35) are equal. This was the last missing piece of the puzzle, so the proof of Theorem 4.5 (and therefore also of Theorem 4.1) is complete.

4.7 Rost's particle process

In Chapter 3 we studied random square Young tableaux, which we saw can also be interpreted as a type of "growth process" for Young diagrams growing from the empty diagram to the square diagram of given order n. We then observed that this process can also be interpreted as an interacting particle system. The corner growth process in continuous time has an analogous interpretation as a system of interacting particles on a one-dimensional lattice, which is quite natural and interesting in its own right. In the literature on interacting particle systems this process is known by a somewhat technical name.[3] To simplify the terminology, here we will refer to it as **Rost's particle process**. It can be thought of as a simple model for a traffic jam, or as a queueing system consisting of a line of customers visiting a succession of service stations. In his paper [110] in which he proved the limit shape theorem for the corner growth process, Rost originally proved a result on the limiting behavior of this particle system and the limit shape theorem was derived as an application. Here we take the opposite route.

In Rost's particle process, particles (which can be visualized, for example, as cars backed up along a road) move randomly on the sites of the integer lattice $\mathbb{Z}$ as a function of a continuous time parameter $t \ge 0$. Each site can contain at most one particle. Initially, particles occupy the positions $0, -1, -2, \ldots$ of the lattice, and the positions $1, 2, \ldots$ are vacant; this is the "infinite traffic jam." Subsequently, each particle moves one step to the right at random times. A particle can only move to the right if the space to its right is vacant. Once a particle has reached a given position and the space to the right of that position becomes vacant, the time it takes the particle to move to the right is an Exp(1)-distributed random variable, independent of all other sources of randomness.

Label the particles with the labels $1, 2, \ldots$, as counted from the "front of the line," so that the particle with label i is the particle that started in position $-i+1$. Formally, the randomness input into the process is a family $(\tau_{i,j})_{i,j\geq 1}$ of i.i.d. random variables with distribution Exp(1), where $\tau_{i,j}$ is interpreted as the time it took particle number i to make its jth move to the right (from position $j-i$ to position $j-i+1$) once it arrived in position $j-i$ and position $j-i+1$ has become vacant.

Given the array $(\tau_{i,j})_{i,j\geq 1}$ it is possible to compute where any particle is at any given time. A convenient way to do this, which will immediately reveal the connection to the corner growth process, is as follows. For $i, j \geq 1$ denote by $G(i, j)$ the time (starting from time 0) it will take particle i to reach position $j-i+1$ (i.e., to complete its jth move to the right). By the rules we specified, the random variables $G(i, j)$ satisfy the recurrence

$$G(i, j) = G(i, j-1) \vee G(i-1, j) + \tau_{i,j}, \tag{4.47}$$

since $G(i, j-1)$ represents the time when particle i arrived in position $j-i$, and $G(i-1, j)$ represents the time when particle $i-1$ *departed* from position $j-i+1$ (leaving it vacant); at the latter of these two times, particle i is ready to make its jth move to the right and the (i, j)th random clock "starts ticking." The recurrence (4.47) holds for all $i, j \geq 1$ with the convention that $G(k, 0) = G(0, k) = 0$ for all $k \geq 1$ (it is easy to see why this convention is needed). Since this is the same recurrence as (4.12), we have proved the following result.

Lemma 4.19 *Rost's particle process is equivalent to the corner growth process in continuous time. More precisely, the array $(G(i, j))_{i,j\geq 1}$ as defined earlier is equal in distribution to the family of passage times in the corner growth process.*

It is worth noting that there is a second, equivalent way of defining Rost's particle process that is slightly more elegant and used in much of the literature on the subject. In this version, each lattice position k is equipped with a standard Poisson process of random times on $[0, \infty)$ that dictate when a particle occupying the position will *attempt* to move to the right. Each time this "Poisson clock" rings, if the position is empty, or if the position is occupied by a particle but the position to its right is also occupied, then nothing happens; if the position is occupied by a particle and the position to the right is vacant, the particle moves to the right. The different Poisson processes associated with different lattice positions are all independent.

When defined this way, the process is a special case of a process known as the **totally asymmetric simple exclusion process** (usually abbreviated to **TASEP**), part of an even larger family of models known as **exclusion processes**. We will not go into the more general theory of these processes, but the box on the next page provides some more background.

It is not hard to check that the definition in terms of Poisson processes is equivalent to the first one we gave earlier. The equivalence is related to the standard connection between Poisson processes and the exponential distribution, and to the fact that it does not matter whether we associate the random clock times to lattice positions or to individual particle/vacancy pairs, as long as they are all independent.

Let us now explore some consequences of the equivalence between Rost's particle process and the corner growth process. One natural question, considered by Rost, is how the entire particle configuration, considered as a bulk mass, evolves over time at the so-called macroscopic scale in which looks not at individual particles but rather at average densities of particles over large regions. To formulate this question more precisely, denote by N_t the counting measure of particles at time t (that is, $N_t(E)$ is the number of particles in the set E at time t). What can we say about the behavior of N_t for large t?

First, note that the range of positions over which the interesting part of the measure N_t is spread out is approximately $[-t, t]$ (that is, as t grows larger we want to look at a larger and larger range of positions – time and space must be scaled similarly). The reason for this is as follows: the movement of particle number 1 is particularly simple, since the time for it to make j moves to the right is $G(1, j) = \sum_{k=1}^{j} \tau_{1,j}$, a sum of j i.i.d. Exp(1) random variables. In other words, particle 1 is executing a continuous-time random walk in which with rate 1 it moves one step to the right. By the strong law of large numbers we have $\frac{1}{j}G(1, j) \to \mathbb{E}\tau_{1,1} = 1$ almost surely as $j \to \infty$. Equivalently, the position $Y_1(t)$ of particle 1 at time t, which is a Poi(t) random variable, satisfies $Y_1(t)/t \to 1$ almost surely as $t \to \infty$. So, to the right of position $(1 + o(1))t$ we expect to not find any particles – the

Exclusion processes

The **exclusion process** is the name given to a family of interacting particle systems introduced by Frank Spitzer in 1970 [123]. Given a graph G – usually taken to be a lattice (finite or infinite) such as $\mathbb{Z}$, $\mathbb{Z}^d$ or a cycle graph – an exclusion process on G consists of an evolving configuration of particles occupying at any time some subset of the vertices (or "sites" in the probabilistic parlance) of G and moving from one site to another at random times. The crucial rule that gives the process its name, known as the **exclusion rule**, is that each site can be occupied by at most one particle at a time. In addition, the assumption is that the process of random times at which each particle attempts to jump (succeeding if and only if the site it is trying to jump to is vacant at that time) is a standard Poisson process on $[0, \infty)$, with processes associated with different particles being independent. Furthermore, when a particle decides to jump, it also randomly selects a site to jump to from some specified distribution (which can depend, usually in a translation-equivariant way, on the particle's position).

The different exclusion processes differ in the graph G and in the distribution of positions for jumping. Assume $G = \mathbb{Z}$. In the **totally asymmetric simple exclusion process**, or **TASEP**, all jumps occur one step to the right. In the **partially asymmetric simple exclusion process** (**PASEP**), jumps occur one step to the right with some fixed probability $p \in (0, 1)$, $p \neq 1/2$, or one step to the left with the complementary probability $1 - p$. The case when $p = 1/2$, that is, when jumps in both directions are equally likely, is known as the **symmetric simple exclusion process**. (In all the above cases, the adjective "simple" refers to the fact that only jumps to adjacent sites are permitted, in analogy with the classical "simple random walk" of probability theory.) These processes along with other nonsimple variants in which particles jump some random number of units away from their current position, and generalizations to higher dimensions and other graphs, have been studied extensively and are a fertile ground for the interaction of ideas from probability theory, statistical physics, integrable systems, partial differential equations and other areas. For more details, see [77, 78]

local density of particles will be 0 with high probability – but in position t and slightly to the left we may find a positive density of particles trailing the leading particle.

Similarly, we can interpret a vacant position as a type of particle, known as a **hole**, and consider the position of the first hole (counting from the left, that is, the hole that starts in position 0) – analogously and symmetrically to the first particle, it is traveling *to the left* at random times with rate 1.

In fact, the time for the first hole to move i steps to the left is exactly $G(i, 1) = \sum_{k=1}^{i} \tau_{k,1}$, so by a similar reasoning, the position $Y_1^*(t)$ of the first hole at time t satisfies $Y_1^*(t)/t \to -1$ almost surely as $t \to \infty$. So, to the left of $(-1 + o(1))t$ we expect to find only particles at time t: the "traffic jam" hasn't begun to unwind this far to the left, but a bit further to the right we will find some vacant sites.

Rost's discovery was that as one travels from position $-t$ to position t, the density of particles decreases from 1 to 0 in the simplest way one might imagine, namely as a linear function. The precise result is as follows.

Theorem 4.20 (Limiting density profile in Rost's particle process) *Define a function $h: \mathbb{R} \to [0, 1]$ by*

$$h(x) = \begin{cases} 1 & \text{if } x < -1, \\ \frac{1}{2}(1 - x) & \text{if } -1 \le x \le 1, \\ 0 & \text{if } x > 1. \end{cases} \tag{4.48}$$

Then h describes the limiting particle density profile in Rost's particle process. More precisely, for any $-\infty < a < b < \infty$ we have the almost sure convergence

$$\frac{1}{t} N_t[at, bt] \to \int_a^b h(x)\, dx \quad \text{as } t \to \infty. \tag{4.49}$$

Proof It is easy to see that it is enough to prove (4.49) for arbitrary $a \in \mathbb{R}$ and $b = \infty$. If $a > 1$ then, by the preceding comments on the asymptotic position of the leading particle, with probability 1 we'll have $N_t[at, \infty) = 0$ for large enough t, so indeed (4.49) holds. Similarly, if $a < -1$ then with probability 1, for all t large enough the relation $N_t[at, \infty) = 1 + \lfloor -at \rfloor$ will hold, since no particle to the left of position $\lfloor at \rfloor$ will have moved by time t. So, we get that almost surely,

$$\frac{1}{t} N_t[at, \infty) \xrightarrow[t \to \infty]{} -a = \int_{-a}^{\infty} h(x)\, dx.$$

Finally, consider $-1 \le a \le 1$. Note that for any $k \in \mathbb{Z}$, any given particle $i \ge 1$ is counted in the number $N_t[k, \infty)$ of particles whose position at time t is $\ge k$ if and only if $G(i, k + i - 1) \le t$ (where $G(i, j)$ is interpreted as 0 for $j \le 0$); so, $N_t[k, \infty)$ is the maximal i satisfying this condition. The idea is to translate this into a statement about the corner growth process $(\Lambda(t))_{t \ge 0}$

associated with the passage times $(G(i, j))_{i,j\geq 1}$. The positions $(i, j) \in \mathbb{N}^2$ for which $G(i, j) \leq t$ correspond to the top-right corners of cells in the planar set $\text{set}_{\Lambda(t)}$, so, by Theorem 4.5, we have

$$\frac{1}{t}N_t[at, \infty) \xrightarrow[t\to\infty]{\text{a.s.}} \phi(a) := \max\{0 \leq x \leq 1 \, : \, \Psi(x, x + a) \leq 1\}$$
$$= \max\left\{0 \leq x \leq 1 \, : \, x + a \leq (1 - \sqrt{x})^2\right\}.$$

It is now easy to check that for $-1 \leq a \leq 1$ we have $\phi(a) = \frac{1}{4}(1 - a)^2 = \int_a^1 h(x)\, dx$ where h is defined in (4.48). □

There is another interesting way to look at Theorem 4.20. Instead of looking at the particle density profile at some fixed time, we can look how the profile evolves over time. That is, fixing some scaling factor n, we consider the family of random measures $(\nu_t^{(n)})_{t\geq 0}$ on $\mathbb{R}$ defined by

$$\nu_t^{(n)}(E) = N_{nt}(nE).$$

It can be easily checked (Exercise 4.5) that Theorem 4.20 implies the following result.

Theorem 4.21 (Rost's hydrodynamic limit theorem) *As $n \to \infty$, the random process $(\nu_t^{(n)})_{t\geq 0}$ converges almost surely to the deterministic family of measures $u(x, t)\, dx$, where*

$$u(x, t) = h\left(\frac{x}{t}\right) = \begin{cases} 1 & \text{if } x < -t, \\ \dfrac{t - x}{2t} & \text{if } -t \leq x \leq t, \\ 0 & \text{if } x > t, \end{cases}$$

in the following precise sense: for any $-\infty < a < b < \infty$ and $t \geq 0$ we have

$$\nu_t^{(n)}[a, b] \to \int_a^b u(x, t)\, dx \quad \text{almost surely as } n \to \infty.$$

Conceptually, the meaning of such a "hydrodynamic" limit is that we imagine that as the scaling of time and space becomes finer and finer, the random movement of particles converges to a continuous (and nonrandom) "flow" and the picture resembles that of a continuous fluid. Furthermore, the fluid obeys a continuous dynamical law: it can be checked easily that the function $u(x, t)$ satisfies the partial differential equation (PDE)

$$\frac{\partial u(x,t)}{\partial t} = -\frac{\partial}{\partial x}\left(u(x,t)(1-u(x,t))\right), \tag{4.50}$$

a version of a PDE from fluid dynamics known as the **inviscid Burgers' equation**. (At least, u satisfies the PDE wherever its partial derivatives exist. One needs to be careful around the points of nondifferentiability, but even there one can say the equation (4.50) is satisfied if one assumes the correct notion of a solution.) Note that $u(x, t)$ satisfies the initial condition $u(x, 0) = \mathbf{1}_{(-\infty,0)}(x)$. In the theory of Burgers' equation, an initial condition with this type of jump discontinuity is known as a **shock initial condition** – it is the continuous analogue of the front of a traffic jam.

The reason the PDE (4.50) gives a useful way of looking at the hydrodynamic limit theorem (which is equivalent to the limit shape theorem in the corner growth process) is that the preceding results can be generalized significantly, and in the more general formulation it is the equation (4.50) which plays the central role. We have focused our attention on Rost's particle process, which is an interacting particle system with a particular initial condition. One can consider more generally the totally asymmetric simple exclusion process (or TASEP – see the box on p. 248), which is a system of interacting particles on $\mathbb{Z}$ evolving according to the same stochastic rules as Rost's process, except with an arbitrary initial state (formally, the configuration space from which the initial state can be chosen and in which the system evolves is $\{0, 1\}^{\mathbb{Z}}$). One can then imagine a scenario in which the initial state is chosen as a discrete approximation to some real-valued density profile $u_0 \colon \mathbb{R} \to [0, 1]$. It turns out that, under mild assumptions, the subsequent evolution of the system converges to the solution of the equation (4.50) with initial condition $u(x, 0) = u_{(}x)$. The precise formulation of this elegant result and its proof are beyond the scope of this book; more details can be found in [115].

We conclude this section by posing another natural question about Rost's process which can be easily answered using the tools we developed. This time we focus our attention on the motion of individual particles: for each $k \geq 1$, let $Y_k(t)$ denote the position of particle number k at time t. For fixed k, we refer to the random function $(Y_k(t))_{t\geq 0}$ as the **trajectory of particle** k. It is natural to ask what is the asymptotic behavior of $Y_k(t)$. Because of the scaling relation between the time and space coordinate, we consider this question not for fixed k (where the answer turns out to be trivial: the kth particle eventually moves forward with asymptotic speed 1), but rather for k that scales linearly with the length of the time interval in which we are interested.

More precisely, if n is a discrete parameter that will grow large, we will take $k \approx xn$ for some fixed $x \geq 0$, and simultaneously speed up time by a factor of n, to get a properly scaled particle trajectory $(Y_k(nt))_{t\geq 0}$. This leads to the following result, which is illustrated in Fig. 4.6.

Theorem 4.22 (Limiting particle trajectories in Rost's particle process) *Define a function* $F\colon [0,\infty)^2 \to \mathbb{R}$ *by*

$$F(t,x) = \begin{cases} -x & \text{if } 0 \leq t \leq x, \\ t - 2\sqrt{xt} & \text{if } t \geq x. \end{cases}$$

For any $t, x \geq 0$, *if* $k = k(n)$ *is a sequence of positive integers such that* $k(n)/n \to x$ *as* $n \to \infty$, *then we have the almost sure convergence*

$$\frac{1}{n} Y_{k(n)}(nt) \to F(t,x) \quad \text{as } n \to \infty.$$

Proof Since particle k starts in position $-k+1$, we have the relation

$$Y_k(t) = -k + 1 + \max\{m \geq 0 \,:\, G(k,m) \leq t\},$$

so

$$\begin{aligned} \frac{1}{n} Y_{k(n)}(nt) &= \frac{-k(n)+1}{n} + \frac{1}{n} \max\{m \geq 0 \,:\, G(k(n),m) \leq t\} \\ &= -x + \frac{1}{n} \max\{m \geq 0 \,:\, G(n \cdot k(n)/n, n \cdot m/n) \leq t\} + o(1). \end{aligned} \tag{4.51}$$

If $0 \leq t < x$ then using Theorem 4.13, for any $m \geq 1$ we have

$$\frac{1}{n} G(n \cdot k(n)/n, n \cdot m/n) \geq \frac{1}{n} G(n \cdot k(n)/n, 1) \xrightarrow[n\to\infty]{} \Psi(x,0) > t,$$

so almost surely, for large enough values of n the maximum on the right-hand side of (4.51) will be attained when $m = 0$, and we get that $\frac{1}{n} Y_{k(n)}(nt) \to -x = F(t,x)$ almost surely.

In the other case where $t \geq x$, by Theorem 4.13 and obvious monotonicity considerations which we leave to the reader to make precise, the last

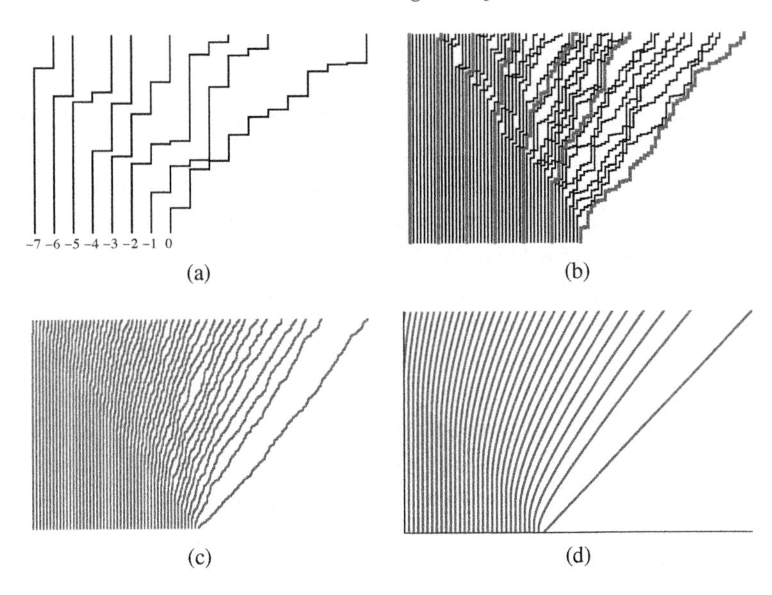

Figure 4.6 Simulated particle trajectories in Rost's particle process at several scales: (a) 8 particles; (b) 50 particles; (c) 400 particles (every eighth particle trajectory is shown). (d) The limiting trajectories. (In all four figures the vertical axis represents time.)

expression is seen to converge almost surely as $n \to \infty$ to

$$-x + \max\{y \geq 0 \,:\, \Psi(x, y) \leq t\} = -x + (\sqrt{t} - \sqrt{x})^2,$$

which is again equal to $F(t, x)$. □

4.8 The multicorner growth process

We now turn to another direction in which the results of the previous sections can be extended. The last-passage percolation formalism we developed suggests a large family of processes in which the random clock times $\tau_{i,j}$ can be a family of i.i.d. random variables sampled from some arbitrary distribution on $[0, \infty)$. Some results, and in particular an asymptotic shape theorem analogous to Theorem 4.13 based on a use of the subadditive ergodic theorem, can be proved in this generality (see [115],

Theorem 2.1). Unfortunately, the exponential distribution is almost the only case for which the limit shape can be found explicitly (and, as we see in the next chapter, for which many other quantities of interest can be computed precisely).

It turns out however that there is another family of distributions other than the exponential distributions for which the last-passage percolation model can be analyzed in a similar level of detail and precision. These are the geometric distributions, the discrete-time analogues of the exponential distribution. Furthermore, the last-passage percolation process associated with the geometric distributions has an interpretation as a growth process of random Young diagrams that is itself very natural. It is also satisfying that this process can be analyzed in a manner almost identical to the analysis of the original corner growth process – all the proofs remain valid after making a few small modifications that we will point out.

Let us start by describing the random growth process, which we call the **multicorner growth process**, and then explain the connection to geometric clock times. Fix a parameter $0 < p < 1$ that will remain constant throughout the discussion, and will be referred to as the **growth rate parameter**. The multicorner growth process is a growing family $(\lambda^{(n)})_{n=0}^{\infty}$ of random Young diagrams that evolves (in discrete time) according to the following rules: first, $\lambda^{(0)} = \emptyset$, the empty diagram; second, for each n, the $(n+1)$th diagram $\lambda^{(n+1)}$ is obtained from $\lambda^{(n)}$ by tossing a coin with bias p independently for each external corner (i, j) of $\lambda^{(n)}$ and adding the cell in position (i, j) if the coin toss was successful. Thus, at any transition from time n to $n+1$, the number of new cells that get added to the diagram is random and varies between 0 and the number of external corners of the nth diagram.

Formally, we define the process as a Markov chain with an explicitly given transition rule. For two Young diagrams μ, λ, denote $\mu \nearrow\!\!\!\nearrow \lambda$ if $\mu \subseteq \lambda \subseteq \mu \cup \text{ext}(\mu)$, that is, if λ can be obtained from μ by adding to μ a set of new cells consisting of some subset of its external corners. The transition rule for the process $(\lambda^{(n)})_{n=0}^{\infty}$ is given by

$$\mathbb{P}\left(\lambda^{(n+1)} = \mu_{n+1} \,\middle|\, \lambda^{(0)} = \mu_0, \ldots, \lambda^{(n)} = \mu_n\right) = p^k (1-p)^{e-k} \tag{4.52}$$

for any Young diagrams $\mu_0, \ldots, \mu_{n+1}$ satisfying $\emptyset = \mu_0 \nearrow\!\!\!\nearrow \ldots \nearrow\!\!\!\nearrow \mu_{n+1}$, provided that the conditioning event has positive probability, and where we denote $e = |\operatorname{ext}(\mu_n)|$ (the number of external corners of μ_n) and $k = |\mu_{n+1}|-|\mu_n|$ (the number of cells that need to be added to get from μ_n to μ_{n+1}).

Note that in the special case $p = 1/2$, an equivalent description of the growth rule would be that a subset of the external corners of $\lambda^{(n)}$ is chosen uniformly at random, and those cells get added to the diagram. This corresponds to a random walk on a graph that extends the Young graph by adding self-loops and additional edges to account for the possibility of adding more than one box at a time (i.e., the directed graph on Young diagrams with adjacency relation $\mu \nearrow\!\!\nearrow \lambda$). In the case $p \neq 1/2$, the growth process can be interpreted as a *weighted* random walk on the same graph. In any case, the description using coin-tossing seems more elegant.

The definition of the multicorner growth process given above is quite intuitive, but as with the case of the corner growth process, to get a good understanding of the process we need to represent it in terms of last-passage percolation. In this case there is no need to change the time parametrization, since the discrete time parametrization we used in the definition is already the correct one needed for the analysis. The connection to last-passage percolation is explained in the following theorem.

Theorem 4.23 *For each* $(i, j) \in \mathbb{N}^2$, *denote*

$$G(i, j) = \min\left\{n \geq 0 \ :\ (i, j) \in \lambda^{(n)}\right\}.$$

The array of random variables $(G(i, j))_{i,j\geq 1}$ *is equal in distribution to the array of passage times in last-passage percolation defined by the recurrence* (4.12) *from a family of i.i.d. clock times* $(\tau_{i,j})_{i,j\geq 1}$ *with the geometric distribution* Geom(p).

Proof Start with the passage times $(G(i, j))_{i,j\geq 1}$ defined by (4.12) from i.i.d. Geom(p)-distributed clock times $(\tau_{i,j})_{i,j\geq 1}$. We can associate with these passage times a growing family $(\nu^{(n)})_{n=0}^{\infty}$ of Young diagrams, defined by

$$\nu^{(n)} = \left\{(i, j) \in \mathbb{N}^2 \ :\ G(i, j) \leq n\right\}.$$

To prove the theorem it will be enough to show that the family $(\nu^{(n)})_{n=0}^{\infty}$ is equal in distribution to the multicorner growth process as defined previously, that is, that it satisfies $\nu^{(0)} = \emptyset$ (which it does, trivially, since $G(i, j) \geq 1$ for all i, j) and the transition rule (4.52).

First, as a trivial case of (4.52) we need to check that growth from $\lambda^{(n)} =$

μ to $\lambda^{(n+1)} = \lambda$ can occur only if $\mu \nearrow \lambda$. This is true since the fact that $\tau_{i,j} \geq 1$ for all i, j implies that the array of passage times $G(i, j)$ is strictly increasing by integer increments along rows and columns, so if $G(i, j) \leq n + 1$ then $G(i-1, j) \vee G(i, j-1) \leq n$ and therefore (i, j) is an external corner of $\nu^{(n)}$.

Second, fix deterministic Young diagrams $\emptyset = \mu_0 \nearrow \ldots \nearrow \mu_n$. We proceed as in the proof of Lemma 4.3. Define an event E by

$$E = \left\{ \nu^{(j)} = \mu_j,\ j = 0, \ldots, n \right\},$$

and assume that E has positive probability. Denote $K = \operatorname{ext}(\mu_n)$. Note that the occurrence of the event E can be decided by looking at the family of clock times $\boldsymbol{\tau} = (\tau_{i,j})_{(i,j) \in \mu_n \cup K}$. We again consider the vector $\boldsymbol{\tau}$ to be made up of two vector components $\boldsymbol{\tau}_{\text{in}} = (\tau_{p,q})_{(p,q) \in \mu_n}$ and $\boldsymbol{\tau}_{\text{out}} = (\tau_{i,j})_{(i,j) \in K}$. For each $i, j \geq 1$ denote by $\kappa(i, j)$ the minimal index k for which $(i, j) \in \lambda^{(k)}$, and note that inside the event E, $\kappa(i, j)$ is constant for each $(i, j) \in \mu_n \cup K$.

Now observe that E can be represented in the form

$$E = \{\boldsymbol{\tau}_{\text{in}} \in A\} \cap \{\tau_{i,j} > n - \kappa(i, j) \text{ for all } (i, j) \in K\}$$

for some nonempty set $A \subset \mathbb{N}^{\mu_n}$. The reason why such a representation is possible is easy to see: the condition in the first event in the intersection is an event designed to guarantee that for any $0 \leq k \leq n$ and $(p, q) \in \mu_n$, $G(p, q) = k$ if and only if $(p, q) \in \mu_k \setminus \mu_{k-1}$, and the condition in the second event ensures that $G(i, j) > n$ for all $(i, j) \in K$. For similar reasons, we have a representation for the event $E \cap \{\lambda^{(n+1)} = \mu_{n+1}\}$ in the form

$$\begin{aligned} E \cap \{\nu^{(n+1)} = \mu_{n+1}\} = \{\boldsymbol{\tau}_{\text{in}} \in A\} & \\ \cap \{\tau_{i,j} &= n + 1 - \kappa(i, j) \text{ for all } (i, j) \in \mu_{n+1} \setminus \mu_n\} \\ \cap \{\tau_{i,j} &> n + 1 - \kappa(i, j) \text{ for all } (i, j) \in K \setminus \mu_{n+1}\} \end{aligned}$$

(where, importantly, the set A is the same in both representations). We therefore have that

$$\mathbb{P}\left(\nu^{(n+1)} = \mu_{n+1} \,|\, E\right) = \frac{\mathbb{P}\left(E \cap \left\{\nu^{(n+1)} = \mu_{n+1}\right\}\right)}{\mathbb{P}(E)}$$

$$= \frac{\prod_{(i,j)\in\mu_{n+1}\setminus\mu_n} \mathbb{P}(\tau_{i,j} = n+1-\kappa(i,j)) \prod_{(i,j)\in K\setminus\mu_{n+1}} \mathbb{P}(\tau_{i,j} > n+1-\kappa(i,j))}{\prod_{(i,j)\in K} \mathbb{P}(\tau_{i,j} > n-\kappa(i,j))}$$

$$= \frac{\prod_{(i,j)\in\mu_{n+1}\setminus\mu_n} p(1-p)^{n-\kappa(i,j)} \prod_{(i,j)\in K\setminus\mu_{n+1}} (1-p)^{n+1-\kappa(i,j)}}{\prod_{(i,j)\in K} (1-p)^{n-\kappa(i,j)}} = p^{|\mu_{n+1}\setminus\mu_n|}(1-p)^{|K\setminus\mu_{n+1}|},$$

which is exactly (4.52). □

Our final goal for this chapter is to prove the following limit shape theorem for the multicorner growth process, which is analogous to Theorem 4.5.

Theorem 4.24 (Limit shape theorem for the multicorner growth process[4]) *Let* $(\lambda^{(n)})_{n=0}^{\infty}$ *denote as above the multicorner growth process with growth rate parameter* $0 < p < 1$. *Define a set* $\Delta_{\mathrm{MCG}}^p \subset \mathbb{R}^2$ *by*

$$\Delta_{\mathrm{MCG}}^p = \left\{(x,y) \,:\, x, y \geq 0,\ x + y + 2\sqrt{(1-p)xy} \leq p\right\}. \tag{4.53}$$

As $n \to \infty$, *the scaled planar set* $\frac{1}{n}\mathrm{set}_{\lambda^{(n)}}$ *converges in probability to* Δ_{MCG}^p. *That is, for any* $\epsilon > 0$, *we have*

$$\mathbb{P}\left((1-\epsilon)\Delta_{\mathrm{MCG}}^p \subset \frac{1}{n}\mathrm{set}_{\lambda^{(n)}} \subset (1+\epsilon)\Delta_{\mathrm{MCG}}^p\right) \to 1 \text{ as } n \to \infty.$$

The curved part of the boundary of the shape Δ_{MCG}^p is an arc of an ellipse whose principal axes are parallel to the u–v rotated coordinate axes. Fig. 4.7 shows the limit shapes and the associated ellipses for various values of p. Exercises 4.12–4.13 sketch a delightful and surprising application of this result to the study of random domino tilings.

The approach to proving Theorem 4.24 follows exactly the same steps that were used in our proof of Theorem 4.5: first, look at generalized passage times that correspond to a slowed down version of the process for which exact computations are possible. This provides bounds on the passage times in the original process, from which we can deduce that the pas-

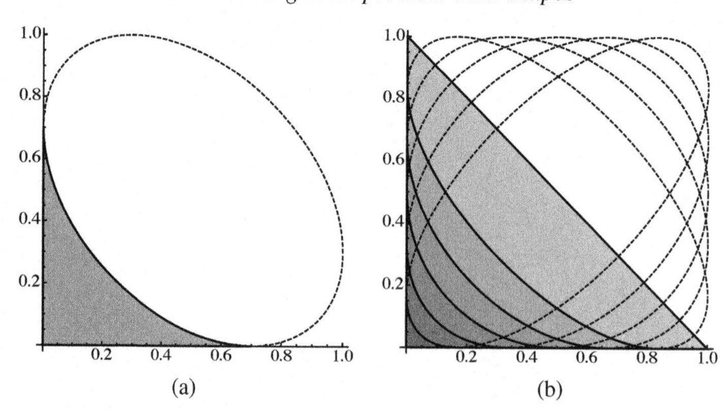

Figure 4.7 (a) The limit shape Δ^p_{MCG} when $p = 0.7$ and the ellipse whose arc forms the curved part of the boundary of the limit shape. (b) The limit shapes (and associated ellipses) for $p = 1/6, 1/3, 1/2, 2/3, 5/6$. The triangle shape corresponds to the limit as $p \nearrow 1$ of the shapes. For $p = 1/2$ the ellipse is the circle of radius $1/2$ centered at $(1/2, 1/2)$.

sage times converge in a suitable scaling limit to an unknown limiting function $\Phi_p(x, y)$. Finally, the connection between the slowed down process and the original process can be used to derive an equation relating $\Phi_p(x, y)$ to the analogous, explicitly known, function for the slowed down process. Solving this equation for $\Phi_p(x, y)$ turns out to be equivalent to inverting a Legendre transform.

We start with the definition of the slowed down version of the process. Fix a parameter r satisfying $0 < r < p$ (which will be analogous to the parameter α we used for the slowed down corner growth process). Analogously to the case of the exponential clock times used before, we augment the array of i.i.d. geometric clock times $(\tau_{i,j})_{i,j\geq 1}$ with additional clock times for the row and column with index 0, which are taken to be random variables $\tau_{0,0}, \tau_{k,0}, \tau_{0,k}$ for $k \geq 1$ (independent of each other and of the existing clock times) whose distributions are given by

$$\tau_{0,0} = 0, \tag{4.54}$$

$$\tau_{i,0} \sim \text{Geom}(r) \qquad (i \geq 1), \tag{4.55}$$

$$\tau_{0,j} \sim \text{Geom}((p - r)/(1 - r)) \qquad (j \geq 1). \tag{4.56}$$

Denote as before in the augmented model

$$G_r(i, j) = G(0, 0; i, j)$$

and refer to $G_r(i, j)$ as the **slowed down passage time to** (i, j) (with parameter r). A key property that makes the analysis of the slowed down model possible is the following lemma, which is analogous to Lemma 4.25.

Lemma 4.25 *Let $0 < r < p < 1$. Let X, Y, Z be independent random variables such that $X \sim \text{Geom}(r)$, $Y \sim \text{Geom}((p-r)/(1-r))$ and $Z \sim \text{Geom}(p)$. Define random variables U, V, W by*

$$U = Z + (X - Y)^+,$$
$$V = Z + (Y - X)^+,$$
$$W = X \wedge Y.$$

Then we have the equality in distribution $(U, V, W) \stackrel{d}{=} (X, Y, Z)$.

Proof Let T be the transformation defined in (4.22). If $u, v, w \geq 1$ are integers, then, by the properties of T discussed in the proof of Lemma 4.10, when $u \leq v$ we have that

$$\begin{aligned}
\mathbb{P}(U = u, V = v, W = w) &= \mathbb{P}(Z = u, X = w, Y = v + w - u) \\
&= p(1-p)^{u-1} \cdot r(1-r)^{w-1} \cdot \frac{p-r}{1-r}\left(\frac{1-p}{1-r}\right)^{v+w-u-1} \\
&= r(1-r)^{u-1} \cdot \frac{p-r}{1-r}\left(\frac{1-p}{1-r}\right)^{v-1} \cdot p(1-p)^{w-1} \\
&= \mathbb{P}(X = u, Y = v, Z = w).
\end{aligned}$$

Similarly, when $u > v$, we have

$$\begin{aligned}
\mathbb{P}(U = u, V = v, W = w) &= \mathbb{P}(Z = v, Y = w, X = u + w - v) \\
&= p(1-p)^{v-1} \cdot r(1-r)^{u+w-v-1} \cdot \frac{p-r}{1-r}\left(\frac{1-p}{1-r}\right)^{w-1} \\
&= r(1-r)^{u-1} \cdot \frac{p-r}{1-r}\left(\frac{1-p}{1-r}\right)^{v-1} \cdot p(1-p)^{w-1} \\
&= \mathbb{P}(X = u, Y = v, Z = w).
\end{aligned}$$

□

Consider as before the x- and y-increments of the slowed down passage times, defined by

$$\begin{aligned} X(i,j) &= G_r(i,j) - G_r(i-1,j), \\ Y(i,j) &= G_r(i,j) - G_r(i,j-1), \end{aligned} \qquad (i,j \geq 0).$$

Lemma 4.26 *For any down-right path $((i_n, j_n))_{n\in\mathbb{Z}}$, let the associated family of random variables $(W_n)_{n\in\mathbb{Z}}$ be defined as in (4.23) in terms of the increment random variables $X(i,j), Y(i,j)$. Then $(W_n)_{n\in\mathbb{Z}}$ is a family of independent random variables, and for each n the distribution of W_n is given by*

$$W_n \sim \begin{cases} \mathrm{Geom}(r) & \text{if } (i_n, j_n) - (i_{n-1}, j_{n-1}) = (1,0), \\ \mathrm{Geom}((p-r)/(1-r)) & \text{if } (i_n, j_n) - (i_{n-1}, j_{n-1}) = (0,-1). \end{cases}$$

Proof The proof is identical to the proof of Lemma 4.11, except that it uses the definition (4.54)–(4.56) of the slowed down process that is associated with the multicorner growth process, and uses Lemma 4.25 where the proof of Lemma 4.11 made use of Lemma 4.10. □

From these results we get the following conclusions analogous to the ones we showed in Section 4.4. First, we get a formula for the expected value of the slowed down passage times:

$$\mathbb{E}G_r(i,j) = \frac{i}{r} + \frac{(1-r)j}{p-r}. \tag{4.57}$$

Second, by choosing an arbitrary value of r (for example $r = p/2$) in (4.57) and using monotonicity we get an upper bound for the original passage times which has the form

$$\mathbb{E}G(i,j) \leq C_p(i+j) \qquad (i,j \geq 1), \tag{4.58}$$

where $C_p > 0$ is a constant that depends on p. This will be needed for the application of the subadditive ergodic theorem. Third, we have a result on the asymptotic behavior of the slowed down passage times, whose proof is completely analogous to Theorem 4.12 and is omitted.

Theorem 4.27 *For any $x, y > 0$ we have the almost sure convergence*

$$\frac{1}{n} G_r(\lfloor nx \rfloor, \lfloor ny \rfloor) \to \Phi_{p,r}(x,y) := \frac{x}{r} + \frac{(1-r)y}{p-r} \quad \text{as } n \to \infty. \tag{4.59}$$

Next, with the aid of these results, we can prove a result on the asymptotics of the passage times in the original process.

Theorem 4.28 (Asymptotics of the passage times in the multicorner growth process) *There exists a function* $\Phi_p \colon [0,\infty)^2 \to [0,\infty)$ *such that for all* $x, y \geq 0$, *if* $((i_n, j_n))_{n=1}^{\infty}$ *is a sequence of positions in* $\mathbb{N}^2$ *such that* $\frac{1}{n}(i_n, j_n) \to (x, y)$ *as* $n \to \infty$, *then we have the almost sure convergence*

$$\frac{1}{n} G(i_n, j_n) \to \Phi_p(x, y) \text{ as } n \to \infty. \tag{4.60}$$

The function Φ_p *satisfies the same list of properties as the function* Ψ *in Theorem 4.13.*

Proof The proof is identical to the proof of Theorem 4.13, except for the following modification: to prove that Φ_p can be extended to a continuous function on the x- and y-axes, observe that Φ_p satisfies the upper bound

$$\Phi_p(x, y) \leq \Phi_{p,r}(x, y) = \frac{x}{r} + \frac{y(1-r)}{p-r},$$

analogously to (4.32). To get the best bound for a given value of p, take $r_* = p\sqrt{x}/(\sqrt{x} + \sqrt{(1-p)y})$. For this value, we get after a short computation that

$$\Phi_p(x, y) \leq \frac{1}{p}\left(x + y + 2\sqrt{(1-p)xy}\right) \tag{4.61}$$

(As in the discussion of the limiting function Ψ, it will turn out at the end of the analysis that the right-hand side of (4.61) is the correct expression for Φ_p.) On the other hand, we have the fairly trivial lower bound

$$\Phi_p(x, y) \geq \frac{x}{p} \vee \frac{y}{p}$$

that follows from the strong law of large numbers in an analogous manner to (4.34). The two bounds are asymptotically sharp as (x, y) approaches the axes, so Φ_p extends to a continuous function on $[0,\infty)^2$ that satisfies

$$\Phi_p(z, 0) = \Phi_p(0, z) = \frac{z}{p}. \qquad \square$$

Using Theorem 4.28, one can prove the existence of a limit shape as we did in Theorem 4.15. The proof of the following result is similar to the proof of that result and is omitted.

Theorem 4.29 *Define a set*

$$D_{\mathrm{MCG},p} = \Big\{(x,y) \in [0,\infty)^2 \ :\ \Phi_p(x,y) \le 1\Big\}. \tag{4.62}$$

As $n \to \infty$, the scaled planar set $n^{-1}\mathrm{set}_{\lambda^{(n)}}$ converges in probability to the set $D_{\mathrm{MCG},p}$.

It remains to find the explicit form of the function Φ_p. As before, the idea is to use the relationship between the slowed down and original passage times to obtain an equation relating Φ_p to the functions $\Phi_{p,r}$, which can then be solved by inverting a Legendre transform. The equation, which is analogous to (4.38) and is proved in exactly the same manner, is as follows.

Theorem 4.30 *For any $x, y \ge 0$ and $0 < \alpha < 1$ we have*

$$\Phi_{p,r}(x,y) = \max_{0\le u\le x}\left\{\frac{u}{r} + \Phi_p(x-u,y)\right\} \\ \vee \max_{0\le v\le x}\left\{\frac{(1-r)v}{p-r} + \Phi_p(x,y-v)\right\}. \tag{4.63}$$

Once again, it will be enough to consider (4.63) in the case $x = y = 1$, where it takes the form

$$\frac{1}{r} + \frac{1-r}{p-r} = \Phi_{p,r}(1,1) \\ = \max_{0\le u\le 1}\left\{\frac{u}{r} + \Phi_p(1-u,1)\right\} \vee \max_{0\le u\le 1}\left\{\frac{(1-r)u}{p-r} + \Phi_p(1,1-u)\right\}.$$

Because of the symmetry of Φ_p, it is sufficient to consider the range of values $0 < r \le 1 - \sqrt{1-p}$, which are the values (check!) where $1/r \ge (1-r)/(p-r)$ and therefore the first maximum dominates the second. Denoting a function $f_p(u) = -\Phi_p(1-u,1)$ and setting $s = 1/r$, we get the equation

$$s + \frac{s-1}{ps-1} = \max_{0\le u\le 1}\left\{su - f_p(u)\right\} \qquad \left(s \ge \frac{1}{p}\left(1+\sqrt{1-p}\right)\right), \tag{4.64}$$

which we recognize as having an obvious connection to a statement about Legendre transforms.

The function f_p is a monotone nondecreasing, continuous and convex function on $[0,1]$ that satisfies $f_p(1) = -1/p$. Extend it to a function on $\mathbb{R}$

by setting

$$\tilde{f}_p(u) = \begin{cases} f_p(u) & \text{if } 0 \le u \le 1, \\ \infty & \text{otherwise.} \end{cases}$$

Then we can rewrite (4.64) in the from

$$s + \frac{s-1}{ps-1} = \sup_{u\in\mathbb{R}} \left\{ su - \tilde{f}_p(u) \right\} \qquad \left(s \ge \frac{1}{p}\left(1 + \sqrt{1-p}\right) \right).$$

The extended function $\tilde{f}_p$ is lower-semicontinuous and convex. Its Legendre transform $h_p(s) = \tilde{f}_p^*(s) = \sup_{u\in\mathbb{R}}\{su - \tilde{f}_p(u)\}$ is a convex function that satisfies $h_p(s) = s + (s-1)/(ps-1)$ for $s \ge \frac{1}{p}\left(1 + \sqrt{1-p}\right)$. Furthermore, it is monotone nondecreasing (since the supremum in its definition is always attained for nonnegative values of u), so, since its right-derivative at the point $s_0 = \frac{1}{p}\left(1 + \sqrt{1-p}\right)$ satisfies

$$h_p'(s_0+) = \frac{d}{ds}\bigg|_{s=s_0} \left(s + \frac{s-1}{ps-1} \right) = 1 - \frac{1-p}{(ps_0-1)^2} = 0,$$

it follows by an argument similar to that used in Section 4.6 that $h_p(s) = h_p(s_0) = 2s_0$ for $s \le s_0$. That is, we have

$$h_p(s) = \begin{cases} s + \dfrac{s-1}{ps-1} & \text{if } s \ge \frac{1}{p}\left(1 + \sqrt{1-p}\right), \\ \frac{2}{p}\left(1 + \sqrt{1-p}\right) & \text{if } s \le \frac{1}{p}\left(1 + \sqrt{1-p}\right). \end{cases}$$

Applying Theorem 4.17, we can recover f_p as the Legendre transform of h_p. This leads (after some trivial examination of end cases which we leave to the reader) to the calculus problem of finding the value of s that maximizes $us - s - \frac{s-1}{ps-1}$. The solution is

$$s_* = \frac{1}{p}\left(1 + \frac{\sqrt{1-p}}{\sqrt{1-u}} \right)$$

and the maximum value is $us_* - s_* - \frac{s_*-1}{ps_*-1}$, which after a short computation results in

$$f_p(u) = -\frac{1}{p}\left(2 - u + 2\sqrt{(1-p)(1-u)}\right).$$

From this result, we can obtain, arguing exactly as in Section 4.6, the following result.

Theorem 4.31 *The function Φ_p is given by*

$$\Phi_p(x, y) = \frac{1}{p}\left(x + y + 2\sqrt{(1-p)xy}\right). \tag{4.65}$$

Combining (4.65) with Theorem 4.29 finishes the proof of the limit shape theorem for the multicorner growth process, Theorem 4.24, which was our goal.[5]

Exercises

4.1 (☕☕) Given numbers $\alpha_1, \ldots, \alpha_k > 0$, let $X_1, \ldots, X_k$ be independent random variables such that $X_j \sim \text{Exp}(\alpha_j)$ for $1 \le j \le k$. Define random variables

$$Y = \operatorname{argmin}(X_1, \ldots, X_k),$$
$$Z = \min(X_1, \ldots, X_k).$$

Show that Y and Z are independent, Z has distribution $\text{Exp}(\alpha_1 + \ldots + \alpha_k)$, and the distribution of Y is given by

$$\mathbb{P}(Y = j) = \frac{\alpha_j}{\alpha_1 + \ldots + \alpha_k} \qquad (1 \le j \le k).$$

4.2 (☕☕) Prove that if $a \in \mathbb{R}$ and $(X_n)_{n=1}^{\infty}$ is a sequence of random variables, defined on the same probability space, such that $X_n \to a$ in probability, then there is a subsequence $(X_{n_k})_{k=1}^{\infty}$ such that $X_{n_k} \to a$ almost surely.

4.3 (☕☕) Let $X_1, X_2, \ldots,$ be a sequence of i.i.d. random variables such that $\mathbb{E}X_1 = 0$ and $\mathbb{E}X_1^4 < \infty$. Prove that there exists a constant $C > 0$ such that for any $a > 0$ we have

$$\mathbb{P}\left(\left|\sum_{k=1}^{n} X_k\right| > a\right) \le \frac{C}{n^2 a^4}.$$

4.4 (☕☕) Using the notation of Section 4.4, define $Z_{i,j} = \min(X_{i+1,j}, Y_{i,j+1})$. By strengthening the reasoning employed in Lemma 4.11, prove that the random variables $(Z_{i,j})_{i,j=0}^{\infty}$ form a family of i.i.d. variables with distribution Exp(1).

4.5 (☕☕) Prove that Theorem 4.20 implies Theorem 4.21.

4.6 (☕) Prove that in the superadditivity inequality (4.28), we have equality if and only if the vectors (x_1, y_1) and (x_2, y_2) are linearly dependent.

4.7 (☕☕☕☕) (Mountford–Guiol [90]) Let $G(i, j)$ denote the passage times in the corner growth process, and let $\Psi(x, y) = (\sqrt{x} + \sqrt{y})^2$. Prove the existence of constants $C, c, \epsilon > 0$ such that for all $i, j \ge 1$ we have that

$$\mathbb{P}\left(\frac{1}{i+j}\left|G(i,j) - \Psi\left(\frac{i}{i+j}, \frac{j}{i+j}\right)\right| \ge (i+j)^{-\epsilon}\right) \le Ce^{-c(i+j)^{\epsilon}}.$$

4.8 Given positions $(a, b), (i, j) \in \mathbb{N}^2$ with $(a, b) \preceq (i, j)$, the (random) unique path in $\mathcal{Z}(a, b; i, j)$ for which the maximum in the definition (4.13) of the passage times in last-passage percolation is attained is called the **geodesic from** (a, b) **to** (i, j).

(a) (☕☕☕) Fix $x, y \geq 0$. Let $((i_n, j_n))_{n=1}^{\infty}$ be a sequence of positions in $\mathbb{N}^2$ such that $i_n/n \to x$, $j_n/n \to y$ as $n \to \infty$. Use the results of Exercises 4.6 and 4.7 above to show that as $n \to \infty$, the geodesics from $(1, 1)$ to (i_n, j_n) converge almost surely after scaling to a straight line from $(0, 0)$ to (x, y), in the following precise sense: if we denote the geodesic from $(1, 1)$ to (i_n, j_n) by $(p_k^n, q_k^n)_{k=0}^{a_n}$ (where clearly $a_n = i_n + j_n - 1$), then we have that

$$\max_{0 \leq k \leq a_n} \left\| \frac{1}{n}(p_k^n, q_k^n) - \frac{k}{a_n}(x, y) \right\| \xrightarrow[n\to\infty]{\text{a.s.}} 0.$$

(b) (☕☕☕) Formulate and prove an analogous statement for the limit shape of the geodesics from $(0, 0)$ to (i_n, j_n) in the slowed down last-passage percolation model with parameter $\alpha > 0$, defined in (4.15)–(4.17).

4.9 (☕☕) (Fristedt [46]) Given an integer partition $\lambda = (\lambda_1, \ldots, \lambda_m)$, for each $k \geq 1$ we denote by $\nu_k = \nu_k(\lambda)$ the number of parts of λ equal to k, and refer to ν_k as the **multiplicity of** k **in** λ. The numbers $(\nu_k)_{k=1}^{\infty}$ satisfy $\nu_k \geq 0$ for all k and $\sum_{k=1}^{\infty} k\nu_k = |\lambda|$. Conversely, it is easy to see that any sequence of numbers satisfying these conditions is the sequence of multiplicities for a unique partition.

Fix a number $0 < x < 1$. Let $N_1, N_2, \ldots$ be a sequence of independent random variables such that $N_k \sim \mathrm{Geom}_0(1 - x^k)$, that is, $\mathbb{P}(N_k = m) = (1 - x^k)x^{km}$ for $m \geq 0$, and denote $N = \sum_{k=0}^{\infty} kN_k$.

(a) Prove that $N < \infty$ almost surely.

(b) Let Λ be the random integer partition for which the part multiplicities are given by $N_1, N_2, \ldots$. Show that for any partition $\mu \in \mathcal{P}^*$ we have that

$$\mathbb{P}_x(\Lambda = \mu) = \frac{x^{|\mu|}}{F(x)},$$

where $F(x) = \prod_{k=1}^{\infty}(1 - x^m)^{-1}$. (The notation $\mathbb{P}_x$ emphasizes the dependence of the probability on the parameter x.) Deduce that

$$\mathbb{P}_x(N = n) = \frac{p(n)x^n}{F(x)} \qquad (n \geq 0), \tag{4.66}$$

where $p(n)$ denotes the number of partitions of n, and that for any $n \geq 0$, conditioned on the event $\{N = n\}$, Λ is uniformly distributed over the set $\mathcal{P}(n)$ of partitions of order n.

(c) Deduce from parts (a) and (b) above the identity

$$F(x) = 1 + \sum_{n=1}^{\infty} p(n)x^n.$$

(This is Euler's product formula, which was also proved in Exercise 1.16.)

(d) Show that the parts of the conjugate partition Λ', in the usual decreasing order, can be recovered using the relation

$$\Lambda'_k = \sum_{m=k}^{\infty} N_m \qquad (m \geq 1)$$

(with the convention that $\Lambda'_k = 0$ if k is greater than the number of parts of Λ').

4.10 (a) (☕☕☕) Let $x_n = e^{-\pi/\sqrt{6n}} = 1 - \frac{\pi}{\sqrt{6n}} + O\left(\frac{1}{n}\right)$. Using the notation of Exercise 4.9 above, show that when the parameter x in the definition of the random partition Λ is taken as x_n, when $n \to \infty$ we have that

$$\mathbb{E}_{x_n}(N) = n\left(1 + O\left(\frac{1}{\sqrt{n}}\right)\right), \tag{4.67}$$

$$\mathrm{Var}_{x_n}(N) = (1 + o(1))\frac{2\sqrt{6}}{\pi}n^{3/2}. \tag{4.68}$$

(b) (☕☕☕) Denote by $\Lambda^{(n)}$ the random partition Λ with the choice of parameter $x = x_n$. Show that $\Lambda^{(n)}$ satisfies the following limit shape result: if we define a function $\phi_n\colon [0, \infty) \to [0, \infty)$ encoding Λ as in (1.17) by

$$\phi_n(x) = n^{-1/2}\Lambda^{(n)}_{\lfloor n^{1/2}x+1\rfloor},$$

then we have the convergence in probability

$$\phi_n(x) \to -\frac{1}{c}\log\left(1 - e^{-cx}\right) \text{ as } n \to \infty,$$

where $c = \pi/\sqrt{6}$. Note that this result can be interpreted in terms of the convergence in probability of the planar set $\mathrm{set}_{\Lambda^{(n)}}$ to a limit set Δ_{Unif}, which has the symmetric description

$$\Delta_{\mathrm{Unif}} = \{(x, y) \,:\, x, y \geq 0, e^{-cx} + e^{-cy} = 1\}.$$

However, the convergence will be in a weaker sense than that of the limit shape theorems we proved for the Plancherel growth process and corner growth process, since the set Δ_{Unif} has horizontal and vertical asymptotes so is unbounded; see Fig. 4.8.

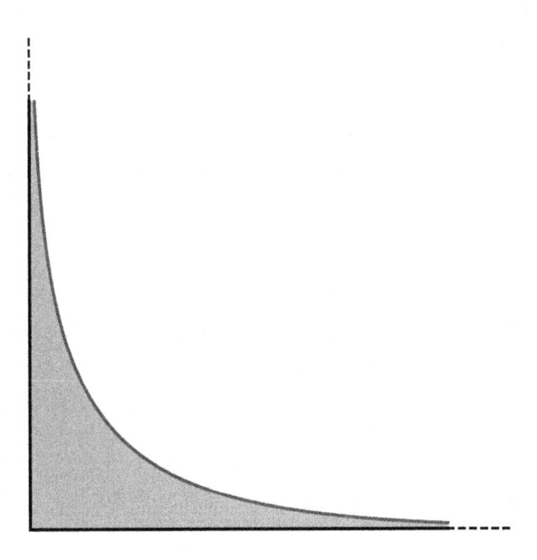

Figure 4.8 The limit shape Δ_{Unif} of uniformly random partitions.

(c) (☕☕☕) Show that the partition generating function $F(x)$ has the asymptotic expansion

$$\log F(e^{-s}) = \frac{\pi^2}{6s} + \frac{1}{2}\log s - \frac{1}{2}\log(2\pi) + o(1) \tag{4.69}$$

as $s \searrow 0$. (A proof can be found in [91].)

(d) (☕☕☕) Show that as $n \to \infty$ we have that

$$\mathbb{P}_{x_n}(N = n) = (1 + o(1))\frac{1}{\sqrt{2\pi \operatorname{Var}_{x_n}(N)}}. \tag{4.70}$$

Note that, in view of the relations (4.67) and (4.68), this can be thought of as the statement that the random variable N, defined as the sum of many independent components, satisfies a "local central limit theorem at 0"; the technique of proof should use the standard Fourier-analytic method of proving local central limit theorems – see, e.g., chapter 3 of [33] – but is technically hard because the variables being summed are not identically distributed, which turns the problem into a somewhat challenging exercise in complex analysis.

(e) (☕☕) Combine the relations (4.66), (4.69), and (4.70) to obtain a proof of the Hardy–Ramanujan asymptotic formula (1.16) for $p(n)$.

(f) (☕☕☕☕) (Temperley [133], Szalay–Turán [130], Vershik [141]) For each $n \geq 1$, let $\lambda^{(n)}$ denote a uniformly random partition of n. Prove that the random partitions $\lambda^{(n)}$ have the same limit shape Δ_{Unif} as the random partitions $\Lambda^{(n)}$; that is, they satisfy the same result as in part (b) above.

4.11 (☕☕☕☕☕) A **plane partition** is the three-dimensional analogue of a Young diagram. Formally, if $n \geq 0$ is an integer, a plane partition of order n is an infinite array $(a_{i,j})_{i,j\geq 1}$ of nonnegative integers such that $n = \sum_{i,j} a_{i,j}$ and such that $a_{i,j} \geq a_{i+1,j} \vee a_{i,j+1}$ for all $i, j \geq 1$. Graphically one can visualize a plane partition as a stack of n unit cubes stacked against a corner of three large walls in a monotonically decreasing fashion; $a_{i,j}$ corresponds to the height of the stack of cubes above the unit square $[i-1, i] \times [j-1, j] \times \{0\}$.
Find a formula for the limit shape for the corner growth process in *three* dimensions, corresponding to a random walk on the graph of plane partitions in which one starts with the empty plane partition of order 0 and successively adds cubes, each time choosing the position of where to add the new cube uniformly at random from the available positions.

4.12 A **domino** is a rectangle in the plane, aligned with the x- and y-axes and having dimensions 2×1 or 1×2. A representation of a set $S \subset \mathbb{R}^2$ as a union of dominos with disjoint interiors is called a **domino tiling** of S.

(a) (☕) Show that the number of domino tilings of the $2\times n$ rectangle $[0, n]\times [0, 2]$ is F_{n+1}, the $(n+1)$th Fibonacci number.

(b) (☕☕☕) (Elkies–Kuperberg–Larsen–Propp [37]) Let $n \geq 1$. The **Aztec diamond of order** n, denoted AD_n, is the union of lattice squares $[j, j+1] \times [k, k+1]$, $(a, b \in \mathbb{Z})$, that are contained in the rotated square $\{(x, y) : |x| + |y| \leq n+1\}$ (see Fig. 4.9). Prove that the number of domino tilings of AD_n is $2^{n(n+1)/2}$.

(c) (☕☕☕) (Elkies et al. [37]) In a domino tiling of AD_n, a **horizontal** domino is a domino of the form $[a, a+2] \times [b, b+1]$ for some $a, b \in \mathbb{Z}$, and is called **north-going** if $a + b + n$ is even or **south-going** if $a + b + n$ is odd. A **vertical** domino has the form $[a, a+1] \times [b, b+2]$ for some $a, b \in \mathbb{Z}$, and is called **west-going** if $a + b + n$ is even or **east-going** if $a + b + n$ is odd.
Prove that the following algorithm, known as the **domino shuffling algorithm**, will recursively generate a uniformly random domino tiling of AD_n:

Step 1. Start by choosing at random one of the two tilings of AD_1, each with equal probability.

Step 2. For each $1 \leq k \leq n-1$, transform the tiling of AD_k into a tiling of AD_{k+1} by carrying out the following steps:

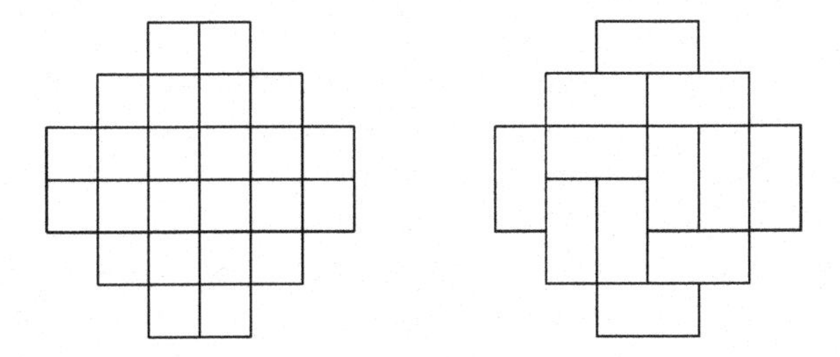

Figure 4.9 The Aztec diamond of order 3 and one of its 64 domino tilings.

- **Destruction:** Any pair of dominos consisting of a north-going domino directly below a south-going domino is removed; any pair of dominos consisting of an east-going domino directly to the left of a west-going domino is removed.
- **Sliding:** Each remaining domino slides one step in the direction it is heading, that is, north-going dominos to the north, and so on. (It needs to be shown that after the sliding no two dominos attempt to occupy the same square.)
- **Creation:** The dominos now occupy a region of the form $\mathrm{AD}_{k+1} \setminus E$, where (it needs to be shown) the closure of E has the property that it can be represented in a unique way as the union of 2×2 blocks with disjoint interiors. For each such 2×2 block, fill it with two newly created dominos which are chosen as either two horizontal or two vertical dominos, according to the result of a fair coin toss, performed independently of all other random choices.

Note: As part of the proof of correctness of the algorithm, one can obtain a proof of the claim of part (b) above.

4.13 (Jockusch-Propp-Shor [60]) Given a domino tiling of AD_n, two dominos are called adjacent if they share a lattice edge of $\mathbb{Z}^2$. The **north arctic region** of the tiling is the union of all north-going dominos that are connected to a domino touching the "north pole" $(0, n)$ via a chain of adjacent north-going dominos (the reason for the name is that the dominos in this region are "frozen" into a rigid brickwork pattern). Similarly, the **south arctic region** is the union of south-going dominos connected to the south pole $(0, -n)$, the **east arctic region** is the union of east-going dominos connected to the east pole $(-n, 0)$ and the **west arctic region** is the union of west-going dominos connected to the west pole $(n, 0)$. These four arctic regions are collectively referred to as the **arctic region** of the tiling. The region tiled by dominos not in the arctic region is called the **temperate region**.

(a) (☕☕) Show that the north arctic region $\mathcal{N}$ can be encoded in terms of a Young diagram whose box positions (i, j) are the addresses of the north-going dominos in $\mathcal{N}$ in a rotated coordinate system based at the north pole. More precisely, position (i, j) corresponds to the 2×1 rectangle $[j - i - 1, j - i + 1] \times [n - i - j + 1, n - i - j + 2]$ (that is, a north-going domino in the top row of AD_n would have position $(1, 1)$; the two possible rectangles that may contain north-going dominos in the row below to the top row are associated with positions $(2, 1)$ and $(1, 2)$; etc.). Prove that the set of positions $(i, j) \in \mathbb{N}^2$ thus defined from the dominos in $\mathcal{N}$ is the set of cells of a Young diagram.

(b) (☕☕) Show that whenever Step 2 in the domino shuffling algorithm described above is executed to "grow" a domino tiling of AD_n into a domino tiling of AD_{n+1}, the Young diagram encoded by the north arctic region as described above grows exactly according to the growth rule (4.52) of the multicorner growth process with $p = 1/2$.

(c) (☕☕) With the help of Theorem 4.24, deduce from the result of part (d) above the following result, known as the "arctic circle" theorem (see Fig. 4.10).

Theorem 4.32 (Arctic circle theorem for domino tilings of the Aztec diamond) *For each $n \geq 1$, let T_n denote a uniformly random domino tiling of the Aztec diamond* AD_n, *and denote its temperate region by* Temp_n. *As $n \to \infty$, the shape of the temperate region* Temp_n *converges in probability after scaling to the disk $D = \{(x, y) : x^2 + y^2 \leq 1/2\}$ inscribed within the diamond (the scaling limit of the discrete Aztec diamonds)* $\{(x, y) : |x| + |y| = 1\}$. *More precisely, for any $\epsilon > 0$ we have that*

$$\mathbb{P}\left((1 - \epsilon)D \subset \frac{1}{n}\mathrm{Temp}_n \subset (1 + \epsilon)D\right) \to 1 \quad \text{as } n \to \infty.$$

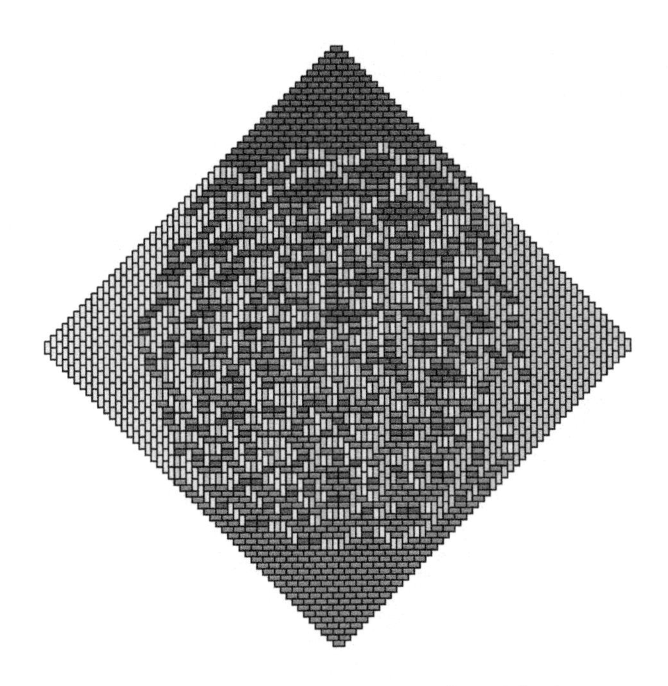

Figure 4.10 A uniformly random domino tiling of the Aztec diamond AD_{50}, sampled by domino shuffling.

5

The corner growth process: distributional results

Chapter summary. In 1997, Kurt Johansson discovered that the corner growth process we studied in the previous chapter is directly related to longest increasing subsequences in **generalized permutations**. This connection can be studied via the **RSK algorithm**, which is an extension of the Robinson–Schensted algorithm discussed in Chapter 1, leading to a remarkable explicit representation for the distribution of the passage times, that is itself related to **Wishart matrices** from random matrix theory. Applying ideas from the theory of **orthogonal polynomials** and asymptotic analysis techniques, we prove Johansson's result that the distribution of the passage times converges to the Tracy–Widom distribution F_2.

5.1 The fluctuations of $G(m, n)$ and the Tracy–Widom distribution

In previous chapters we studied two natural processes of randomly growing Young diagrams, and derived the limiting shapes for both: the Plancherel growth process, which was used in Chapter 1 to solve the Ulam–Hammersley problem of deriving the (first-order) asymptotics of the maximal increasing subsequence length in a random permutation; and the corner growth process, which we analyzed in Chapter 4, where we also saw it bears an interesting relation to other natural random processes such as the totally asymmetric simple exclusion process and random domino tilings.

Although the conceptual resemblance between these two processes is satisfying, our analysis of the corner growth process involved probabilistic techniques with very little combinatorial content, and, in particular, had nothing to do with the mathematics of longest increasing subsequences, which is ostensibly the subject of this work.

It may thus come as a pleasant surprise that there is a much deeper connection, discovered by Kurt Johansson in 1997, between the corner growth process, longest increasing subsequences, and many of the ideas and concepts discussed in earlier chapters of this book. Our goal in this chapter is to study this connection and pursue it to its (rather beautiful) set of logical conclusions, following Johansson's landmark paper [62]. Ultimately, we will be led to a much deeper understanding of the passage times associated with the corner growth process and the multicorner growth process. The culmination of our efforts will be in the proof of the following result about the limiting distribution of the passage times.

Theorem 5.1 (Limit law for the passage times) *For $x, y > 0$ define functions*

$$\Psi(x,y) = (\sqrt{x} + \sqrt{y})^2, \tag{5.1}$$

$$\sigma(x,y) = (xy)^{-1/6}(\sqrt{x} + \sqrt{y})^{4/3}. \tag{5.2}$$

Let $(m_k)_{k=1}^{\infty}, (n_k)_{k=1}^{\infty}$ be sequences of positive integers with the properties that

$$m_k, n_k \to \infty \text{ as } k \to \infty, \tag{5.3}$$

$$0 < \liminf_{k\to\infty} \frac{m_k}{n_k} < \limsup_{k\to\infty} \frac{m_k}{n_k} < \infty. \tag{5.4}$$

Then as $k \to \infty$, the passage times $G(m_k, n_k)$ associated with the corner growth process converge in distribution after rescaling to the Tracy–Widom distribution F_2 defined in (2.3). *More precisely, we have*

$$\mathbb{P}\left(\frac{G(m_k,n_k) - \Psi(m_k,n_k)}{\sigma(m_k,n_k)} \le t\right) \xrightarrow[k\to\infty]{} F_2(t) \qquad (t \in \mathbb{R}). \tag{5.5}$$

Because of the connection between the corner growth process and Rost's particle process described in the previous chapter (Section 4.7), Theorem 5.1 can be formulated equivalently as a statement about the limiting distribution as $n, m \to \infty$ of the time when the mth leading particle has made its nth move in Rost's particle process. A far-reaching generalization of this result for the case of the Asymmetric Simple Exclusion Process (ASEP), a generalization of Rost's particle process in which particles can move both to the left and to the right, was proved by Tracy and Widom [137].

5.2 Longest increasing subsequences in generalized permutations

We start with a few fairly simple combinatorial observations. One clue that the corner growth process and its cousin, the multicorner growth process, may be related to longest increasing subsequences is that the definition of the passage times in (4.13) involves taking a maximum. Johansson noticed that, in the case of the multicorner growth process, the passage times can in fact be interpreted literally as maximal (weakly) increasing subsequence lengths in certain sequences of numbers that (unlike permutations) may involve repetitions.

To make this observation precise, we make the following definitions. A **generalized permutation of length** k **and row bounds** (m, n) is a two-line array of integers that has the form

$$\sigma = \begin{pmatrix} i_1 & i_2 & \cdots & i_k \\ j_1 & j_2 & \cdots & j_k \end{pmatrix},$$

where $1 \le i_1, \ldots, i_k \le m$, $1 \le j_1, \ldots, j_k \le n$, and where the columns are ordered lexicographically, in the sense that if $s < t$ then either $i_s < i_t$, or $i_s = i_t$ and $j_s \le j_t$. Denote by $\mathcal{P}^k_{m,n}$ the set of generalized permutations of length k and row bounds (m, n). If $\sigma = \left(\begin{smallmatrix} i_1 & \cdots & i_k \\ j_1 & \cdots & j_k \end{smallmatrix}\right)$ is a generalized permutation and $1 \le s_1 < \ldots < s_d \le k$ is a sequence of column positions, we refer to the generalized permutation $\left(\begin{smallmatrix} i_{s_1} & \cdots & i_{s_d} \\ j_{s_1} & \cdots & j_{s_d} \end{smallmatrix}\right)$ as a **subsequence of** σ, and call such a subsequence **increasing** if $j_{s_1} \le \ldots \le j_{s_d}$. Note that generalized permutations are indeed generalizations of ordinary permutations (interpreted as two-line arrays in the usual way), and the definition of an increasing subsequence generalizes that concept for an ordinary permutation. If $\sigma \in \mathcal{P}^k_{m,n}$, as for ordinary permutations let $L(\sigma)$ denote the maximal length of an increasing subsequence of σ. Equivalently, $L(\sigma)$ is the maximal length of a weakly increasing subsequence of the bottom row of σ.

One last definition we will need is an alternative way to encode elements of $\mathcal{P}^k_{m,n}$ as matrices. Let $\mathcal{M}^k_{m,n}$ denote the set of $m \times n$ matrices $(a_{i,j})_{1 \le i \le m, 1 \le j \le n}$ with nonnegative integer entries satisfying $\sum_{i,j} a_{i,j} = k$. We can associate with a generalized permutation $\sigma \in \mathcal{P}^k_{m,n}$ a matrix $M_\sigma = (a_{i,j})_{i,j} \in \mathcal{M}^k_{m,n}$ by setting $a_{i,j}$ to be the number of columns in σ equal to $\binom{i}{j}$. For example, the matrix associated with the generalized permutation

$$\sigma = \begin{pmatrix} 1 & 1 & 1 & 1 & 2 & 2 & 2 & 2 & 3 & 3 \\ 1 & 1 & 4 & 5 & 3 & 3 & 3 & 5 & 2 & 5 \end{pmatrix}, \tag{5.6}$$

(considered as an element of $\mathcal{P}_{3,5}^{10}$) is

$$M_\sigma = \begin{pmatrix} 2 & 0 & 0 & 1 & 1 \\ 0 & 0 & 3 & 0 & 1 \\ 0 & 1 & 0 & 0 & 1 \end{pmatrix}.$$

Conversely, given the matrix M_σ we can recover σ by writing a two-line array containing $a_{i,j}$ copies of $\binom{i}{j}$ for each $1 \le i \le m,\ 1 \le j \le n$ and sorting the columns lexicographically. Thus, we have the following easy result.

Lemma 5.2 *The map $\sigma \mapsto M_\sigma$ defines a bijection between $\mathcal{P}_{m,n}^k$ and $\mathcal{M}_{m,n}^k$.*

It is natural to ask what the function $\sigma \mapsto L(\sigma)$ looks like when we interpret generalized permutations as matrices, that is, how is $L(\sigma)$ computed in terms of the associated matrix M_σ. The answer to this question involves a familiar expression from our study of last passage percolation in the previous chapter.

Lemma 5.3 *If $M = (a_{i,j})_{i,j} \in \mathcal{M}_{m,n}^k$ and $\sigma \in \mathcal{P}_{m,n}^k$ is the generalized permutation associated to it via the bijection of Lemma 5.2, then*

$$L(\sigma) = \max\left\{ \sum_{\ell=0}^{d} a_{p_\ell,q_\ell} \ :\ (p_\ell, q_\ell)_{\ell=0}^{d} \in \mathcal{Z}(1,1;m,n) \right\} \tag{5.7}$$

where $\mathcal{Z}(1,1;m,n)$ is the set of up-right paths from $(1,1)$ to (m,n), defined in Section 4.3.

Proof Denote the expression on the right-hand side of (5.7) by $\tilde{L}(M)$. First, we show that $\tilde{L}(M) \le L(\sigma)$. Let $(p_\ell, q_\ell)_{\ell=0}^{d} \in \mathcal{Z}(1,1;m,n)$. Let μ be the generalized permutation

$$\mu = \left(\begin{array}{ccccccc} \overbrace{p_0 \cdots p_0}^{a_{p_0,q_0} \text{ columns}} & \overbrace{p_1 \cdots p_1}^{a_{p_1,q_1} \text{ columns}} & \cdots & \overbrace{p_d \cdots p_d}^{a_{p_d,q_d} \text{ columns}} \\ q_0 \cdots q_0 & q_1 \cdots q_1 & \cdots & q_d \cdots q_d \end{array} \right).$$

Then clearly μ is a subsequence of σ, and because $(p_\ell, q_\ell)_{\ell=0}^{d}$ is an up-right path, it is also easy to see that it is an increasing subsequence, and has length $\sum_{\ell=0}^{d} a_{p_\ell,q_\ell}$. It follows that $\sum_{\ell=0}^{d} a_{p_\ell,q_\ell} \le L(\sigma)$. Since this inequality holds for an arbitrary up-right path in $\mathcal{Z}(1,1;m,n)$, we have shown that $\tilde{L}(M) \le L(\sigma)$.

To see that the opposite inequality $\tilde{L}(M) \ge L(\sigma)$ holds, let $\mu = \begin{pmatrix} i_1 & \cdots & i_s \\ j_1 & \cdots & j_s \end{pmatrix}$

be an increasing subsequence of σ of maximal length $s = L(\sigma)$. Let $\binom{\gamma_1}{\delta_1}$, $\ldots$, $\binom{\gamma_r}{\delta_r}$ be a list of the *distinct* columns of μ, in the order in which they appear in μ; recall that each column $\binom{\gamma_t}{\delta_t}$ appears at most a_{γ_t,δ_t} times in μ, so we have that $\sum_{t=1}^{r} a_{\gamma_t,\delta_t} \geq s$. But now note that by the definition of $(\gamma_t, \delta_t)_{t=1}^{r}$, for each $1 \leq t \leq r-1$ the differences $\gamma_{t+1} - \gamma_t$ and $\delta_{t+1} - \delta_t$ are both nonnegative and at least one of them is positive. It follows that there is an up-right path $(p_\ell, q_\ell)_{\ell=0}^{d} \in \mathcal{Z}(1, 1; m, n)$ that contains all the pairs (γ_t, δ_t). From this we get the chain of inequalities

$$\tilde{L}(M) \geq \sum_{\ell=0}^{d} a_{p_\ell,q_\ell} \geq \sum_{t=1}^{r} a_{\gamma_t,\delta_t} \geq s = L(\sigma),$$

which proves the claim. □

From now on, we also use the notation $L(M)$ to denote the right-hand side of (5.7), which by the above result is equal to the maximal increasing subsequence length in σ, where σ and M are the equivalent ways of thinking about generalized permutations. Comparing (5.7) with (4.13), we have the following immediate corollary.

Corollary 5.4 *Let $0 < p < 1$, and let $(\tau_{i,j})_{i,j\geq 1}$ be an array of i.i.d. random variables with the geometric distribution* Geom(p). *Let $(G(i, j))_{i,j\geq 1}$ be the passage times in last-passage percolation with the array $(\tau_{i,j})_{i,j\geq 1}$ of clock times, associated with the multicorner growth process (see Theorem 4.23). For each $m, n \geq 1$ let $M_{m,n}$ be the matrix $M_{m,n} = (\tau_{i,j})_{1\leq i\leq m,\, 1\leq j\leq n}$. Then the passage times have the representation*

$$G(m, n) = L(M_{m,n}) \qquad (m, n \geq 1). \tag{5.8}$$

The problem of understanding the passage times in the multicorner growth process has thus been reduced to the study of the generalized permutation statistic $L(\cdot)$ of a random rectangular matrix of i.i.d. geometrically distributed random variables. It is of course far from obvious that this reduction makes the problem any easier or more tractable; however, similarly to the ideas we encountered in Chapter 1, it turns out that maximal increasing subsequence lengths of generalized permutations are related to the mathematics of Young tableaux. By using an extension of the Robinson–Schensted algorithm discussed in Chapter 1 we will be able to derive a striking explicit formula for the distribution function of the passage times (not only for the multicorner growth process, but also for the corner growth process associated with exponentially distributed clock times), which will be the starting point of a powerful analysis. We discuss the relevant combinatorial ideas in the next two sections.

5.3 Semistandard Young tableaux and the Robinson–Schensted–Knuth algorithm

Recall from Section 1.6 that the Robinson–Schensted algorithm maps a permutation $\sigma \in S_k$ to a triple (λ, P, Q), where λ is a Young diagram of order k, and P and Q are Young tableaux of shape λ. Donald E. Knuth discovered [70] an extension of the algorithm to the case of generalized permutations. The extended algorithm has become known as the **Robinson–Schensted–Knuth (RSK) algorithm**. Starting from a generalized permutation $\sigma \in \mathcal{P}^k_{m,n}$, its output will still be a triple (λ, P, Q) where λ is a Young diagram of order k, but now P and Q will be λ-shaped arrays of integers that are generalizations of the "standard" kind of Young tableaux we are used to, and are known as **semistandard Young tableaux**.

The basic idea is the same and involves "growing" the diagram λ and the tableaux P and Q starting from an empty diagram. Let us start with the tableau P: as before, it plays the role of the **insertion tableau**, and is formed by the application of a sequence of **insertion steps** in which the entries of the *bottom row* of the two-line array σ are inserted one by one into the existing tableau, enlarging it – and the associated diagram λ – by one cell. The insertion is performed using the same rules as described on p. 16 (originally in the context of the patience sorting algorithm) and in Section 1.6; the only difference is that the sequence of numbers being inserted may contain repetitions. For clarity, let us state how this works more explicitly. As with the Robinson–Schensted algorithm, each entry from the bottom row of σ is inserted into the first row of the growing tableau P, which leads to a cascade of "bumping" events and recursive insertions into rows below the first row, according to the following rules:

1. A number x being inserted into a row of a tableau will be placed in the leftmost position of the row whose current entry is (strictly) bigger than x. If there are no such positions, x is placed in an unoccupied position to the right of all currently occupied positions.

2. If x was placed in a position already occupied by a number y, then y gets bumped down to the next row where it will be inserted recursively according to rule 1 above.

As an illustration, inserting the number 2 into the tableau

1	1	2	2	4	4
3	3	4	6		
4	5				

results in the new tableau

1	1	2	2	2	4
3	3	4	4		
4	5	6			

It is useful to associate with each insertion step a **bumping sequence**, which is the sequence of positions at which the entry of the tableau P changed. In the preceding example, the bumping sequence is $(5, 1)$, $(4, 2)$, $(3, 3)$.

Next, the tableau Q plays the role of the **recording tableau** just as in the Robinson–Schensted case; after each insertion step, we add a single entry to Q by setting the entry in the just-added cell of the shape λ to the number taken from the *top row* of σ directly above the bottom-row entry that was just inserted into P. As an example, readers can verify that applying the algorithm to the generalized permutation σ in (5.6) produces the following insertion and recording tableaux:

1	1	2	3	3	5	5
3	5					
4						

1	1	1	1	2	2	3
2	2					
3						

From the example we see that P and Q are not arbitrary arrays of numbers but still have some useful monotonicity properties, although they are not standard Young tableaux. Given a Young diagram λ, define a **semistandard Young tableau** (or **semistandard tableau**) to be a filling of the cells of λ with positive integers, such that rows are weakly increasing and columns are *strictly* increasing.

Lemma 5.5 (a) *When a number x is inserted into a semistandard tableau T, the bumping sequence $(a_1, 1), \dots (a_k, k)$ is weakly monotone, that is, $a_i \ge a_{i+1}$ for all i.*

(b) *Given a semistandard tableau T and numbers $x \ge y$, let $(a_1, 1), \dots, (a_k, k)$ and $(b_1, 1), \dots, (b_m, m)$ be the bumping sequences associated with first inserting x into T and then inserting y into the tableau obtained from T by the first insertion. Then we have $m \le k$ and $b_i \ge a_i$ for $1 \le i \le m$.*

(c) *The insertion tableau P computed by the RSK algorithm is a semistandard tableau.*

(d) *The recording tableau Q computed by the RSK algorithm is a semistandard tableau.*

(e) *During the application of the RSK algorithm, equal entries in the recording tableau Q are added from left to right.*

Claim (a) of Lemma 5.5 is trivial, and each of the subsequent claims is straightforward to prove from the definitions and from the previous claims. The formal proof is left to the reader (Exercise 5.2); for more details see [125], Section 7.11.

Note that the last claim in the lemma is particularly useful, since it implies that, as with the simpler case of the Robinson–Schensted algorithm, the original generalized permutation σ can be recovered from the tableaux P and Q. As before, this is done by applying a sequence of **deletion steps** to the tableau P, with the starting point for each deletion being the leftmost cell from among the cells containing the maximal entry of the recording tableau Q. After applying the deletion to P, the cell from which the deletion began is removed also from Q, and the process is repeated until both tableaux have been "emptied." Each deletion step also yields as a byproduct a column $\binom{i}{j}$ of σ, and Lemma 5.5(e) guarantees that this choice for the order in which to apply the deletions is the correct one to recover the columns of σ precisely in reverse order.

One final property of the RSK algorithm is its connection to longest increasing subsequences, which is similar to the case of the Robinson–Schensted algorithm.

Lemma 5.6 *If the RSK algorithm associates the triple (λ, P, Q) with the generalized permutation σ, then we have*

$$L(\sigma) = \lambda_1 \quad \textit{(the length of the first row of } \lambda\textit{).}$$

Proof If $\sigma = \left(\begin{smallmatrix} i_1 & \cdots & i_k \\ j_1 & \cdots & j_k \end{smallmatrix}\right)$, recall that $L(\sigma)$ is the maximal length of a weakly increasing subsequence in the sequence $J = (j_1, \ldots, j_k)$. Following the evolution of the first row of the insertion tableau, we see that we are effectively performing a patience sorting procedure on J (see Section 1.5). The proof is now identical to the proof of Lemma 1.7. □

Summarizing the preceding discussion, we have outlined the proof of the following important result.

Theorem 5.7 *The RSK algorithm is a bijection between the set $\mathcal{P}^k_{m,n}$ of generalized permutations of length k and row bounds (m, n) and the set of triples (λ, P, Q) where λ is a Young diagram, P is a semistandard Young tableau of shape λ and entries from $\{1, \ldots, m\}$, and Q is a semistandard Young tableau of shape λ and entries from $\{1, \ldots, n\}$. Under this bijection, $L(\sigma)$ is equal to the length of the first row of λ.*

Exercises 5.3 and 5.4 provide some simple numerical examples, which are highly recommended as a way to get a better intuitive feel of this elegant but nontrivial construction.

5.4 An enumeration formula for semistandard tableaux

Another tool we need for our analysis of the passage times in the multi-corner growth process is the following enumeration formula for semistandard Young tableaux.[1]

Theorem 5.8 *Let $\lambda = (\lambda_1, \ldots, \lambda_m)$ be a Young diagram. For any $n \geq m$, the number of semistandard Young tableaux of shape λ with entries from $\{1, \ldots, n\}$ is given by*

$$\prod_{1 \leq i < j \leq n} \frac{\lambda_i - \lambda_j + j - i}{j - i}, \tag{5.9}$$

where λ_i is interpreted as 0 for i greater than the number of parts of λ.

Note that in the case $n < m$ the requirement of strict monotonicity along columns means that there are no semistandard tableaux of shape λ.

Our approach to the proof of Theorem 5.8 will be to transform the problem into that of enumerating a different family of combinatorial objects known as **Gelfand–Tsetlin patterns**. Formally, a Gelfand–Tsetlin (GT) pattern with n rows is a triangular array of positive integers

$$G = (g_{i,j})_{1 \le j \le i \le n}$$

such that the inequalities $g_{i+1,j} \le g_{i,j} < g_{i+1,j+1}$ hold for all applicable indices i, j. Here is an example of a pattern with five rows:

$$\begin{array}{ccccccccc} & & & & 3 & & & & \\ & & & 3 & & 4 & & & \\ & & 1 & & 4 & & 6 & & \\ & 1 & & 2 & & 6 & & 7 & \\ 1 & & 2 & & 5 & & 7 & & 8 \end{array}$$

Given a GT pattern with n rows, we can use it to construct a semistandard Young tableau T according to the following recipe: for each $1 \le i \le n$, the ith row of the pattern encodes a Young diagram $\lambda^{(i)}$ whose parts are

$$(g_{i,i} - i, \ldots, g_{i,2} - 2, g_{i,1} - 1)$$

with any trailing zeroes removed. With this definition, it is easy to see that the properties of the GT pattern imply that for any $2 \le i \le n$, $\lambda^{(i-1)}$ is contained in $\lambda^{(i)}$. The semistandard Young tableau T is then defined as the array of shape $\lambda^{(n)}$ obtained by writing for each $1 \le i \le n$ the entry i in all the cells of the "skew-shape" which is the difference $\lambda^{(i)} \setminus \lambda^{(i-1)}$ (where $\lambda^{(0)}$ is interpreted as the empty diagram $\emptyset$) between two successive diagrams on the list.

To illustrate this construction, the GT pattern in the preceding example leads to the Young diagrams

$$\begin{aligned} \lambda^{(1)} &= (2), \\ \lambda^{(2)} &= (2, 2), \\ \lambda^{(3)} &= (3, 2), \\ \lambda^{(4)} &= (3, 3), \\ \lambda^{(5)} &= (3, 3, 2), \end{aligned}$$

and the associated semistandard tableau is therefore

1	1	3
2	2	4
5	5	

We omit the easy verification of the fact that the array of numbers resulting from this construction is in general a semistandard Young tableau. We also leave to the reader (Exercise 5.5) to verify that the construction can be inverted, in the following precise sense.

Lemma 5.9 *Let λ be a Young diagram with m parts, and let $n \geq m$. Define numbers $x_1, \ldots, x_n$ by*

$$x_j = \lambda_{n+1-j} + j \qquad (1 \leq j \leq n).$$

Then the construction above defines a bijection between the set of Gelfand–Tsetlin patterns with n rows and bottom row $(x_1, \ldots, x_n)$ and the set of semistandard Young tableaux of shape λ with entries from $\{1, \ldots, n\}$.

By the lemma, the proof of Theorem 5.8 has been reduced to proving the following result on the enumeration of GT patterns with given bottom row.

Theorem 5.10 *For any $n \geq 1$ and integers $1 \leq x_1 < \ldots < x_n$, the number of Gelfand–Tsetlin patterns with bottom row $(x_1, \ldots, x_n)$ is given by*

$$\prod_{1 \leq i < j \leq n} \frac{x_j - x_i}{j - i}. \tag{5.10}$$

Proof Denote by $W_n(x_1, \ldots, x_n)$ the number of GT patterns with bottom row $(x_1, \ldots, x_n)$. By partitioning the patterns enumerated by $W_n(x_1, \ldots, x_n)$ according to their $(n-1)$th row $(y_1, \ldots, y_{n-1})$, we see that the family of functions W_n satisfies the recurrence

$$\begin{aligned} W_n(x_1, \ldots, x_n) &= \sum_{\substack{(y_1, \ldots, y_{n-1}) \\ \forall j \; x_j \leq y_j < x_{j+1}}} W_{n-1}(y_1, \ldots, y_{n-1}) \\ &= \sum_{y_1 = x_1}^{x_2 - 1} \sum_{y_2 = x_2}^{x_3 - 1} \cdots \sum_{y_{n-1} = x_{n-1}}^{x_n - 1} W_{n-1}(y_1, \ldots, y_{n-1}), \end{aligned}$$

or equivalently

$$W_n(x_1,\ldots,x_n) = \mathop{\mathrm{S}}_{y_1=x_1}^{x_2} \mathop{\mathrm{S}}_{y_2=x_2}^{x_3} \cdots \mathop{\mathrm{S}}_{y_{n-1}=x_{n-1}}^{x_n} W_{n-1}(y_1,\ldots,y_{n-1}), \tag{5.11}$$

where we introduce a modified summation operator

$$\mathop{\mathrm{S}}_{u=a}^{b} f(u) = \sum_{u=a}^{b-1} f(u) \qquad (a,b \in \mathbb{Z},\ a < b).$$

It is natural to extend the definition of $\mathrm{S}_{u=a}^{b} f(u)$ by setting $\mathrm{S}_{u=a}^{b} f(u) = 0$ if $a = b$ and $\mathrm{S}_{u=a}^{b} f(u) = -\mathrm{S}_{u=b}^{a} f(u)$ if $a > b$. With this extended definition, the recurrence (5.11) gives meaning to $W_n(x_1,\ldots,x_n)$ for arbitrary integer values $x_1,\ldots,x_n$. Note also the following easy facts. First, the modified summation operator satisfies

$$\mathop{\mathrm{S}}_{u=a}^{c} f(u) = \mathop{\mathrm{S}}_{u=a}^{b} f(u) + \mathop{\mathrm{S}}_{u=b}^{c} f(u)$$

for any integers a, b, c. Second, if $f(u_1,\ldots,u_d)$ is a polynomial in $u_1,\ldots,u_d$ of total degree m with a (not necessarily unique) highest-order monomial $c\prod_{j=1}^{d} u_j^{\alpha_j}$, where $m = \sum_j \alpha_j$, then $\mathrm{S}_{u_1=a}^{b} f(u_1,\ldots,u_d)$ is a polynomial in $a, b, u_2,\ldots,u_d$ of total degree $m+1$ which contains the two monomials $\frac{c}{\alpha_1+1} b^{\alpha_1+1}\prod_{j=2}^{d} u_j^{\alpha_j}$ and $-\frac{c}{\alpha_1+1} a^{\alpha_1+1}\prod_{j=2}^{d} u_j^{\alpha_j}$ of degree $m+1$.

Third, we claim that if $f(y_1,\ldots,y_{n-1})$ is an antisymmetric polynomial in $y_1,\ldots,y_{n-1}$ then

$$g(x_1,\ldots,x_n) = \mathop{\mathrm{S}}_{y_1=x_1}^{x_2} \mathop{\mathrm{S}}_{y_2=x_2}^{x_3} \cdots \mathop{\mathrm{S}}_{y_{n-1}=x_{n-1}}^{x_n} f(y_1,\ldots,y_{n-1})$$

is in turn antisymmetric in $x_1,\ldots,x_n$. To check this, fix $2 \le j \le n-2$, and write

$$\begin{aligned} -g(\ldots,x_{j+1},x_j,\ldots) &= - \mathop{\mathrm{S}}_{y_{j-1}=x_{j-1}}^{x_{j+1}} \mathop{\mathrm{S}}_{y_j=x_{j+1}}^{x_j} \mathop{\mathrm{S}}_{y_{j+1}=x_j}^{x_{j+2}} \mathop{\mathrm{S}}_{[\text{all other } y_i\text{'s}]} f(y_1,\ldots,y_{n-1}) \\ &= \mathop{\mathrm{S}}_{[\text{other } y_i\text{'s}]} \left(\mathop{\mathrm{S}}_{y_{j-1}=x_{j-1}}^{x_j} + \mathop{\mathrm{S}}_{y_{j-1}=x_j}^{x_{j+1}} \right) \mathop{\mathrm{S}}_{y_j=x_j}^{x_{j+1}} \left(\mathop{\mathrm{S}}_{y_{j+1}=x_j}^{x_{j+1}} + \mathop{\mathrm{S}}_{y_{j+1}=x_{j+1}}^{x_{j+2}} \right) f(y_1,\ldots,y_{n-1}) \end{aligned}$$

$$= \mathop{\mathrm{S}}_{[\text{other } y_i\text{'s}]} \Bigg[\mathop{\mathrm{S}}_{y_{j-1}=x_{j-1}}^{x_j} \mathop{\mathrm{S}}_{y_j=x_j}^{x_{j+1}} \mathop{\mathrm{S}}_{y_{j+1}=x_j}^{x_{j+1}} f(y_1,\ldots,y_{n-1})$$

$$+ \mathop{\mathrm{S}}_{y_{j-1}=x_j}^{x_{j+1}} \mathop{\mathrm{S}}_{y_j=x_j}^{x_{j+1}} \mathop{\mathrm{S}}_{y_{j+1}=x_j}^{x_{j+1}} f(y_1,\ldots,y_{n-1})$$

$$+ \mathop{\mathrm{S}}_{y_{j-1}=x_j}^{x_{j+1}} \mathop{\mathrm{S}}_{y_j=x_j}^{x_{j+1}} \mathop{\mathrm{S}}_{y_{j+1}=x_{j+1}}^{x_{j+2}} f(y_1,\ldots,y_{n-1})$$

$$+ \mathop{\mathrm{S}}_{y_{j-1}=x_{j-1}}^{x_j} \mathop{\mathrm{S}}_{y_j=x_j}^{x_{j+1}} \mathop{\mathrm{S}}_{y_{j+1}=x_{j+1}}^{x_{j+2}} f(y_1,\ldots,y_{n-1}) \Bigg].$$

Here, we see that in the bracketed expression the first three triple summations vanish by the antisymmetry of f. This leaves

$$\mathop{\mathrm{S}}_{[\text{other } y_i\text{'s}]} \mathop{\mathrm{S}}_{y_{j-1}=x_{j-1}}^{x_j} \mathop{\mathrm{S}}_{y_j=x_j}^{x_{j+1}} \mathop{\mathrm{S}}_{y_{j+1}=x_{j+1}}^{x_{j+2}} f(y_1,\ldots,y_{n-1}) = g(x_1,\ldots,x_n).$$

We have shown that $g(x_1,\ldots,x_{j+1},x_j,\ldots,x_n) = -g(x_1,\ldots,x_j,x_{j+1},\ldots,x_n)$. The verification of antisymmetry for the remaining cases $j = 1$ and $j = n-1$ is similar and is left to the reader.

We can now prove by induction on n that $W_n(x_1,\ldots,x_n)$ is given by (5.10). For $n = 1$ the claim is trivial. Assuming the claim is true for W_{n-1}, the recurrence (5.11) gives that

$$W_n(x_1,\ldots,x_n) = \prod_{j=1}^{n-2} \frac{1}{j!} \cdot \mathop{\mathrm{S}}_{y_1=x_1}^{x_2} \mathop{\mathrm{S}}_{y_2=x_2}^{x_3} \cdots \mathop{\mathrm{S}}_{y_{n-1}=x_{n-1}}^{x_n} \prod_{1\le i<j\le n-1} (y_j - y_i).$$

By the above observations, this is an antisymmetric polynomial of total degree $n(n-1)/2$ in $x_1,\ldots,x_n$. By considering the monomial $y_1^0 y_2^1 \ldots y_{n-1}^{n-2}$ of the Vandermonde product being summed, we see that $W_n(x_1,\ldots,x_n)$ has a highest-order monomial $x_1^0 x_2^1 \ldots x_n^{n-1}$ with leading coefficient $\prod_{j=1}^{n-1}(j!)^{-1}$.

To conclude, now use a well-known property of the Vandermonde product $\prod_{1\le i<j\le n}(x_j - x_i)$, namely that it is the unique (up to multiplication by a constant) antisymmetric polynomial in n variables of total degree $n(n-1)/2$ (see Exercise 5.6). Together with the above observations this proves that $W_n(x_1,\ldots,x_n) = \prod_{j=1}^{n}(j!)^{-1} \prod_{1\le i<j\le n}(x_j - x_i)$. □

5.5 The distribution of $G(m, n)$

Using the combinatorial tools developed in the previous sections, we are now ready to prove an important explicit formula for the distribution of the passage times in the corner growth process and the multicorner growth process.

Theorem 5.11 *For $0 < p < 1$ and $k \geq 0$ let*

$$w_k^{(p)}(x) = \binom{x+k}{k}(1-p)^x,$$
$$W_k(x) = x^k e^{-x}.$$

For any $m \geq n \geq 1$, define numerical factors

$$C_{m,n,p} = \frac{p^{mn} \cdot ((m-n)!)^n}{(1-p)^{mn+n(n-1)/2} \cdot n!} \cdot \prod_{j=0}^{n-1} \frac{1}{j!(m-n+j)!}, \tag{5.12}$$

$$D_{m,n} = \frac{1}{n!} \prod_{j=0}^{n-1} \frac{1}{j!(m-n+j)!}. \tag{5.13}$$

Then the distribution function of the passage times associated with the multicorner growth process with parameter p is given for any $m \geq n \geq 1$ by

$$\mathbb{P}(G(m,n) \leq t) = C_{m,n,p} \sum_{x_1,\ldots,x_n=0}^{\lfloor t \rfloor + n-1} \prod_{1 \leq i < j \leq n} (x_i - x_j)^2 \prod_{i=1}^{n} w_{m-n}^{(p)}(x_i). \tag{5.14}$$

The distribution function of the passage times associated with the corner growth process is given by

$$\mathbb{P}(G(m,n) \leq t) = D_{m,n} \int_0^t \ldots \int_0^t \prod_{1 \leq i < j \leq n} (x_i - x_j)^2 \prod_{j=1}^{n} W_{m-n}(x_j)\, dx_1 \ldots dx_n. \tag{5.15}$$

Proof Denote $S_{m,n} = \sum_{i=1}^{m} \sum_{j=1}^{n} \tau_{i,j}$, where $\tau_{i,j}$ are the Geom(p)-distributed i.i.d. clock times associated with the multicorner growth process. Let $k \geq 0$. Conditioned on the event $\{S_{m,n} = k\}$, the matrix $M_{m,n} = (\tau_{i,j})_{1 \leq i \leq m,\, 1 \leq j \leq n}$ is in $\mathcal{M}_{m,n}^k$. For any matrix $M = (a_{i,j})_{i,j} \in \mathcal{M}_{m,n}^k$, we have

$$\mathbb{P}(M_{m,n} = M) = \prod_{i=1}^{m} \prod_{j=1}^{n} \mathbb{P}(\tau_{i,j} = m_{i,j}) = \prod_{i,j} \left(p(1-p)^{a_{i,j}-1} \right)$$
$$= \left(\frac{p}{1-p} \right)^{mn} (1-p)^k.$$

It follows that

$$\mathbb{P}(S_{m,n} = k) = \sum_{M \in \mathcal{M}_{m,n}^k} \mathbb{P}(M_{m,n} = M) = \left(\frac{p}{1-p} \right)^{mn} (1-p)^k \left| \mathcal{M}_{m,n}^k \right|,$$

and therefore also

$$\mathbb{P}(M_{m,n} = M \mid S_{m,n} = k) = \left| \mathcal{M}_{m,n}^k \right|^{-1}.$$

That is, the conditional distribution of $M_{m,n}$ given the event $\{S_{m,n} = k\}$ is the uniform distribution on $\mathcal{M}_{m,n}^k$. (The cardinality of $\mathcal{M}_{m,n}^k$ is unimportant since it will end up cancelling out by the end of the computation, but in case you are curious about its value, see Exercise 5.7.) We now proceed to evaluate the probability $\mathbb{P}(G(m, n) \leq t)$ by conditioning on $S_{m,n}$ and making use of (5.8). Assume for convenience that t is an integer (clearly, knowing (5.14) for integer t implies the general case). We have

$$\mathbb{P}(G(m, n) \leq t) = \sum_{k=0}^{\infty} \mathbb{P}(S_{m,n} = k) \mathbb{P}(G(m, n) \leq t \mid S_{m,n} = k)$$
$$= \sum_{k=0}^{\infty} \mathbb{P}(S_{m,n} = k) \mathbb{P}(L(M_{m,n}) \leq t \mid S_{m,n} = k)$$
$$= \sum_{k=0}^{\infty} \left(\frac{p}{1-p} \right)^{mn} (1-p)^k \left| \left\{ M \in \mathcal{M}_{m,n}^k \mid L(M) \leq t \right\} \right|. \quad (5.16)$$

Now the RSK algorithm can be brought to bear: by the correspondence between matrices and triples (λ, P, Q), the cardinality of the set of matrices in $\mathcal{M}_{m,n}^k$ satisfying $L(M) \leq t$ can be represented as

$$\left| \left\{ M \in \mathcal{M}_{m,n}^k \mid L(M) \leq t \right\} \right| = \sum_{\lambda \vdash k, \lambda_1 \leq t} N(\lambda, m, n), \quad (5.17)$$

where for a Young diagram λ of order k, $N(\lambda, m, n)$ denotes the number of pairs (P, Q) of semistandard Young tableaux of shape λ such that P has entries from $\{1, \ldots, m\}$ and Q has entries from $\{1, \ldots, n\}$. Recalling that we assumed that $m \geq n$, note that $N(\lambda, m, n) = 0$ if λ has more than n parts, by the strict monotonicity of semistandard tableaux along columns. For λ with at most n parts, $\lambda_j = 0$ if $j > n$, so, by Theorem 5.8, we have

$$\begin{aligned} N(\lambda, m, n) &= \prod_{1 \leq i < j \leq m} \frac{\lambda_i - \lambda_j + j - i}{j - i} \cdot \prod_{1 \leq i < j \leq n} \frac{\lambda_i - \lambda_j + j - i}{j - i} \\ &= \frac{U_{m-n}}{U_n U_m} \prod_{1 \leq i < j \leq n} \left(\lambda_i - \lambda_j + j - i\right)^2 \cdot \prod_{i=1}^{n} \prod_{j=n+1}^{m} (\lambda_i + j - i), \end{aligned}$$

where we denote $U_k = \prod_{1 \leq i < j \leq k} (j - i) = 1!2! \ldots (k-1)!$. Note that a simple rearrangement of terms shows that $U_{m-n}/U_m U_n = \left(\prod_{j=0}^{n-1} j!(m - n + j)!\right)^{-1}$. Now, denoting $x_j = \lambda_j + n - j$ for $j = 1, \ldots, n$, the x_j's satisfy $x_1 > x_2 > \ldots > x_n \geq 0$, so we can rewrite this as

$$\begin{aligned} N(\lambda, m, n) &= \frac{U_{m-n}}{U_n U_m} \prod_{1 \leq i < j \leq n} (x_i - x_j)^2 \cdot \prod_{i=1}^{n} \prod_{j=n+1}^{m} (x_i + j - n) \\ &= \frac{U_{m-n}}{U_n U_m} \prod_{1 \leq i < j \leq n} (x_i - x_j)^2 \cdot \prod_{j=1}^{n} \frac{(x_j + m - n)!}{x_j!} \\ &= \frac{U_{m-n}}{U_n U_m} ((m - n)!)^n \prod_{1 \leq i < j \leq n} (x_i - x_j)^2 \cdot \prod_{j=1}^{n} \binom{x_j + m - n}{m - n}. \end{aligned} \tag{5.18}$$

The condition $\sum_{j=1}^{n} \lambda_j = k$ translates to $\sum_{j=1}^{n} x_j = k + \frac{n(n-1)}{2}$, and $\lambda_1 \leq t$ translates to $x_1 \leq t + n - 1$. So, combining the relations (5.16)–(5.18), we get that

$$\begin{aligned} &\mathbb{P}(G(m, n) \leq t) \\ &= \left(\frac{p}{1 - p}\right)^{mn} \sum_{k=0}^{\infty} (1 - p)^k \sum_{\lambda \vdash k, \lambda_1 \leq t} N(\lambda, m, n) \\ &= p^{mn} (1 - p)^{-mn - n(n-1)/2} \frac{U_{m-n}}{U_n U_m} ((m - n)!)^n \\ &\quad \times \sum_{k=0}^{\infty} \left[\sum_{\substack{0 \leq x_n < \ldots < x_1 \leq t+n-1 \\ \sum_j x_j = k + n(n-1)/2}} \prod_{1 \leq i < j \leq n} (x_i - x_j)^2 \prod_{j=1}^{n} \binom{x_j + m - n}{m - n} \cdot (1 - p)^{\sum_{j=1}^{n} x_j} \right] \\ &= n! \, C_{m,n,p} \sum_{0 \leq x_n < \ldots < x_1 \leq t+n-1} \prod_{1 \leq i < j \leq n} (x_i - x_j)^2 \prod_{j=1}^{n} \binom{x_j + m - n}{m - n} \cdot (1 - p)^{\sum_{j=1}^{n} x_j}. \end{aligned}$$

In this expression the summand is symmetric in $x_1, \ldots, x_n$. Replacing the sum by a summation over *unordered* sequences $0 \le x_1, \ldots, x_n \le t + n - 1$ exactly cancels out the $n!$ factor (and does not add any new contributions coming from vectors $(x_1, \ldots, x_n)$ with nondistinct coordinates, because of the $(x_i - x_j)^2$ factors), giving

$$C_{m,n,p} \sum_{x_1,\ldots,x_n=0}^{t+n-1} \prod_{1\le i<j\le n} (x_i - x_j)^2 \prod_{j=1}^{n} \left((1-p)^{x_j} \binom{x_j + m - n}{m - n} \right)$$
$$= C_{m,n,p} \sum_{x_1,\ldots,x_n=0}^{t+n-1} \prod_{1\le i<j\le n} (x_i - x_j)^2 \prod_{j=1}^{n} w_{m-n}^{(p)}(x_j).$$

This finishes the proof of (5.14).

Next, we prove (5.15) by taking a limit of (5.14) as $p \searrow 0$ and using the well-known fact from elementary probability that if $(X_p)_{0<p<1}$ is a family of random variables where $X_p \sim \text{Geom}(p)$ then $pX_p \xrightarrow{d} \text{Exp}(1)$ as $p \searrow 0$.

Redenote the geometric clock times $\tau_{i,j}$ discussed in the first part of the proof by $\tau_{i,j}^{(p)}$, to emphasize their dependence on p, and redenote the corresponding passage times by $G^{(p)}(m, n)$. Let $(T_{i,j})_{i,j\ge1}$ be an array of i.i.d. random times with the Exp(1) distribution, and let

$$H(m, n) = \max \left\{ \sum_{\ell=0}^{k} T_{p_\ell,q_\ell} \,:\, (p_\ell, q_\ell)_{\ell=0}^{k} \in \mathcal{Z}(1, 1; m, n) \right\}$$

be the associated passage times. Fix $m \ge n \ge 1$. By the relationship mentioned above between the geometric and exponential distributions, we have the convergence in distribution $(p\tau_{i,j}^{(p)})_{1\le i\le m,\, 1\le j\le n} \xrightarrow{d} (T_{i,j})_{1\le i\le m,\, 1\le j\le n}$ as $p \searrow 0$. Since the summation and maximum operations are continuous, we also get that

$$p\, G^{(p)}(m, n) \xrightarrow{d} H(m, n) \;\text{ as } p \searrow 0.$$

It follows that $\mathbb{P}(H(m, n) \le t) = \lim_{p\searrow0} \mathbb{P}(G^{(p)}(m, n) \le t/p)$. To evaluate the limit, note first that, for $u \ge 0$, we have that as $p \searrow 0$,

$$w_k^{(p)}(u/p) = \frac{1}{k!} \prod_{j=1}^{k} \left(\frac{u}{p} + j \right) (1-p)^{u/p} = \frac{1}{k!} \frac{u^k + O(p)}{p^k} e^{-u+O(p)}$$
$$= (1 + O(p)) \frac{1}{p^k k!} (W_k(u) + O(p)),$$

with the constant implicit in the big-O notation being uniform as u ranges over a compact interval $[0, T]$. Hence we can write $\mathbb{P}\left(G^{(p)}(m,n) \le \frac{t}{p}\right)$ as an approximate multidimensional Riemann sum that is seen to converge to a corresponding multidimensional integral. That is, we have

$$
\begin{aligned}
&\mathbb{P}\left(G^{(p)}(m,n) \le \frac{t}{p}\right) \\
&= C_{m,n,p} \sum_{x_1,\ldots,x_n=0}^{\lfloor \frac{t}{p} \rfloor + n-1} \prod_{1\le i<j\le n} (x_i - x_j)^2 \prod_{i=1}^{n} w_{m-n}^{(p)}(x_i) \\
&= C_{m,n,p}\, p^{-n(n-1)} \sum_{x_1,\ldots,x_n=0}^{\lfloor \frac{t}{p} \rfloor + n-1} \prod_{1\le i<j\le n} (px_i - px_j)^2 \prod_{i=1}^{n} w_{m-n}^{(p)}\left(\frac{px_i}{p}\right) \\
&= (1 + O(p)) C_{m,n,p} p^{-n(n-1)} \\
&\quad \times \int_0^t \cdots \int_0^t \prod_{1\le i<j\le n} (u_i - u_j)^2 \prod_{i=1}^{n} \left(\frac{W_{m-n}(u_i) + O(p)}{p^{m-n}(m-n)!}\right) \frac{du_1 \ldots du_n}{p^n} \\
&= (1 + O(p)) \frac{C_{m,n,p} p^{-mn}}{((m-n)!)^n} \\
&\quad \times \int_0^t \cdots \int_0^t \prod_{1\le i<j\le n} (u_i - u_j)^2 \prod_{i=1}^{n} W_{m-n}(u_i)\, du_1 \ldots du_n.
\end{aligned}
$$

Taking the limit as $p \searrow 0$ gives (5.15) with the normalization constant $D_{m,n}$ being given by (5.13), as claimed. □

5.6 The Fredholm determinant representation

For most of the remainder of the chapter, we focus on the passage times $G(m,n)$ in the case of the corner growth process. The case of the multi-corner growth process can be treated using the same techniques; see Section 5.13.

The next major step in our analysis will be to transform the representation (5.15) for the distribution function of $G(m,n)$ into a new form involving a Fredholm determinant, which is more suitable for asymptotic analysis. The determinant involves a family of special functions called the **Laguerre polynomials**, which we define now. For an integer $\alpha \ge 0$, the Laguerre polynomials with parameter α are the family of polynomials

$(\ell_n^\alpha(x))_{n=0}^\infty$ defined by

$$\ell_n^\alpha(x) = \left(\frac{n!}{(n+\alpha)!}\right)^{1/2} \sum_{k=0}^{n} \frac{(-1)^{n+k}}{k!} \binom{n+\alpha}{k+\alpha} x^k. \tag{5.19}$$

(The Laguerre polynomials are also defined more generally for arbitrary real $\alpha \geq 0$ by replacing factorials with gamma-function factors in the obvious way, but we will need only the case of integer α.) We also define the associated **Laguerre kernels** with parameter α to be the sequence of functions $\mathbf{L}_n^\alpha \colon \mathbb{R} \times \mathbb{R} \to \mathbb{R}$, $(n \geq 0)$, given by

$$\begin{aligned} \mathbf{L}_n^\alpha(x,y) = {} & \sqrt{n(n+\alpha)} \cdot x^{\alpha/2} y^{\alpha/2} e^{-x/2} e^{-y/2} \\ & \times \begin{cases} \dfrac{\ell_n^\alpha(x)\ell_{n-1}^\alpha(y) - \ell_{n-1}^\alpha(x)\ell_n^\alpha(y)}{x-y} & \text{if } x \neq y, \\ (\ell_n^\alpha)'(x)\ell_{n-1}(x) - (\ell_{n-1}^\alpha)'(x)\ell_n^\alpha(x) & \text{if } x = y. \end{cases} \end{aligned} \tag{5.20}$$

Theorem 5.12 *For any $m \geq n \geq 1$, the distribution of the passage times $G(m,n)$ in the corner growth process is given by the Fredholm determinant*

$$\begin{aligned} \mathbb{P}(G(m,n) \leq t) &= \det\left(\mathbf{I} - (\mathbf{L}_n^{m-n})_{[t,\infty)}\right) \\ &= 1 + \sum_{k=1}^{n} \frac{(-1)^k}{k!} \int_t^\infty \ldots \int_t^\infty \det_{i,j=1}^{k} \left(\mathbf{L}_n^{m-n}(x_i,x_j)\right) dx_1 \ldots dx_k. \end{aligned} \tag{5.21}$$

The leap from (5.15) to (5.21) is admittedly a rather nonobvious one, and relies on techniques from the theory of **orthogonal polynomials**. We develop the necessary ideas in the next two sections, and use them to prove Theorem 5.12 in Section 5.9.

5.7 Orthogonal polynomials

Our discussion of orthogonal polynomials will be a minimal one tailored to our present needs; in particular, for simplicity we focus on the so-called absolutely continuous case. See the box on p. 294 for more background on the general theory. The construction of a family of orthogonal polynomials starts with a **weight function**, which is a measurable function $w \colon \mathbb{R} \to$

$[0, \infty)$ with the properties that

$$\int_{\mathbb{R}} w(x)\,dx > 0, \qquad \int_{\mathbb{R}} w(x)|x|^n\,dx < \infty \quad (n \geq 0).$$

We associate with such a function the inner product $\langle \cdot, \cdot \rangle_w$ defined for functions $f, g\colon \mathbb{R} \to \mathbb{R}$ by

$$\langle f, g \rangle_w = \int_{\mathbb{R}} f(x)g(x)\,w(x)\,dx$$

whenever the integral converges absolutely. (As usual, the inner product induces a norm $\|f\|_w = \langle f, f \rangle_w^{1/2}$; the set of functions for which $\|f\|_w < \infty$, with the convention that two functions f, g are considered equal if $\|f - g\|_w = 0$, is the Hilbert space $L^2(\mathbb{R}, w(x)\,dx)$.) A sequence of polynomials $(p_n(x))_{n=0}^{\infty}$ is called **the family of orthogonal polynomials associated with** $w(x)$ if the p_n's satisfy the following properties:

1. p_n is a polynomial of degree n.
2. p_n has a positive leading coefficient.
3. The p_n are orthonormal with respect to the measure $w(x)\,dx$. That is, for all $m, n \geq 0$ we have

$$\langle p_n, p_m \rangle_w = \int_{\mathbb{R}} p_n(x)p_m(x)\,w(x)\,dx = \delta_{mn} = \begin{cases} 1 & \text{if } m = n, \\ 0 & \text{otherwise.} \end{cases} \tag{5.22}$$

Note the use of the definite article in the preceding definition, suggesting uniqueness of such a family. Indeed, we have the following fundamental result.

Theorem 5.13 (Existence and uniqueness of orthogonal polynomial families) *Given a weight function $w(x)$, there exists a unique family of orthogonal polynomials associated with it.*

Proof The proof of existence is essentially the Gram–Schmidt orthogonalization procedure applied to the sequence of monomials $1, x, x^2, \ldots$ in the inner product space $L^2(\mathbb{R}, w(x)\,dx)$. We take $p_0(x) = \langle 1, 1 \rangle_w^{1/2}$, and for $n \geq 1$ define $p_n(x)$ inductively in terms of $p_0, \ldots, p_{n-1}$ as

$$p_n(x) = \frac{x^n - \sum_{k=0}^{n-1} \langle x^n, p_k(x) \rangle_w \, p_k(x)}{\left\| x^n - \sum_{k=0}^{n-1} \langle x^n, p_k(x) \rangle_w \, p_k(x) \right\|_w}.$$

Note that, by induction, assuming that p_k is a polynomial of degree k for all $0 \le k < n$, we have that $x^n - \sum_{k=0}^{n-1} \langle x^n, p_k(x) \rangle_w \, p_k(x)$ is a monic polynomial of degree n. This implies that the denominator in the definition of p_n is positive and that p_n is a polynomial of degree n with positive leading coefficient. The orthonormality is also easy to verify inductively just as with the standard Gram–Schmidt procedure.

To verify uniqueness, assume that $(q_n)_{n=0}^{\infty}$ is another family satisfying properties 1–3 in the preceding definition. An easy corollary of properties 1–2 is that for any $n \ge 0$ we have that

$$\mathrm{span}\{p_0, \ldots, p_n\} = \mathrm{span}\{q_0, \ldots, q_n\} = \mathrm{span}\{1, \ldots, x^n\} =: \mathrm{Poly}_n(\mathbb{R}).$$

It is easy to see that we must have $q_0 = p_0 = \left(\int_{\mathbb{R}} w(x)\, dx\right)^{-1/2}$. Assume by induction that we proved that $q_k = p_k$ for all $0 \le k \le n-1$. Then by property 3, both p_n and q_n are orthogonal to $p_0, \ldots, p_{n-1}$ and are therefore in the orthogonal complement of (the n-dimensional space) $\mathrm{Poly}_{n-1}(\mathbb{R})$ considered as a vector subspace of (the $(n+1)$-dimensional space) $\mathrm{Poly}_n(\mathbb{R})$. This orthogonal complement is one-dimensional, so p_n and q_n are proportional to each other. By properties 1–2 and the case $m = n$ of property 3, it is easy to see that they must be equal. □

Various specific choices of the weight function $w(x)$ lead to interesting families of polynomials, the most important of which are named after famous 18th- and 19th-century mathematicians who pioneered the field. For example, when $w(x) = \mathbf{1}_{[0,1]}(x)$ one obtains the **Legendre polynomials**; $w(x) = e^{-x^2}$ leads to the **Hermite polynomials**; $w(x) = 1/\sqrt{1-x^2}$ gives the **Chebyshev polynomials**, etc. These so-called classical families of orthogonal polynomials often arise naturally as answers to concrete questions and have found applications in many areas of pure and applied mathematics. See the box on the next page for a list of the more common families and some of their uses.

Continuing with the general theory, denote the leading coefficient of p_n by κ_n. For convenience, we also denote $p_{-1} \equiv 0$ and let κ_{-1} denote an arbitrary real number.

Lemma 5.14 (Three-term recurrence) *Let $(A_n)_{n=1}^{\infty}$ and $(C_n)_{n=1}^{\infty}$ be given by*

$$A_n = \frac{\kappa_n}{\kappa_{n-1}}, \qquad C_n = \frac{\kappa_n \kappa_{n-2}}{\kappa_{n-1}^2}.$$

Orthogonal polynomials and their applications

Orthogonal polynomials were first studied by the Russian mathematician Pafnuty Chebyshev in the 19th century in connection with the study of continued fraction expansions. The **classical orthogonal polynomials**, which are specific families of orthogonal polynomials studied by and named after well-known mathematicians of the era, have found many applications in diverse areas of mathematics. Since they satisfy fairly simple linear second-order differential equations, they arise naturally in connection with many important problems of analysis and mathematical physics. The table below lists the families of classical orthogonal polynomials and a few of their applications. For more information on the general theory, refer to [5], [131].

Family	Weight function	Applications
Hermite	e^{-x^2}	Quantum harmonic oscillator, diagonalization of the Fourier transform, GUE random matrices
Chebyshev	$1/\sqrt{1-x^2}$	Polynomial interpolation
Legendre	$\mathbf{1}_{[-1,1]}(x)$	Spherical harmonics, multipole expansions
Jacobi	$(1-x)^\alpha(1+x)^\beta\mathbf{1}_{[-1,1]}(x)$	Representations of $SU(2)$
Laguerre	$e^{-x}x^\alpha\mathbf{1}_{[0,\infty)}(x)$	the hydrogen atom, Wishart random matrices

Orthogonal polynomial ensembles are random point processes that arise in various settings, most notably as eigenvalue distributions in random matrix theory. The key property that makes their detailed analysis possible is that they are determinantal point processes whose correlation kernel is the reproducing kernel of an orthogonal polynomial family. See the survey [72] for more information and many examples. For extensive information about applications of orthogonal polynomials to random matrix theory, see [4], [30], [42], [88].

There are real numbers $(B_n)_{n=1}^\infty$ such that the orthogonal polynomials $(p_n)_{n=0}^\infty$ satisfy the recurrence relation

$$p_n(x) = (A_n x + B_n)p_{n-1}(x) - C_n p_{n-2}(x) \qquad (n \geq 1). \tag{5.23}$$

Proof Define $q_n(x) = p_n(x) - A_n x p_{n-1}(x)$. The subtraction cancels the highest-order term x^n, so we get that $\deg q_n < n$. It follows that $q_n(x)$ can be expressed as a linear combination of $p_0, \ldots, p_{n-1}$. By (5.22) and standard linear algebra, the coefficients in this linear combination are the inner products of q_n with $p_0, \ldots, p_{n-1}$:

$$q_n(x) = \sum_{k=0}^{n-1} \langle q_n, p_k \rangle_w p_k(x).$$

But note that

$$\begin{aligned}
\langle q_n, p_k \rangle_w &= \langle p_n, p_k \rangle - A_n \langle x p_{n-1}, p_k \rangle_w = -A_n \int_{\mathbb{R}} x p_{n-1}(x) p_k(x)\, w(x)\, dx \\
&= -A_n \langle p_{n-1}, x p_k \rangle_w,
\end{aligned}$$

and this is equal to 0 if $0 \leq k \leq n-3$, since in that case the polynomial $x p_k(x)$ is of degree $k+1 \leq n-2$ and is therefore expressible as a linear combination of $p_0, \ldots, p_{n-2}$, and in particular is orthogonal to p_{n-1}. So, we have shown that actually q_n is expressible as

$$q_n(x) = \langle q_n, p_{n-1} \rangle_w p_{n-1}(x) + \langle q_n, p_{n-2} \rangle_w p_{n-2}(x) =: B_n p_{n-1}(x) - D_n p_{n-2}(x),$$

or equivalently that

$$p_n(x) = (A_n x + B_n) p_{n-1}(x) - D_n p_{n-2},$$

for some numbers B_n, D_n. The value of D_n can be found by assuming inductively that (5.23) holds for $n \geq 1$, and writing

$$\begin{aligned}
0 = \langle p_n, p_{n-2} \rangle_w &= \langle A_n x p_{n-1} + B_n p_{n-1} - D_n p_{n-2}, p_{n-2} \rangle_w \\
&= A_n \langle x p_{n-1}, p_{n-2} \rangle_w + B_n \langle p_{n-1}, p_{n-2} \rangle_w - D_n \langle p_{n-2}, p_{n-2} \rangle_w \\
&= A_n \langle p_{n-1}, x p_{n-2} \rangle_w - D_n \\
&= A_n \langle p_{n-1}, (\kappa_{n-2}/\kappa_{n-1}) p_{n-1} + \text{[lin. combination of } p_0, \ldots, p_{n-2}] \rangle_w - D_n \\
&= A_n \frac{\kappa_{n-2}}{\kappa_{n-1}} - D_n,
\end{aligned}$$

which gives that $D_n = A_n \frac{\kappa_{n-2}}{\kappa_{n-1}} = \frac{\kappa_n \kappa_{n-2}}{\kappa_{n-1}^2} = C_n$, as claimed. □

Next, for any $n \geq 0$ define a kernel $K_n : \mathbb{R} \times \mathbb{R} \to \mathbb{R}$ by

$$\mathbf{K}_n(x, y) = \sum_{k=0}^{n-1} p_k(x) p_k(y).$$

$\mathbf{K}_n$ is called the nth **reproducing kernel** associated with the family of orthogonal polynomials $(p_n)_{n=0}^{\infty}$. It is easy to see that if we interpret $\mathbf{K}_n$ in the usual way as a linear operator acting on functions $f \in L^2(\mathbb{R}, w(x)\,dx)$ by

$$(\mathbf{K}_n f)(x) = \int_{\mathbb{R}} \mathbf{K}_n(x,y) f(y)\,dy,$$

then $\mathbf{K}_n$ is simply the orthogonal projection operator onto the linear subspace $\operatorname{span}\{p_0,\ldots,p_{n-1}\} = \operatorname{span}\{1,\ldots,x^{n-1}\}$ of $L^2(\mathbb{R}, w(x)\,dx)$. In particular, if f is a polynomial of degree $\le n-1$ then $\mathbf{K}_n f = f$.

It is also useful to note that $\mathbf{K}_n(x,y) = \mathbf{K}_n(y,x)$, that is, $\mathbf{K}_n$ is symmetric, and furthermore it is a positive-semidefinite kernel, in the sense that for any numbers $x_1,\ldots,x_m \in \mathbb{R}$, the matrix $K_{x_1,\ldots,x_m} = (\mathbf{K}_n(x_i,x_j))_{i,j=1}^m$ is positive-semidefinite. To check this, take a vector $\mathbf{u} = (u_1,\ldots,u_m)$ (considered as a column vector) and note that

$$\begin{aligned} \mathbf{u}^\top K_{x_1,\ldots,x_m}\mathbf{u} &= \sum_{i,j=1}^m u_i u_j \mathbf{K}_n(x_i,x_j) \\ &= \sum_{k=0}^{n-1}\sum_{i,j=1}^m u_i u_j p_k(x_i) p_k(x_j) = \sum_{k=0}^{n-1}\left(\sum_{i=1}^m u_i u_j p_k(x_i)\right)^2 \ge 0. \quad (5.24) \end{aligned}$$

Theorem 5.15 (The Christoffel–Darboux formula) *The reproducing kernel can be expressed in terms of just two successive polynomials $p_{n-1}(x), p_n(x)$, as follows:*

$$\mathbf{K}_n(x,y) = \begin{cases} \frac{\kappa_{n-1}}{\kappa_n} \cdot \dfrac{p_n(x)p_{n-1}(y) - p_{n-1}(x)p_n(y)}{x-y} & \text{if } x \ne y, \\ \frac{\kappa_{n-1}}{\kappa_n}(p_n'(x)p_{n-1}(x) - p_n(x)p_{n-1}'(x)) & \text{if } x = y. \end{cases}$$

Proof The case $x = y$ is proved by taking the limit as $y \to x$ in the case $x \ne y$, using L'Hôpital's rule. For the case $x \ne y$, use (5.23) to write

$$\begin{aligned} &p_k(x)p_{k-1}(y) - p_{k-1}(x)p_k(y) \\ &\quad = \Big((A_k x + B_k)p_{k-1}(x) - C_k p_{k-2}(x)\Big)p_{k-1}(y) \\ &\qquad\qquad - p_{k-1}(x)\Big((A_k y + B_k)p_{k-1}(y) - C_k p_{k-2}(y)\Big) \\ &\quad = A_k(x-y)p_{k-1}(x)p_{k-1}(y) + C_k(p_{k-1}(x)p_{k-2}(y) - p_{k-2}(x)p_{k-1}(y)). \end{aligned}$$

Equivalently,

$$p_{k-1}(x)p_{k-1}(y) = \frac{\kappa_{k-1}}{\kappa_k} \cdot \frac{p_k(x)p_{k-1}(y) - p_{k-1}(x)p_k(y)}{x-y} - \frac{\kappa_{k-2}}{\kappa_{k-1}} \cdot \frac{p_{k-1}(x)p_{k-2}(y) - p_{k-2}(x)p_{k-1}(y)}{x-y}.$$

Summing this equation over $k = 1, \ldots, n$ gives a telescopic sum that evaluates to

$$\mathbf{K}_n(x,y) = \sum_{k=1}^{n} p_{k-1}(x)p_{k-1}(y) = \frac{\kappa_{n-1}}{\kappa_n} \cdot \frac{p_n(x)p_{n-1}(y) - p_{n-1}(x)p_n(y)}{x-y},$$

as claimed. □

5.8 Orthogonal polynomial ensembles

Orthogonal polynomials arise in certain probabilistic settings in connection with a class of random processes known as **orthogonal polynomial ensembles**, which we describe next. Given a weight function $w(x)$, we associate with it a family of probability densities $f_w^{(n)} : \mathbb{R}^n \to [0, \infty)$, $n = 1, 2, \ldots$. For each n, $f_w^{(n)}$ is given by

$$f_w^{(n)}(x_1, \ldots, x_n) = \frac{1}{Z_n} \prod_{1 \le i < j \le n} (x_i - x_j)^2 \prod_{j=1}^{n} w(x_j), \tag{5.25}$$

where

$$Z_n = \int \cdots \int_{\mathbb{R}^n} \prod_{1 \le i < j \le n} (x_i - x_j)^2 \prod_{j=1}^{n} w(x_j)\, dx_1 \ldots dx_n. \tag{5.26}$$

If $\mathbf{X}_w^{(n)} = (X_{w,1}^{(n)}, \ldots, X_{w,n}^{(n)})$ is a random vector that has $f_w^{(n)}$ as its probability density function, we say that $\mathbf{X}_w^{(n)}$ is the nth **orthogonal polynomial ensemble** associated with the weight function w. Note that the coordinates of $\mathbf{X}_w^{(n)}$ have a random order, but since the density of $\mathbf{X}_w^{(n)}$ is a symmetric function of $x_1, \ldots, x_n$, their order structure is a uniformly random permutation in S_n and is statistically independent of the values of the coordinates, so it is natural to ignore the ordering and interpret $\mathbf{X}_w^{(n)}$ as a random set of n real numbers $\xi_{w,1}^{(n)} < \ldots < \xi_{w,n}^{(n)}$, that is, as a random point process (see Chapter 2). Now, if the density (5.25) consisted of just the factor $\prod_{j=1}^{n} w(x_j)$, we would have a rather uninteresting point process of n i.i.d. points sampled according to the weight function $w(x)$ (normalized to be a probability density). What makes things more interesting is the addition of the Vandermonde factor $\prod_{1\le i<j\le n}(x_i - x_j)^2$, which effectively adds a "repulsion" between the random points $X^{(n)}_{w,1},\dots,X^{(n)}_{w,n}$. We see below that this results in the existence of nice determinantal formulas for various quantities associated with the process $\mathbf{X}^{(n)}_w$, and that the orthogonal polynomial family $(p_n(x))_{n=0}^\infty$ associated with the weight function $w(x)$ enters the picture in a natural way.

Lemma 5.16 *Let (Z,Σ,μ) be a finite measure space. For any $n\ge 1$ and for any bounded measurable functions $f_1,\dots,f_n,g_1,\dots,g_n\colon Z\to\mathbb{R}$, we have*

$$\det_{i,j=1}^{n}\left(\int_Z f_i(x)g_j(x)\,d\mu(x)\right) = \frac{1}{n!}\int\cdots\int_{Z^n}\det_{i,j=1}^{n}\left(f_i(x_j)\right)\det_{i,j=1}^{n}\left(g_i(x_j)\right)\prod_{j=1}^{n}d\mu(x_j).$$

Proof

$$\begin{aligned}
&\int\cdots\int_{Z^n}\det_{i,j=1}^{n}\left(f_i(x_j)\right)\det_{i,j=1}^{n}\left(g_i(x_j)\right)\prod_{j=1}^{n}d\mu(x_j)\\
&\qquad=\sum_{\sigma,\tau\in S_n}\operatorname{sgn}(\sigma)\operatorname{sgn}(\tau)\prod_{j=1}^{n}\int_Z f_{\sigma(j)}(x)g_{\tau(j)}(x)\,d\mu(x)\\
&\qquad=\sum_{\pi\in S_n}\sum_{\substack{\sigma,\tau\in S_n\\ \sigma\circ\tau^{-1}=\pi}}\operatorname{sgn}(\sigma\circ\tau^{-1})\prod_{j=1}^{n}\int_Z f_{\sigma(j)}(x)g_{\tau(j)}(x)\,d\mu(x)\\
&\qquad=\sum_{\pi\in S_n}\sum_{\substack{\sigma,\tau\in S_n\\ \sigma\circ\tau^{-1}=\pi}}\operatorname{sgn}(\pi)\prod_{k=1}^{n}\int_Z f_{\pi(k)}(x)g_{k}(x)\,d\mu(x)\\
&\qquad=n!\sum_{\pi\in S_n}\operatorname{sgn}(\pi)\prod_{k=1}^{n}\int_Z f_{\pi(k)}(x)g_{k}(x)\,d\mu(x)\\
&\qquad=n!\det_{i,j=1}^{n}\left(\int_Z f_i(x)g_j(x)\,d\mu(x)\right).
\end{aligned}$$

□

Lemma 5.17 *If $f\colon\mathbb{R}\to\mathbb{R}$ is a bounded measurable function, then we have*

$$\mathbb{E}\left(\prod_{j=1}^{n}f(X^{(n)}_{w,j})\right)=\det_{i,j=0}^{n-1}\left(\int_{\mathbb{R}}p_i(x)p_j(x)f(x)\,w(x)\,dx\right). \tag{5.27}$$

Proof Note that the Vandermonde product $\prod_{1\le i<j\le n}(x_j-x_i)$ can be rewrit-

ten as

$$\prod_{1\le i<j\le n}(x_j-x_i)=\det_{i,j=1}^{n}\left(x_i^{j-1}\right)=\frac{1}{\prod_{j=0}^{n-1}\kappa_j}\det_{i,j=1}^{n}\left(p_{j-1}(x_i)\right).$$

(This does not use the orthonormality property but simply the fact that p_j is a polynomial of degree j with leading coefficient κ_j, so the matrix on the right-hand side is obtained from the standard Vandermonde matrix by applying a triangular sequence of elementary row operations.) We therefore get, using Lemma 5.16 (with the obvious parameters $Z=\mathbb{R}$ and $d\mu(x)=w(x)\,dx$), that

$$\begin{aligned}\mathbb{E}\left(\prod_{j=1}^{n}f(X_{w,j}^{(n)})\right)&=\frac{1}{Z_n}\int\cdots\int_{\mathbb{R}^n}\prod_{1\le i<j\le n}(x_i-x_j)^2\prod_{j=1}^{n}(f(x_j)w(x_j)\,dx_j)\\&=\frac{1}{Z_n\prod_{j=0}^{n-1}\kappa_j^2}\int\cdots\int_{\mathbb{R}^n}\det_{i,j=1}^{n}\left(f(x_i)p_{j-1}(x_i)\right)\det_{i,j=1}^{n}\left(p_{j-1}(x_i)\right)\\&\qquad\times\prod_{j=1}^{n}w(x_j)\,dx_1\ldots dx_n\\&=\frac{n!}{Z_n\prod_{j=0}^{n-1}\kappa_j^2}\det_{i,j=1}^{n}\left(\int_{\mathbb{R}}p_{i-1}(x)p_{j-1}(x)f(x)\,w(x)\,dx\right).\end{aligned}$$

This is similar to (5.27) except for the presence of the numerical factor $\frac{n!}{Z_n\prod_{j=0}^{n-1}\kappa_j^2}$. However, we showed that this relation holds for an arbitrary bounded measurable function f. Specializing to the case $f\equiv 1$, and now finally using the orthonormality relation (5.22), shows that this factor is equal to 1, which proves the result for general f. □

Corollary 5.18 *The normalization constant Z_n defined in* (5.26) *is given by*

$$Z_n=\frac{n!}{\prod_{j=0}^{n-1}\kappa_j^2}.$$

Theorem 5.19 *For any Borel set $E\subset\mathbb{R}$, the probability that all the points $X_{w,1}^{(n)},\ldots,X_{w,n}^{(n)}$ in the orthogonal polynomial ensemble $\mathbf{X}_w^{(n)}$ lie outside of E*

can be represented as a Fredholm determinant, namely

$$\begin{aligned}\mathbb{P}(X^{(n)}_{w,1}, \ldots, X^{(n)}_{w,n} \notin E) &= \det\left(\mathbf{I} - (\mathbf{K}_n)_{|L^2(E,w(x)\,dx)}\right) \\ &= 1 + \sum_{m=1}^{n} \frac{(-1)^m}{m!} \int \cdots \int_{E^m} \det_{i,j=1}^{m}\left(\mathbf{K}_n(x_i, x_j)\right) \prod_{j=1}^{m} w(x_j)\,dx_1 \ldots dx_m. \end{aligned} \tag{5.28}$$

Proof Let $f\colon \mathbb{R} \to \mathbb{R}$ be the function $f(x) = \mathbf{1}_{\mathbb{R}\setminus E}(x) = 1 - \mathbf{1}_E(x)$. Applying Lemma 5.17, we have

$$\begin{aligned}\mathbb{P}(X^{(n)}_{w,1}, \ldots, X^{(n)}_{w,n} \notin E) &= \mathbb{E}\left[\prod_{j=1}^{n} f(X^{(n)}_{w,j})\right] = \det_{i,j=0}^{n-1}\left(\int_{\mathbb{R}} p_i(x)p_j(x)(1 - \mathbf{1}_E(x))w(x)\,dx\right) \\ &= \det_{i,j=0}^{n-1}\left(\delta_{ij} - \int_{\mathbb{R}} p_i(x)p_j(x)\mathbf{1}_E(x)w(x)\,dx\right) \\ &= 1 + \sum_{m=1}^{n} \frac{(-1)^m}{m!} \sum_{0\le i_1,\ldots,i_m\le n-1} \det_{k,\ell=1}^{m}\left(\int_{\mathbb{R}} p_{i_k}(x)p_{i_\ell}(x)\mathbf{1}_E(x)w(x)\,dx\right)\end{aligned}$$

(if you find the last transition confusing, refer to equation (2.21) and the remark immediately following it). Applying Lemma 5.16 and interchanging the summation and integration operations, this can be rewritten as

$$\begin{aligned}1 + \sum_{m=1}^{n} &\frac{(-1)^m}{m!} \int \cdots \int_{\mathbb{R}^m} \left[\frac{1}{m!} \sum_{0\le i_1,\ldots,i_m\le n-1} \det_{k,\ell=1}^{m}\left(p_{i_k}(x_\ell)\right) \right.\\ &\qquad \left. \times \det_{k,\ell=1}^{m}\left(p_{i_k}(x_\ell)\mathbf{1}_E(x_\ell)\right)\right] \prod_{k=1}^{m} (w(x_k)\,dx_k) \\ = 1 + \sum_{m=1}^{n} &\frac{(-1)^m}{m!} \int \cdots \int_{\mathbb{R}^m} \left[\frac{1}{m!} \sum_{0\le i_1,\ldots,i_m\le n-1} \det_{k,\ell=1}^{m}\left(p_{i_k}(x_\ell)\right) \det_{k,\ell=1}^{m}\left(p_{i_k}(x_\ell)\right)\right] \\ &\qquad \times \prod_{k=1}^{m} (\mathbf{1}_E(x_k)\, w(x_k)\,dx_k).\end{aligned}$$

In this last expression, the inner m-fold summation can be interpreted as an m-fold integration with respect to the counting measure on $\{0, \ldots, n-1\}$, so Lemma 5.16 can be applied again, transforming our expression into

$$1+\sum_{m=1}^{n}\frac{(-1)^m}{m!}\int\dots\int_{\mathbb{R}^m}\det_{k,\ell=1}^{m}\left(\sum_{i=0}^{n-1}p_{i-1}(x_k)p_{j-1}(x_\ell)\right)\prod_{k=1}^{m}(\mathbf{1}_E(x_k)\,w(x_k)\,dx_k)$$

$$=1+\sum_{m=1}^{n}\frac{(-1)^m}{m!}\int\dots\int_{E^m}\det_{i,j=1}^{m}\left(\mathbf{K}_n(x_i,x_j)\right)\prod_{j=1}^{m}w(x_j)\,dx_1\dots dx_m,$$

which is precisely the right-hand side of (5.28). □

Closely related to the existence of the identity (5.28) is the fact that the orthogonal polynomial ensemble $\mathbf{X}_w^{(n)}$ is a determinantal point process (of the absolutely continuous variety; see the discussion in Chapter 2) whose correlation kernel is related to the reproducing kernel $\mathbf{K}_n(x,y)$. The following result makes this precise.

Theorem 5.20 *Define a modified version $\widetilde{\mathbf{K}}_n(x,y)$ of the reproducing kernel by*

$$\widetilde{\mathbf{K}}_n(x,y)=\sqrt{w(x)w(y)}\mathbf{K}(x,y).$$

For any $1\le m\le n$, the marginal joint density function of the first m coordinates $(X_{w,1}^{(n)},\dots,X_{w,m}^{(n)})$ of $\mathbf{X}_w^{(n)}$ is given by

$$f_w^{(n,m)}(x_1,\dots,x_m)=\frac{(n-m)!}{n!}\det_{i,j=1}^{m}\left(\widetilde{\mathbf{K}}_n(x_i,x_j)\right).$$

We will not need Theorem 5.20 for our purposes. See Exercises 5.9 and 5.10 for a proof idea.

5.9 The Laguerre polynomials

We collect here a few elementary properties of the Laguerre polynomials $\ell_n^\alpha(x)$ defined in (5.19). For convenience, define a simpler variant

$$L_n^\alpha(x)=\sum_{k=0}^{n}\frac{(-1)^k}{k!}\binom{n+\alpha}{k+\alpha}x^k$$

so that $\ell_n^\alpha(x)=(-1)^n\sqrt{\frac{n!}{(n+\alpha)!}}L_n^\alpha(x)$. Note that $L_n^\alpha(x)$ is an nth degree polynomial with leading coefficient $(-1)^n/n!$. It is also immediate to check that $(L_n^\alpha)'(x)=-L_{n-1}^{\alpha+1}(x)$, or equivalently in terms of $\ell_n^\alpha(x)$ we have

$$(\ell_n^\alpha)'(x)=\sqrt{n}\,\ell_{n-1}^{\alpha+1}(x). \tag{5.29}$$

Lemma 5.21 *The polynomials $L_n^\alpha(x)$ satisfy*

$$L_n^\alpha(x) = \frac{1}{n!} e^x x^{-\alpha} \frac{d^n}{dx^n} \left(e^{-x} x^{n+\alpha}\right) \qquad (n \geq 0). \tag{5.30}$$

Proof By the Leibniz rule $\frac{d^n}{dx^n}(fg) = \sum_{k=0}^n \binom{n}{k} f^{(k)} g^{(n-k)}$, we have

$$\begin{aligned} \frac{d^n}{dx^n}\left(e^{-x}x^{n+\alpha}\right) &= \sum_{k=0}^n \binom{n}{k} (-1)^k e^{-x}(n+\alpha)(n+\alpha-1)\dots(k+\alpha+1)x^{k+\alpha} \\ &= n! e^{-x} x^\alpha \sum_{k=0}^n \frac{(-1)^k}{k!} \binom{n+\alpha}{k+\alpha} x^k. \end{aligned}$$

□

Lemma 5.22 *The polynomials $L_n^\alpha(x)$ satisfy*

$$\int_0^\infty L_n(x) L_m(x) e^{-x} x^\alpha \, dx = \frac{(n+\alpha)!}{n!} \delta_{nm} \qquad (n, m \geq 0). \tag{5.31}$$

Proof Assume without loss of generality that $n \geq m$. By (5.30), the left-hand side of (5.31) can be written as

$$\frac{1}{n!} \int_0^\infty \frac{d^n}{dx^n} \left(e^{-x} x^{n+\alpha}\right) L_m^\alpha(x) \, dx.$$

Integrating by parts n times repeatedly gives

$$\frac{(-1)^n}{n!} \int_0^\infty e^{-x} x^{n+\alpha} \frac{d^n L_m^\alpha(x)}{dx^n} \, dx = \begin{cases} 0 & \text{if } n > m, \\ \frac{1}{n!} \int_0^\infty e^{-x} x^{n+\alpha} \, dx = \frac{(n+\alpha)!}{n!} & \text{if } n = m, \end{cases}$$

as claimed. □

Corollary 5.23 *The Laguerre polynomials $\ell_n^\alpha(x)$ are the orthogonal polynomials associated with the weight function $W_\alpha(x) = x^\alpha e^{-x} \mathbf{1}_{[0,\infty)}(x)$.*

Using the Christoffel–Darboux formula (Theorem 5.15) we now see that the Laguerre kernel $\mathbf{L}_n^\alpha(x, y)$ defined in (5.20) is a slightly modified version of the reproducing kernel $\mathbf{K}_n^\alpha$ associated with the Laguerre polynomials $\ell_n^\alpha(x)$. More precisely, we have the relation

$$\mathbf{L}_n^\alpha(x, y) = \sqrt{W_\alpha(x) W_\alpha(y)} \mathbf{K}_n^\alpha(x, y). \tag{5.32}$$

Proof of Theorem 5.12 By Theorem 5.11, the probability $\mathbb{P}(G(m, n) \leq t)$ is equal to the probability that all the points of the orthogonal polynomial ensemble $\mathbf{X}_{W_{m-n}}^{(n)}$ associated with the weight function $W_{m-n}(x) = e^{-x} x^{m-n} \mathbf{1}_{[0,\infty)}(x)$ lie in the interval $[0, t]$. By Theorem 5.19, this probability is expressed by the Fredholm determinant (5.28), where $w(x) = W_{m-n}(x)$, the set E is taken as $[0,t]$ and $\mathbf{K}_n = \mathbf{K}_n^{m-n}$ is the nth reproducing kernel associated with the Laguerre polynomials $(\ell_n^{m-n})_{n=0}^{\infty}$ (which we proved above are the orthogonal polynomials for the weight function $W_{m-n}(x)$). Making the appropriate substitutions according to (5.32), the weight factors $W_{m-n}(x_j)$ cancel out and we get precisely the identity (5.21). □

The preceding observations relating the distribution of the passage times $G(m,n)$ to the orthogonal polynomial ensemble $\mathbf{X}_{W_{m-n}}^{(n)}$, in combination with Corollary 5.18, also give an alternative way to derive the value of the normalization constant $D_{m,n}$ in (5.13), (5.15). Readers should check that the value obtained from this approach is in agreement with (5.13) (Exercise 5.12).

A final property of the Laguerre polynomials we will make use of is a contour integral representation, which will be the starting point of an asymptotic analysis we undertake in the next few sections.

Lemma 5.24 *The polynomial $\ell_n^{\alpha}(x)$ has the contour integral representation*

$$\ell_n^{\alpha}(x) = \left(\frac{n!}{(n+\alpha)!}\right)^{1/2} \frac{1}{2\pi i} \oint_{|z|=r} \frac{e^{xz}(1-z)^{n+\alpha}}{z^{n+1}}\, dz. \tag{5.33}$$

where r is an arbitrary positive number.

Proof The function $e^{xz}(1-z)^{n+\alpha}$ is (for fixed $x \in \mathbb{R}$ and integers $n, \alpha \geq 0$) an entire function of a complex variable z. Its power series expansion can be computed as

$$\begin{aligned} e^{xz}(1-z)^{n+\alpha} &= \sum_{k=0}^{\infty} \frac{x^k z^k}{k!} \cdot \sum_{j=0}^{n+\alpha} (-1)^j \binom{n+\alpha}{j} z^j \\ &= \sum_{m=0}^{\infty} (-1)^m \left(\sum_{k=0}^{m} \frac{(-1)^k}{k!} \binom{n+\alpha}{m-k} x^k \right) z^m. \end{aligned}$$

In particular, the coefficient of z^n is $(-1)^n L_n^{\alpha}(x)$, so by the residue theorem we have

$$L_n^{\alpha}(x) = \frac{(-1)^n}{2\pi i} \oint_{|z|=r} \frac{e^{xz}(1-z)^{n+\alpha}}{z^{n+1}}\, dz. \tag{5.34}$$

which is equivalent to (5.33). □

5.10 Asymptotics for the Laguerre kernels

Having derived the Fredholm determinant representation (5.21) for the distribution function of the passage times, the next step toward our goal of proving Theorem 5.1 is to study the asymptotics of the Laguerre kernels $\mathbf{L}_n^\alpha(x, y) = \mathbf{L}_n^{m-n}(x, y)$ appearing in this representation. We will prove that after appropriate scaling these kernels converge to the Airy kernel $\mathbf{A}(\cdot, \cdot)$ (defined in (2.1)); together with some auxiliary bounds, we will be able to deduce convergence of the Fredholm determinants and thus prove Theorem 5.1.[2]

The assumptions in Theorem 5.1 lead us to consider the asymptotic regime in which $m, n \to \infty$ together in such a way that the ratio m/n is bounded away from 0 and ∞. We may assume without of loss of generality that $m \geq n$ (which is required for (5.21) to hold), since the distribution of $G(m, n)$ is symmetric in m and n. Given real variables s, t, we introduce scaled variables x, y (which depend implicitly on m and n) given by

$$x = \Psi(m, n) + \sigma(m, n)t,$$
$$y = \Psi(m, n) + \sigma(m, n)s.$$

It is convenient to eliminate the variable m by introducing the real-valued parameter $\gamma = m/n$, and writing

$$m = \gamma n, \tag{5.35}$$
$$\alpha = m - n = (\gamma - 1)n. \tag{5.36}$$

Denote further

$$\beta = (1 + \sqrt{\gamma})^2, \tag{5.37}$$
$$\sigma_0 = \gamma^{-1/6}(1 + \sqrt{\gamma})^{4/3}, \tag{5.38}$$

so that we may write

$$x = \beta n + \sigma_0 n^{1/3} t, \tag{5.39}$$
$$y = \beta n + \sigma_0 n^{1/3} s. \tag{5.40}$$

Thus, the asymptotics we are interested in involve letting $n \to \infty$ while γ is allowed to range over a compact interval $[1, \Gamma]$ for some arbitrary constant $1 < \Gamma < \infty$, with γn always assumed to take integer values. In other words, we let γ range over the set $[1,\Gamma]_n$, which for convenience is defined by

$$[1,\Gamma]_n = \{\gamma \in [1,\Gamma] \,:\, \gamma n \in \mathbb{N}\}. \tag{5.41}$$

(Note that the choice to use the parameter γ in place of m offers mainly a psychological advantage, since it allows one to think of m, n as tending to infinity together such that their ratio m/n stays approximately constant; indeed, the original formulation of Theorem 5.1 in [62] assumes explicitly that m/n converges to some fixed limit $\gamma \geq 1$.) As a final notational device, let $\mathbf{\Lambda}_n(t,s)$ denote a rescaled version of the Laguerre kernel, defined by

$$\mathbf{\Lambda}_n(t,s) = \sigma_0 n^{1/3} \mathbf{L}_n^\alpha(x,y) = \sigma_0 n^{1/3} \mathbf{L}_n^\alpha(\beta n + \sigma_0 n^{1/3} t, \beta n + \sigma_0 n^{1/3} s). \tag{5.42}$$

Theorem 5.25 (Airy asymptotics for the Laguerre kernels) *The scaled Laguerre kernels satisfy the following properties:*

(a) *For any $\Gamma > 1$ and $s, t \in \mathbb{R}$, $\mathbf{\Lambda}_n(s,t) \to \mathbf{A}(s,t)$ as $n \to \infty$, uniformly as γ ranges over $[1,\Gamma]_n$.*
(b) *For any $T > 0$, we have*

$$\sup_{n\geq 1} \sup_{\gamma\in[1,\Gamma]_n} \sup_{s\in[-T,T]} \mathbf{\Lambda}_n(s,s) < \infty. \tag{5.43}$$

(c) *For any $\epsilon > 0$, there is a number T such that*

$$\sup_{n\geq 1} \sup_{\gamma\in[1,\Gamma]_n} \int_T^\infty \mathbf{\Lambda}_n(s,s)\,ds < \epsilon\,. \tag{5.44}$$

Proof of Theorem 5.1 assuming Theorem 5.25 Using the notation above,

$$\begin{aligned}
&\mathbb{P}\left(\frac{G(m,n)-\Psi(m,n)}{\sigma(m,n)} \leq t\right) \\
&\quad = \mathbb{P}\left(G(m,n) \leq \Psi(m,n) + \sigma(m,n)t\right) = \mathbb{P}\left(G(m,n) \leq \beta n + \sigma_0 n^{1/3} t\right) \\
&\quad = 1 + \sum_{k=1}^\infty \frac{(-1)^k}{k!} \int_{\beta n+\sigma_0 n^{1/3} t}^\infty \ldots \int_{\beta n+\sigma_0 n^{1/3} t}^\infty \det_{i,j=1}^{k} \left(\mathbf{L}_n^\alpha(x_i,x_j)\right) dx_1 \ldots dx_k \\
&\quad = 1 + \sum_{k=1}^\infty \frac{(-1)^k}{k!} \int_t^\infty \ldots \int_t^\infty \det_{i,j=1}^{k} \left(\mathbf{\Lambda}_n(t_i,t_j)\right) dt_1 \ldots dt_k,
\end{aligned}$$

where the last transition involved a change of variables $x_j = \beta n + \sigma_0 n^{1/3} t_j$, $(j = 1,\ldots,k)$. Our goal is to show that as $n \to \infty$, this Fredholm determinant converges to

$$F_2(t) = 1 + \sum_{k=1}^\infty \frac{(-1)^k}{k!} \int_t^\infty \ldots \int_t^\infty \det_{i,j=1}^{n} \left(\mathbf{A}(t_i,t_j)\right) dt_1 \ldots dt_n,$$

uniformly as γ ranges over $[1,\Gamma]_n$. Let $\epsilon > 0$ be given. Fix some large number $T > |t|$ whose value will be specified later. We have the bound

$$\left|\mathbb{P}\left(\frac{G(m,n)-\Psi(m,n)}{\sigma(m,n)} \le t\right) - F_2(t)\right| \le \sum_{k=1}^{\infty} \frac{1}{k!}\left(I_{k,T}^{(n)} + J_{k,T}^{(n)} + M_{k,T}\right), \tag{5.45}$$

where we define

$$I_{k,T}^{(n)} = \int_t^T \cdots \int_t^T \left| \det_{i,j=1}^{k}\left(\mathbf{\Lambda}_n(t_i,t_j)\right) - \det_{i,j=1}^{k}\left(\mathbf{A}(t_i,t_j)\right)\right| dt_1 \ldots dt_k,$$

$$J_{k,T}^{(n)} = k\int_T^\infty \int_t^\infty \cdots \int_t^\infty \left| \det_{i,j=1}^{k}\left(\mathbf{\Lambda}_n(t_i,t_j)\right)\right| dt_1 \ldots dt_k,$$

$$M_{k,T} = k\int_T^\infty \int_t^\infty \cdots \int_t^\infty \left| \det_{i,j=1}^{k}\left(\mathbf{A}(t_i,t_j)\right)\right| dt_1 \ldots dt_k.$$

We now estimate each of the quantities $I_{k,T}^{(n)}, J_{k,T}^{(n)}, M_{k,T}$ separately. We use Hadamard's inequality (2.29) from Chapter 2; note that this inequality is applicable to both the matrices $(\mathbf{A}(t_i,t_j))_{i,j=1}^k$ and $(\mathbf{\Lambda}_n(t_i,t_j))_{i,j=1}^k$ since they are symmetric and positive-semidefinite (see Exercise 2.18 and (5.24)). First, for $M_{k,T}$ we have

$$\begin{aligned} M_{k,T} &\le k\int_T^\infty \int_t^\infty \cdots \int_t^\infty \prod_{i=1}^{k} \mathbf{A}(t_i,t_i)\, dt_1 \ldots dt_k \\ &= k\int_T^\infty \mathbf{A}(s,s)\,ds \cdot \left(\int_t^\infty \mathbf{A}(s,s)\,ds\right)^{k-1}, \end{aligned}$$

By Lemma 2.24, $\int_T^\infty \mathbf{A}(s,s)\,ds$ can be made arbitrarily small by choosing T large enough. In particular, for large enough T we will have that

$$\begin{aligned} \sum_{k=1}^{\infty} \frac{1}{k!} M_{k,T} &\le \sum_{k=1}^{\infty} \frac{k}{k!}\left(\int_t^\infty \mathbf{A}(s,s)\,ds\right)^{k-1} \cdot \int_T^\infty \mathbf{A}(s,s)\,ds \\ &= \exp\left(\int_t^\infty \mathbf{A}(s,s)\,ds\right) \cdot \int_T^\infty \mathbf{A}(s,s)\,ds < \epsilon. \end{aligned} \tag{5.46}$$

Second, $J_{k,T}^{(n)}$ can be bounded using similar reasoning, by noting that

$$\begin{aligned} J_{k,T}^{(n)} &\le k\int_T^\infty \int_t^\infty \cdots \int_t^\infty \prod_{i=1}^{k} \mathbf{\Lambda}_n(t_i,t_i)\, dt_1 \ldots dt_k \\ &= k\int_T^\infty \mathbf{\Lambda}_n(s,s)\,ds \cdot \left(\int_t^\infty \mathbf{\Lambda}_n(s,s)\,ds\right)^{k-1}. \end{aligned}$$

By part (c) of Theorem 5.25, by choosing T large enough we can again ensure that

$$\sum_{k=1}^{\infty} \frac{1}{k!} J_{k,T}^{(n)} \leq \sum_{k=1}^{\infty} \frac{k}{k!} \left(\int_t^{\infty} \boldsymbol{\Lambda}_n(s,s)\,ds \right)^{k-1} \cdot \int_T^{\infty} \boldsymbol{\Lambda}_n(s,s)\,ds < \epsilon \qquad (5.47)$$

for all $n \geq 1$ and all $\gamma \in [1,\Gamma]_n$. Let T be chosen so that both (5.46) and (5.47) are satisfied. Next, by part (a) of Theorem 5.25, we have for each $k \geq 1$ that

$$\left| \det_{i,j=1}^{k} \left(\boldsymbol{\Lambda}_n(t_i,t_j) \right) - \det_{i,j=1}^{k} \left(\mathbf{A}(t_i,t_j) \right) \right| \xrightarrow[n\to\infty]{} 0 \qquad (t_1,\ldots,t_k \in [t,T]),$$

uniformly as γ ranges over $[1,\Gamma]_n$, and furthermore, by part (b) and (2.29), we have for $(t_1,\ldots,t_k) \in [t,T]^k$ that

$$\left| \det_{i,j=1}^{k} \left(\boldsymbol{\Lambda}_n(t_i,t_j) \right) - \det_{i,j=1}^{k} \left(\mathbf{A}(t_i,t_j) \right) \right|$$
$$\leq \left(\sup_{n\geq 1} \sup_{\gamma\in[1,\Gamma]_n} \sup_{s\in[t,T]} \boldsymbol{\Lambda}_n(s,s) \right)^k + \left(\sup_{s\in[t,T]} \mathbf{A}(s,s)\,ds \right)^k \leq C^k k^{k/2} \qquad (5.48)$$

for some constant $C > 0$. It follows by the bounded convergence theorem that $I_{k,T}^{(n)} \to 0$ as $n \to \infty$, *for any fixed* $k \geq 1$. Therefore also for any fixed integer $N \geq 1$ we have that

$$\sum_{k=1}^{N} \frac{1}{k!} I_{k,T}^{(n)} \xrightarrow[n\to\infty]{} 0.$$

To finish the proof, we claim that N can be chosen large enough so that (for all $n \geq 1$ and all $\gamma \in [1,\Gamma]_n$)

$$\sum_{k=N+1}^{\infty} \frac{1}{k!} I_{k,T}^{(n)} < \epsilon. \qquad (5.49)$$

Indeed, from (5.48) we get that $I_{k,T}^{(n)} \leq (T-t)^k C^k k^{k/2}$, so that

$$\sum_{k=N+1}^{\infty} \frac{1}{k!} I_{k,T}^{(n)} < \sum_{k=N+1}^{\infty} \frac{(T-t)^k C^k k^{k/2}}{k!},$$

which decays to 0 as $N \to \infty$.

Combining the above results (5.45), (5.46), (5.47), (5.48), and (5.49) gives finally that

$$\limsup_{n\to\infty} \left| \mathbb{P}\left(\frac{G(m,n)-\Psi(m,n)}{\sigma(m,n)} \le t \right) - F_2(t) \right|$$
$$\le \sum_{k=1}^{N} \frac{1}{k!} I_{k,T}^{(n)} + \sum_{k=N+1}^{\infty} \frac{1}{k!} I_{k,T}^{(n)} + \sum_{k=1}^{\infty} \frac{1}{k!} (J_{k,T}^{(n)} + M_{k,T})$$
$$\le 0 + \epsilon + \epsilon + \epsilon = 3\epsilon.$$

Since ϵ was an arbitrary positive number, this finishes the proof. □

The analysis leading up to a proof of Theorem 5.25 is presented in the next two sections. The following lemma will provide a helpful entry point.

Lemma 5.26 *Assume the notational conventions* (5.35)–(5.42). *Define functions*

$$A_n(x) = \gamma^{1/4} n^{1/2} \left(\frac{n!}{(\gamma n)!} \right)^{1/2} e^{-x/2} x^{\alpha/2}, \tag{5.50}$$

$$h_1(z) = 1, \tag{5.51}$$

$$h_2(z) = \sqrt{\gamma}\frac{z}{1-z} - 1, \tag{5.52}$$

$$h_3(z) = z - \frac{1}{1+\sqrt{\gamma}}, \tag{5.53}$$

$$h_4(z) = h_2(z) h_3(z), \tag{5.54}$$

and for a function $h(z)$ of a complex variable, denote

$$I_n(h,x) = \frac{1}{2\pi i} \oint_{|z|=r} \frac{e^{xz}(1-z)^{\gamma n}}{z^{n+1}} h(z)\,dz. \tag{5.55}$$

Then the Laguerre kernel $\mathbf{L}_n^\alpha(x,y)$ can be represented as

$$\mathbf{L}_n^\alpha(x,y) = \begin{cases} A_n(x)A_n(y) \cdot \dfrac{I_n(h_1,x)I_n(h_2,y) - I_n(h_2,x)I_n(h_1,y)}{x-y} & \text{if } x \ne y, \\ A_n(x)^2 (I_n(h_2,x)I_n(h_3,x) - I_n(h_1,x)I_n(h_4,x)) & \text{if } x = y. \end{cases} \tag{5.56}$$

Proof Rewrite the case $x \ne y$ of (5.20) as

$$\mathbf{L}_n^\alpha(x,y) = \sqrt{n \cdot \gamma n} \cdot e^{-x/2} x^{\alpha/2} e^{-y/2} y^{\alpha/2}$$
$$\times \frac{\ell_n^\alpha(x)(\ell_{n-1}^\alpha(y) - \ell_n^\alpha(y)) - (\ell_{n-1}^\alpha(x) - \ell_n^\alpha(x))\ell_n^\alpha(y)}{x-y}, \tag{5.57}$$

and note that, by (5.33), we have

$$\ell_n^\alpha(x) = \left(\frac{n!}{(\gamma n)!}\right)^{1/2} \frac{1}{2\pi i} \oint_{|z|=r} \frac{e^{xz}(1-z)^{\gamma n}}{z^{n+1}}\,dz$$

$$= \left(\frac{n!}{(\gamma n)!}\right)^{1/2} I_n(h_1, x), \tag{5.58}$$

$$\ell_{n-1}^\alpha(x) = \left(\frac{(n-1)!}{(\gamma n-1)!}\right)^{1/2} \frac{1}{2\pi i} \oint_{|z|=r} \frac{e^{xz}(1-z)^{\gamma n-1}}{z^{n}}\,dz$$

$$= \left(\frac{n!}{(\gamma n)!}\right)^{1/2} \frac{1}{2\pi i} \oint_{|z|=r} \frac{e^{xz}(1-z)^{\gamma n}}{z^{n+1}} \cdot \sqrt{\gamma}\frac{z}{1-z}\,dz, \tag{5.59}$$

and therefore also

$$\ell_{n-1}^\alpha(x) - \ell_n^\alpha(x) = \left(\frac{n!}{(\gamma n)!}\right)^{1/2} I_n(h_2, z). \tag{5.60}$$

Thus, the right-hand side of (5.57) is equal to

$$A_n(x)A_n(y) \cdot \frac{I_n(h_1,x)I_n(h_2,y) - I_n(h_2,x)I_n(h_1,y)}{x-y},$$

as claimed. Similarly, to prove the claim in the case $x = y$, first rewrite $\mathbf{L}_n^\alpha(x,x)$, using (5.20) and (5.29), in the somewhat convoluted form

$$\mathbf{L}_n^\alpha(x,x) = \sqrt{n\cdot\gamma n}\cdot e^{-x}x^\alpha\Bigg((\ell_{n-1}^\alpha(x) - \ell_n^\alpha(x))\left(\sqrt{n}\ell_{n-1}^{\alpha+1}(x) - \frac{1}{1+\sqrt{\gamma}}\ell_n^\alpha(x)\right)$$
$$- \ell_n^\alpha(x)\Bigg(\sqrt{n-1}\ell_{n-2}^{\alpha+1} - \sqrt{n}\ell_{n-1}^{\alpha+1}(x)$$
$$- \frac{1}{1+\sqrt{\gamma}}\ell_{n-1}^\alpha(x) + \frac{1}{1+\sqrt{\gamma}}\ell_n^\alpha(x)\Bigg)\Bigg). \tag{5.61}$$

Now observe that, similarly and in addition to (5.58)–(5.60), we have the relations

$$\sqrt{n}\ell_{n-1}^{\alpha+1}(x) = \left(\frac{n!}{(\gamma n)!}\right)^{1/2} I_n(z, x),$$

$$\sqrt{n-1}\ell_{n-2}^{\alpha+1}(x) = \left(\frac{n!}{(\gamma n)!}\right)^{1/2} I_n\left(\sqrt{\gamma}\frac{z^2}{1-z}, x\right)$$

which therefore also gives that

$$\sqrt{n}\ell_{n-1}^{\alpha+1}(x) - \frac{1}{1+\sqrt{\gamma}}\ell_n^\alpha(x) = \left(\frac{n!}{(\gamma n)!}\right)^{1/2} I_n(h_3, x). \tag{5.62}$$

and

$$\sqrt{n-1}\,\ell_{n-2}^{\alpha+1} - \sqrt{n}\,\ell_{n-1}^{\alpha+1}(x) - \frac{1}{1+\sqrt{\gamma}}\ell_{n-1}^{\alpha}(x) + \frac{1}{1+\sqrt{\gamma}}\ell_{n}^{\alpha}(x)$$
$$= \left(\frac{n!}{(\gamma n)!}\right)^{1/2} I_n(h_4, x). \qquad (5.63)$$

Combining (5.58), (5.60), (5.61), (5.62), and (5.63) gives the desired representation for $\mathbf{L}_n^{\alpha}(x, x)$. □

Readers may have noticed that the choice of functions h_1, h_2, h_3, h_4 in the representation (5.56) is not unique, and indeed is more complicated than might seem necessary. However, as we will see later, this choice is carefully crafted to bring about a cancellation of leading terms in an asymptotic expansion. The constant $1/(1+\sqrt{\gamma})$ which appears in the definition of h_3 and in some of the formulas in the proof above also plays an important role that will become apparent in the next section.

5.11 Asymptotic analysis of $I_n(h, x)$

We now reach the main part of the analysis, in which we study the asymptotic behavior as $n \to \infty$ of $I_n(h, x)$. (Because of Lemma 5.26, our main interest is of course with $h = h_j$ for $j = 1, 2, 3, 4$.) The main asymptotic result is the following one.

Theorem 5.27 *Let $\Gamma > 1$ and let $t \in \mathbb{R}$. Define*

$$r_0 = \frac{1}{1+\sqrt{\gamma}}, \qquad (5.64)$$

$$g_0 = 1 + \sqrt{\gamma} + \gamma \log\left(1 - \frac{1}{1+\sqrt{\gamma}}\right) + \log(1+\sqrt{\gamma}). \qquad (5.65)$$

Let $h(z)$ be a function of a complex variable z and of the parameter γ, that is analytic in z in the unit disk and depends continuously on $\gamma \in [1, \infty)$. Assume that $h(z)$ has a zero of order k at $z = r_0$, where $0 \le k < 7$; that is, it has the behavior

$$h(z) = a_k(z - r_0)^k + O(|z - r_0|^{k+1}) \qquad (5.66)$$

in a neighborhood of r_0, where $a_k \neq 0$, and a_k may depend on γ but k does *not. Then for any $t \in \mathbb{R}$ and $\Gamma > 1$, we have as $n \to \infty$ that*

$$I_n(h, x) = \left(\mathrm{Ai}^{(k)}(t) + O\left(n^{-1/24}\right)\right) a_k \gamma^{(k+1)/6} (1 + \sqrt{\gamma})^{-(4k+1)/3} \times n^{-(k+1)/3} \exp\left(g_0 n + \sigma_0 t r_0 n^{1/3}\right), \tag{5.67}$$

uniformly as γ ranges over $[1, \Gamma]_n$.

The assumption $0 \le k < 7$ above can be dropped at the cost of a modification to the exponent $-1/24$ in the error term, but the result as formulated above suffices for our needs.

Proof In the proof that follows, we leave to the reader the easy verification that all constants that appear in asymptotic bounds, whether explicit or implicit in the big-O notation, can be chosen in a way that the estimates hold uniformly as γ ranges over $[1, \Gamma]_n$, with $\Gamma > 1$ being an arbitrary fixed number. Aside from the need for such verification, we may now treat γ as a parameter whose value can be thought of as fixed.

Recall that $\mathrm{Ai}(t)$ is the Airy function defined in (2.1), which also has the equivalent definition (2.67). More generally, the derivatives of the Airy function are given by (2.68).

To start the analysis, define a function

$$g(z) = \beta z + \gamma \,\mathrm{Log}(1 - z) - \mathrm{Log}\, z, \tag{5.68}$$

where as before $\mathrm{Log}(w)$ denotes the principal value of the complex logarithm function. Note that $g(z)$ is analytic in $\{z : |z| < 1, -\pi < \arg z < \pi\}$, and that (5.55) can be rewritten as

$$\begin{aligned} I_n(h, x) &= \frac{1}{2\pi i} \oint_{|z|=r} e^{\sigma_0 t n^{1/3} z} \exp\left[n(\beta z + \gamma \,\mathrm{Log}(1 - z) - \mathrm{Log}\, z)\right] h(z) \frac{dz}{z} \\ &= \frac{1}{2\pi i} \oint_{|z|=r} \exp\left(n g(z) + \sigma_0 t n^{1/3} z\right) h(z) \frac{dz}{z}. \end{aligned} \tag{5.69}$$

The form of this integral is suitable for the application of the **saddle point method**, a standard technique of asymptotic analysis. See the box on the next page for more background on this important method. The key idea is to find the correct deformation of the contour of integration such that the main contribution to the integral will come from a small part of the contour around a single point known as the saddle point.

The saddle point method

The **saddle point method** (also known as the **stationary phase method** or **method of steepest descent**) is a powerful technique of asymptotic analysis that is widely applicable to problems in analysis, combinatorics, number theory and many other areas. The idea is to first represent a quantity of interest as a contour integral in the complex plane, then to deform the integration contour in such a way that the bulk of the contribution to the integral comes from the vicinity of a single point (or, in more general versions, a small set of points) known as the **saddle point**. One then estimates the contribution near the saddle point using a Taylor approximation, and separately shows that the contribution away from the saddle point is asymptotically negligible.

Many of the simplest applications of the method involve an integral of the form

$$\frac{1}{2\pi i}\oint_{|z|=r} e^{ng(z)}\frac{dz}{z} \tag{5.70}$$

for some function $g(z)$, where the radius of integration r is arbitrary and n is large. Under certain reasonable assumptions, it can be shown that r should be chosen so that the contour passes through the saddle point, which is the solution to the equation $g'(z) = 0$.

A standard textbook example is to use the saddle point method to prove Stirling's approximation formula for $n!$. Start with the elementary identity

$$\frac{n^n e^{-n}}{n!} = \frac{1}{2\pi i}\oint_{|z|=r}\frac{e^{n(z-1)}}{z^{n+1}}dz,$$

which fits the template (5.70) with $g(z) = z - 1 - \operatorname{Log} z$. Solving the equation $g'(z) = 0$ gives $z = 1$. Setting $r = 1$ and parametrizing the integral gives

$$\frac{n^n e^{-n}}{n!} = \frac{1}{2\pi}\int_{-\pi}^{\pi}\exp(ng(e^{it}))\,dt = I_n + J_n,$$

where $I_n = \frac{1}{2\pi}\int_{-\varepsilon}^{\varepsilon}\exp(ng(e^{it}))\,dt$, $J_n = \frac{1}{2\pi}\int_{(-\pi,\pi)\setminus(-\varepsilon,\varepsilon)}\exp(ng(e^{it}))\,dt$, and where we set $\varepsilon = n^{-2/5}$. Using the Taylor approximation $g(e^{it}) = -\frac{1}{2}t^2 + O(|t|^3)$ that is valid near $t = 0$ and the relation $|e^w| = e^{\operatorname{Re} w}$, it is now straightforward (Exercise 5.19) to show that

$$I_n = (1 + O(n^{-1/5}))(2\pi n)^{-1/2}, \qquad |J_n| = O(e^{-n}), \tag{5.71}$$

which gives the result

$$n! = (1 + O(n^{-1/5}))\sqrt{2\pi n}(n/e)^n, \tag{5.72}$$

a version of Stirling's formula with an explicitly bounded (though suboptimal) error term. See Exercises 5.20 and 5.21 for more examples.

To motivate the computation, consider first the simplest case where $t = 0$ and $h(z) = h_1(z) = 1$, in which the integral (5.69) reduces to the simpler expression

$$\frac{1}{2\pi i} \oint_{|z|=r} \exp\left(ng(z)\right) \frac{dz}{z}.$$

The integration contour is a circle around the origin, and in this case there is no need to consider more general contours, but the radius of integration r is arbitrary, and we will choose it to pass at – or very near to – the saddle point. To find the saddle point, according to the recipe explained in the box on page 312, we have to solve the equation $g'(z) = 0$. This gives the condition

$$g'(z) = \beta - \frac{\gamma}{1-z} - \frac{1}{z} = \frac{\beta z(1-z) - \gamma z - (1-z)}{z(1-z)} = 0, \tag{5.73}$$

or $\beta z^2 - 2(1 + \sqrt{\gamma})z + 1 = 0$, which gives $z = \frac{1}{1+\sqrt{\gamma}} = r_0$; that is, the saddle point lies on the positive real axis, and the "ideal" radius of integration is the constant r_0 defined in (5.64). (Actually, in our particular situation it will turn out to be preferable to take a slightly smaller radius of integration that does not actually pass through the saddle point.) The asymptotics will end up being determined by the behavior of $g(z)$ near the saddle point; in other words, by the first few Taylor coefficients of $g(z)$ at $z = r_0$. Note that the constant coefficient $g(r_0)$ is precisely g_0 defined in (5.65). The next coefficient is $g'(r_0)$, which is equal to 0 by design, and the next coefficient after that is $g''(r_0)/2$, which turns out to also be 0:

$$g''(r_0) = \frac{1}{r_0^2} - \frac{\gamma}{(1-r_0)^2} = (1 + \sqrt{\gamma})^2 - \gamma\left(\frac{1+\sqrt{\gamma}}{\sqrt{\gamma}}\right)^2 = 0.$$

Finally, denote $g_3 = -g'''(r_0)$. We leave to the reader to check that

$$g_3 = \frac{2(1+\sqrt{\gamma})^4}{\sqrt{\gamma}}. \tag{5.74}$$

To summarize, $g(z)$ has the behavior

$$g(z) = g_0 - \tfrac{1}{6} g_3 (z - r_0)^3 + O\left((z - r_0)^4\right) \tag{5.75}$$

for z in a neighborhood of r_0.

Let us get back from the motivating case to the case of general t and general h. The inclusion of the additional factors $h(z) \cdot \exp(\sigma_0 t n^{1/3} z)$ in the integrand does affect the asymptotics, but we shall see that the same radius of integration can still be used. As noted earlier, for the radius of integration we will use a value slightly smaller than r_0, specifically

$$r = r_0 - n^{-1/3}$$

(any constant multiple of $n^{-1/3}$ can be used here). We are ready to start estimating the contour integral (5.69). Parametrizing the contour of integration as $z = re^{i\theta}$, $-\pi \le \theta \le \pi$, gives

$$I_n(h,x) = \frac{1}{2\pi}\int_{-\pi}^{\pi} \exp\left(ng(re^{i\theta}) + \sigma_0 t n^{1/3} re^{i\theta}\right) h(re^{i\theta})\, d\theta.$$

We divide this into two parts, accounting separately for the contributions from the vicinity of the saddle point and away from it, by writing $I_n(h,x) = Q_n + W_n$, where

$$Q_n = \frac{1}{2\pi}\int_{-\varepsilon}^{\varepsilon} \exp\left(ng(re^{i\theta}) + \sigma_0 t n^{1/3} re^{i\theta}\right) h(re^{i\theta})\, d\theta,$$

$$W_n = \frac{1}{2\pi}\int_{(-\pi,\pi)\setminus(-\varepsilon,\varepsilon)} \exp\left(ng(re^{i\theta}) + \sigma_0 t n^{1/3} re^{i\theta}\right) h(re^{i\theta})\, d\theta,$$

and where we set $\varepsilon = n^{-7/24}$.

Analysis of Q_n. Make the change of variables $\theta = n^{-1/3}w$. The range $|\theta| \le \varepsilon$ translates to $|w| \le n^{1/24}$. In this range, using the Taylor expansions (5.66), (5.75), we see that the following estimates hold uniformly in w:

$$\begin{aligned} re^{i\theta} &= (r_0 - n^{-1/3})(1 + iwn^{-1/3} + O(w^2 n^{-2/3})) \\ &= r_0 + (-1 + ir_0 w)n^{-1/3} + O(n^{-7/12}), && (5.76) \\ g(re^{i\theta}) &= g_0 + \tfrac{1}{6} g_3 (r_0 - re^{i\theta})^3 + O(n^{-4/3}) \\ &= g_0 + \tfrac{1}{6} g_3 \left((1 - ir_0 w)n^{-1/3} + O(n^{-7/12})\right)^3 + O(n^{-4/3}) \\ &= g_0 + \tfrac{1}{6} g_3 (1 - ir_0 w)^3 n^{-1} + O(n^{-7/6}), && (5.77) \\ h(re^{i\theta}) &= a_k \left((-1 + ir_0 w)n^{-1/3}\right)^k + O\left(\left(|-1 + ir_0 w|n^{-1/3}\right)^{k+1}\right) \\ &= a_k (-1 + ir_0 w)^k n^{-k/3} + O\left(n^{-\frac{7}{24}(k+1)}\right). && (5.78) \end{aligned}$$

(Note that the assumption $0 \le k < 7$ implies that $7(k+1)/24 > k/3$, so the error term in (5.78) is of smaller order of magnitude than the exact term preceding it.) The integral Q_n therefore becomes

$$\begin{aligned} Q_n &= \frac{n^{-1/3}}{2\pi} \int_{-n^{1/24}}^{n^{1/24}} \exp(g_0 n + \sigma_0 r_0 t n^{1/3}) \\ &\qquad \times \exp\left(\tfrac{1}{6} g_3 (1 - i r_0 w)^3 + \sigma_0 t(-1 + i r_0 w) + O(n^{-1/6})\right) \\ &\qquad \times \left(a_k(-1 + i r_0 w)^k n^{-k/3} + O\left(n^{-\frac{7}{24}(k+1)}\right)\right) dw \\ &= (1 + O(n^{-1/6})) a_k n^{-(k+1)/3} \exp(g_0 n + \sigma_0 r_0 t n^{1/3}) \\ &\quad \times \frac{1}{2\pi} \int_{-n^{1/24}}^{n^{1/24}} \exp\left(\tfrac{1}{6} g_3 (1 - i r_0 w)^3 + \sigma_0 t(-1 + i r_0 w)\right) \\ &\qquad\qquad \times ((-1 + i r_0 w)^k + O(n^{-1/24}))\, dw, \end{aligned} \tag{5.79}$$

where we have $O(n^{-(7-k)/24}) = O(n^{-1/24})$ because of the assumption that $k \le 6$. In the last integral, we consider the main term and the $O(n^{-1/24})$ error term separately. First, the main term (with the $1/2\pi$ factor included) looks like it should converge to the Airy-type integral

$$\frac{1}{2\pi} \int_{-\infty}^{\infty} \exp\left(\tfrac{1}{6} g_3 (1 - i r_0 w)^3 + \sigma_0 t(-1 + i r_0 w))\right) (-1 + i r_0 w)^k dw. \tag{5.80}$$

Indeed, the difference between the two integrals can be estimated by writing

$$\begin{aligned} &\left| \int_{\mathbb{R} \setminus (-n^{1/24}, n^{1/24})} \exp\left(\tfrac{1}{6} g_3 (1 - i r_0 w)^3 + \sigma_0 t(-1 + i r_0 w))\right) (-1 + i r_0 w)^k\, dw \right| \\ &\le 2 \int_{n^{1/24}}^{\infty} \exp\left[\operatorname{Re}\left(\tfrac{1}{6} g_3 (1 - i r_0 w)^3 + \sigma_0 t(-1 + i r_0 w))\right)\right] \cdot |-1 + i r_0 w|^k\, dw \\ &= 2 \int_{n^{1/24}}^{\infty} \exp\left(\tfrac{1}{6} g_3 (1 - 3 r_0^2 w^2) - \sigma_0 t\right) \cdot |-1 + i r_0 w|^k\, dw, \end{aligned}$$

and this is easily seen to be bounded by an expression of the form $Ce^{-cn^{1/12}}$ for some constants $C, c > 0$ (which depend on t and Γ). Furthermore, the integral (5.80) can be evaluated by making the change of variables $v = r_0 (g_3/2)^{1/3} v$. Noting that $\sigma_0 = (g_3/2)^{1/3}$ (see (5.38) and (5.74)), and making use of (2.68), we get that (5.80) is equal to

$$\begin{aligned} &\frac{\sigma_0^{-k}}{2\pi} \int_{-\infty}^{\infty} \exp\left(\tfrac{1}{3}(\sigma_0 - iv)^3 + t(-\sigma_0 + iv)\right)(-\sigma_0 + iv)^k \frac{dv}{r_0 \sigma_0} \\ &\qquad\qquad = \gamma^{(k+1)/6} (1 + \sqrt{\gamma})^{-(4k+1)/3} \operatorname{Ai}^{(k)}(t). \end{aligned}$$

Next, for the $O(n^{-1/24})$ error term, note similarly that

$$\left| \int_{\mathbb{R}\setminus(-n^{1/24},n^{1/24})} \exp\left(\tfrac{1}{6}g_3(1-ir_0w)^3 + \sigma_0 t(-1+ir_0w))\right) O(n^{-1/24})\,dw \right|$$
$$\leq O(n^{-1/24}) \int_{-n^{1/24}}^{n^{1/24}} \exp\left(\tfrac{1}{6}g_3(1-3r_0^2w^2) - \sigma_0 t\right)\,dw = O(n^{-1/24}).$$

Combining the above observations gives that

$$Q_n = \left(\mathrm{Ai}^{(k)}(t) + O\left(n^{-1/24}\right)\right) a_k \gamma^{(k+1)/6}(1+\sqrt{\gamma})^{-(4k+1)/3} \times n^{-(k+1)/3} \exp\left(g_0 n + \sigma_0 t r_0 n^{1/3}\right). \tag{5.81}$$

Analysis of W_n. Denote $M = \sup\left\{|h(z)| \,:\, |z| < 1/(1+\sqrt{\Gamma})\right\}$. Trivially, we have

$$\begin{aligned}|W_n| &\leq \frac{1}{2\pi}\int_{(-\pi,\pi)\setminus(-\varepsilon,\varepsilon)} \left|\exp\left(ng(re^{i\theta}) + \sigma_0 t n^{1/3} re^{i\theta}\right)\right| \cdot |h(re^{i\theta})|\,d\theta \\ &\leq \frac{M}{2\pi}\int_{(-\pi,\pi)\setminus(-\varepsilon,\varepsilon)} \exp\left(\mathrm{Re}\left(ng(re^{i\theta}) + \sigma_0 t n^{1/3} re^{i\theta}\right)\right) d\theta \\ &\leq \frac{M}{2\pi}\exp\left(\sigma_0 t n^{1/3} r\cos\varepsilon\right)\int_{(-\pi,\pi)\setminus(-\varepsilon,\varepsilon)} \exp\left(n\,\mathrm{Re}\,g(re^{i\theta})\right) d\theta.\end{aligned}$$

In the last bound, we claim that the expression $\mathrm{Re}\,g(re^{i\theta})$ in the integrand is an increasing function of $\cos\theta$, and therefore as θ ranges over $(-\pi,\pi)\setminus(-\varepsilon,\varepsilon)$ it takes its maximum precisely at $\theta = \varepsilon$, so we can further write

$$\begin{aligned}|W_n| &\leq M\exp\left(n\,\mathrm{Re}\,g(re^{i\varepsilon}) + \sigma_0 t n^{1/3} r\right) \\ &\leq M\exp\left(n\,\mathrm{Re}\,g(re^{i\varepsilon}) + \sigma_0 t n^{1/3} r_0 - \sigma_0 t\right).\end{aligned} \tag{5.82}$$

To prove this, observe that, from the definition (5.68) of $g(z)$, we have

$$\begin{aligned}\mathrm{Re}\,g(re^{i\theta}) &= \mathrm{Re}\left[\beta re^{i\theta} + \gamma\,\mathrm{Log}(1-re^{i\theta}) - \mathrm{Log}(re^{i\theta})\right] \\ &= \beta r\cos\theta + \tfrac{1}{2}\gamma\log\left((1-r\cos\theta)^2 + r^2\sin^2\theta\right) - \log r \\ &= \beta r\cos\theta + \tfrac{1}{2}\gamma\log(1+r^2-2r\cos\theta) - \log r = h(r,\cos\theta),\end{aligned}$$

where

$$h(r,s) = \beta rs + \tfrac{1}{2}\gamma\log(1+r^2-2rs) - \log r \qquad (-1 \leq s \leq 1).$$

Now differentiate $h(r,s)$ with respect to s; we have

$$\begin{aligned}\frac{\partial h(r,s)}{\partial s} &= \beta r + \tfrac{1}{2}\gamma(-2r)\frac{1}{1+r^2-2rs} = r\left(\beta - \frac{\gamma}{1+r^2-2rs}\right)\\ &= r\left(\beta - \frac{\gamma}{(1-r)^2+2r(1-s)}\right) \geq r\left(\beta - \frac{\gamma}{(1-r)^2}\right)\\ &\geq r\left(\beta - \frac{\gamma}{(1-r_0)^2}\right) = 0,\end{aligned}$$

so $h(r,\cos\theta)$ is indeed an increasing function of $\cos\theta$.

Having proved (5.82), note that the estimates (5.76) and (5.77) apply to $\theta = \varepsilon$, with the corresponding value of w being equal to $n^{1/24}$, so we get that

$$\begin{aligned}n\,\mathrm{Re}\, g(re^{i\varepsilon}) &= ng_0 + \tfrac{1}{6}g_3\,\mathrm{Re}(1 - ir_0n^{1/24})^3 + O(n^{-1/6})\\ &= ng_0 + \tfrac{1}{6}g_3(1 - 3r_0^2n^{1/12}) + O(n^{-1/6}) \leq ng_0 - cn^{1/12} + C\end{aligned}$$

for some constants $C, c > 0$. Combining this with (5.82), we have shown that

$$|W_n| \leq M\exp\left(g_0n + \sigma_0 t r_0 n^{1/3} - \sigma_0 t - cn^{1/12} + C\right). \qquad (5.83)$$

The estimates (5.81) and (5.83) together prove (5.67). □

Theorem 5.28 *Under the same assumptions as in Theorem 5.27, there exist constants $C, c > 0$ that depend on $h(z)$ and Γ, such that for all $n \geq 1$, $\gamma \in [1,\Gamma]_n$, and $t \in \mathbb{R}$, we have*

$$|I_n(h,x)| \leq Cn^{-(k+1)/3}\exp\left(g_0n + \sigma_0 t r_0 n^{1/3} - ct\right). \qquad (5.84)$$

Proof We use the notation and computations from the proof of Theorem 5.27. By (5.79), there exist constants $C', C'', c > 0$, independent of n

and $\gamma \in [1, \Gamma]_n$, such that

$$\begin{aligned}
|Q_n| &\le C' n^{-(k+1)/3} \exp\left(g_0 n + \sigma_0 t r_0 n^{1/3}\right) \\
&\quad \times \int_{-\infty}^{\infty} \exp\left[\mathrm{Re}\left(\tfrac{1}{6} g_3 (1 - i r_0 w)^3 + \sigma_0 t(-1 + i r_0 w)\right)\right] \\
&\qquad\qquad \times (|-1 + i r_0 w|^k + O(1))\, dw \\
&= C' n^{-(k+1)/3} \exp\left(g_0 n + \sigma_0 t r_0 n^{1/3} - \sigma_0 t\right) \\
&\quad \times \int_{-\infty}^{\infty} \exp\left(\tfrac{1}{6} g_3 (1 - 3 r_0^2 w^2)\right) \cdot (|-1 + i r_0 w|^k + O(1))\, dw \\
&\le C'' n^{-(k+1)/3} \exp\left(g_0 n + \sigma_0 t r_0 n^{1/3} - c t\right).
\end{aligned}$$

The fact that $|W_n|$ satisfies a similar bound is immediate from (5.83). Since $I_n(h, x) = Q_n + W_n$, this finishes the proof. □

5.12 Proof of Theorem 5.25

Lemma 5.29 *For fixed $t \in \mathbb{R}$ we have*

$$A_n(x) = (1 + O(n^{-1/5})) n^{1/2} \exp\left(-g_0 n - \sigma_0 r_0 t n^{1/3}\right) \quad \text{as } n \to \infty, \tag{5.85}$$

uniformly as γ ranges over $[1, \Gamma]_n$. Furthermore, there exists a constant $C > 0$ such that for all $t \in \mathbb{R}$, $n \ge 1$ and $\gamma \in [1, \Gamma]_n$, we have

$$A_n(x) \le C n^{1/2} \exp\left(-g_0 n - \sigma_0 r_0 t n^{1/3}\right). \tag{5.86}$$

Proof By the version (5.72) of Stirling's approximation, we have

$$\begin{aligned}
\left(\frac{n!}{(\gamma n)!}\right)^{1/2} &= (1 + O(n^{-1/5})) \left(\frac{\sqrt{2\pi n}(n/e)^n}{\sqrt{2\pi \gamma n}(\gamma n/e)^{\gamma n}}\right)^{1/2} \\
&= (1 + O(n^{-1/5})) \gamma^{-1/4} n^{(1-\gamma)n/2} e^{-(1-\gamma)n/2} \gamma^{-\gamma n/2}.
\end{aligned} \tag{5.87}$$

The factor $x^{\alpha/2} e^{-x/2}$ in $A_n(x)$ behaves asymptotically (for fixed t as $n \to \infty$)

as

$$
\begin{aligned}
x^{\alpha/2}e^{-x/2} &= \left(\beta n + \sigma_0 t n^{1/3}\right)^{(\gamma-1)n/2} \exp\left(-\frac{\beta n}{2} - \frac{\sigma_0 t n^{1/3}}{2}\right) \\
&= (1+\sqrt{\gamma})^{(\gamma-1)n} n^{(\gamma-1)n/2}\left(1 + \frac{\sigma_0 t}{\beta} n^{-2/3}\right)^{(\gamma-1)n/2} \\
&\quad \times \exp\left(-\frac{\beta n}{2} - \frac{\sigma_0 t n^{1/3}}{2}\right) \\
&= (1 + O(n^{-1/3}))(1+\sqrt{\gamma})^{(\gamma-1)n} n^{(\gamma-1)n/2} \\
&\quad \times \exp\left(-\frac{\beta n}{2} - \frac{\sigma_0 t n^{1/3}}{2} + \frac{\sigma_0 t}{\beta}\left(\frac{\gamma-1}{2}\right) n^{1/3}\right). \qquad (5.88)
\end{aligned}
$$

So we get that

$$
\begin{aligned}
A_n(x) &= (1 + O(n^{-1/5}))n^{1/2} \exp\left[\sigma_0 t n^{1/3}\left(-\frac{1}{2} + \frac{\gamma-1}{2\beta}\right)\right] \\
&\quad \times \exp\left[n\left(\frac{\gamma-1}{2} - \frac{\gamma}{2}\log\gamma + (\gamma-1)\log(1+\sqrt{\gamma}) - \frac{\beta}{2}\right)\right].
\end{aligned}
$$

It is easy to check the coefficients of n and $\sigma_0 t n^{1/3}$ in the exponent are equal to $-g_0$ and $-r_0$, respectively, so we get (5.85). A variant of the same computation gives (5.86), by noting that the constant implicit in the error term $O(n^{-1})$ in (5.87) is independent of t, and that in (5.88) the asymptotic estimate

$$
\left(1 + \frac{\sigma_0 t}{\beta} n^{-2/3}\right)^{(\gamma-1)n/2} = (1 + O(n^{-1/3}))\exp\left(\sigma_0 t\left(\frac{\gamma-1}{2\beta}\right)n^{1/3}\right),
$$

in which the constant implicit in the error term *does* depend on t, can be replaced (using the standard fact that $1 + u \le e^u$ for all $u \in \mathbb{R}$) by an inequality

$$
\left(1 + \frac{\sigma_0 t}{\beta} n^{-2/3}\right)^{(\gamma-1)n/2} \le \exp\left(\sigma_0 t\left(\frac{\gamma-1}{2\beta}\right)n^{1/3}\right)
$$

which holds for all n and t. □

Proof Theorem 5.25(a) We use the representation (5.56) for the Laguerre kernel, together with the asymptotic results we proved for above for $I_n(h, x)$ and $A_n(x)$. Specifically, we apply Theorem 5.27 with $h = h_j$, $j = 1, 2, 3, 4$, defined in (5.51)–(5.54). The relevant values of k and a_k for each function

h_j are easy to compute and are as follows:

$$\begin{aligned} h_1(z) &= 1: & k &= 0, & a_k &= 1, \\ h_2(z) &= \sqrt{\gamma}\tfrac{z}{1-z} - 1: & k &= 1, & a_k &= \tfrac{(1+\sqrt{\gamma})^2}{\sqrt{\gamma}}, \\ h_3(z) &= z - r_0: & k &= 1, & a_k &= 1, \\ h_4(z) &= h_2(z)h_3(z): & k &= 2, & a_k &= \tfrac{(1+\sqrt{\gamma})^2}{\sqrt{\gamma}}. \end{aligned}$$

Start with the case $t \neq s$. In this case, combining (5.56), (5.67), and (5.85) gives that

$$\begin{aligned} \mathbf{A}_n(t,s) &= \sigma_0 n^{1/3} A_n(x)A_n(y) \cdot \frac{I_n(h_1,x)I_n(h_2,y) - I_n(h_2,x)I_n(h_1,y)}{x-y} \\ &= n^{1/3} \cdot \frac{1+O(n^{-1/5})}{n^{1/3}(t-s)} \cdot n^{1/2}n^{1/2}\left(\gamma^{1/6}(1+\sqrt{\gamma})^{-1/3}\right)\left(\gamma^{1/3}(1+\sqrt{\gamma})^{-5/3}\right) \\ &\quad \times n^{-1/3}n^{-2/3}\frac{(1+\sqrt{\gamma})^2}{\sqrt{\gamma}}\Big[(\mathrm{Ai}(t)+O(n^{-\frac{1}{24}}))(\mathrm{Ai}'(s)+O(n^{-\frac{1}{24}})) \\ &\qquad\qquad - (\mathrm{Ai}'(t)+O(n^{-\frac{1}{24}}))(\mathrm{Ai}(s)+O(n^{-\frac{1}{24}}))\Big] \\ &= \left(1+O(n^{-1/5})\right)\left(\frac{\mathrm{Ai}(t)\,\mathrm{Ai}'(s) - \mathrm{Ai}(s)\,\mathrm{Ai}'(t)}{t-s} + O(n^{-1/24})\right) \\ &= \mathbf{A}(t,s) + O(n^{-1/24}). \end{aligned}$$

Similarly, in the case $t = s$ we have

$$\begin{aligned} \mathbf{A}_n(t,t) &= \sigma_0 n^{1/3} A_n(x)^2 (I_n(h_2,x)I_n(h_3,x) - I_n(h_1,x)I_n(h_4,y)) \\ &= \left(1+O(n^{-1/5})\right) n^{4/3}\left(\gamma^{-1/6}(1+\sqrt{\gamma})^{4/3}\right)\frac{(1+\sqrt{\gamma})^2}{\sqrt{\gamma}}\left(\gamma^{2/3}(1+\sqrt{\gamma})^{-10/3}\right) \\ &\quad \times \Big(n^{-2/3}n^{-2/3}(\mathrm{Ai}'(t)+O(n^{-\frac{1}{24}}))(\mathrm{Ai}'(t)+O(n^{-\frac{1}{24}})) \\ &\qquad - n^{-1/3}n^{-1}(\mathrm{Ai}(t)+O(n^{-\frac{1}{24}}))(\mathrm{Ai}''(t)+O(n^{-\frac{1}{24}}))\Big) \\ &= \left(1+O(n^{-1/5})\right)\left(\mathrm{Ai}'(t)^2 - \mathrm{Ai}(t)\,\mathrm{Ai}''(t) + O(n^{-1/24})\right) \\ &= \mathbf{A}(t,t) + O(n^{-1/24}). \end{aligned}$$

□

Proof Theorem 5.25(b)–(c) The representation (5.56) combined with the inequalities (5.84) and (5.86) gives that for some constants $c, C' > 0$, we have

$$|\mathbf{A}_n(t,t)| \leq C' e^{-2ct}$$

for all $n \geq 1$, $t \in \mathbb{R}$ and $\gamma \in [1, \Gamma]_n$. This immediately gives (5.43) and (5.44). □

5.13 The passage times in the multicorner growth process

The analysis in Sections 5.6–5.12 carries over in a more or less straightforward way to the case of the multicorner growth process, but the required computations and estimates are more complicated. In this section we formulate the main results for this case. The details of the proofs can be found in Johansson's paper [62] and in chapter 5 of [115].

First, there is a version of the Fredholm determinant representation (5.21) for the case of the multicorner growth process. The determinant involves a discrete family of kernels $\mathbf{M}_n^{(p)} : \mathbb{N}_0 \times \mathbb{N}_0 \to \mathbb{R}$ (where $\mathbb{N}_0 = \mathbb{N} \cup \{0\}$) called the **Meixner kernels**, which is associated with a family of orthogonal polynomials called the **Meixner polynomials**. For any integer $K \geq 0$ and $0 < p < 1$, the Meixner polynomials with parameters K and p are the family of polynomials $(M_n^{(K,p)}(x))_{n=0}^{\infty}$ defined by

$$M_n^{(K,p)}(x) = \frac{p^{K/2}(1-p)^{n/2}}{\binom{n+K-1}{n}^{1/2}} \sum_{k=0}^{n} (1-p)^{-k} \binom{x}{k} \binom{-x-K}{n-k}.$$

The Meixner polynomials satisfy the orthogonality relation

$$\sum_{x=0}^{\infty} M_i^{(K,p)}(x) M_j^{(K,p)}(x) \binom{x+K-1}{K-1} (1-p)^x = \delta_{ij} \qquad (i, j \geq 0),$$

which means they are a family of *discrete orthogonal polynomials*, that is, they are orthogonal with respect to the discrete measure $\mu^{(K,p)}(x)$ defined by $\mu^{(K,p)}(x) = \binom{x+K-1}{K-1}(1-p)^x$ on $\mathbb{N}_0$. Next, define the associated **Meixner kernels** $\mathbf{M}_n^{(K,p)} : \mathbb{N}_0 \times \mathbb{N}_0 \to \mathbb{R}$ by

$$\mathbf{M}_n^{(K,p)}(x,y) = \left(\binom{x+K-1}{K-1} \binom{y+K-1}{K-1} (1-p)^{x+y} \right)^{1/2} \times \begin{cases} \dfrac{M_n^{(K,p)}(x) M_{n-1}^{(K,p)}(y) - M_{n-1}^{(K,p)}(x) M_n^{(K,p)}(y)}{x-y} & \text{if } x \neq y, \\ (M_n^{(K,p)})'(x) M_{n-1}^{(K,p)}(x) - (M_{n-1}^{(K,p)})'(x) M_n^{(K,p)}(x) & \text{if } x = y. \end{cases} \tag{5.89}$$

Theorem 5.30 *For any $m \geq n \geq 1$, the distribution of the passage times $G(m,n)$ in the multicorner growth process is given by the Fredholm determinant*

$$\mathbb{P}(G(m,n) \leq t) = \det\left(\mathbf{I} - (\mathbf{M}_n^{m-n+1})_{\{n+t,n+t+1,\ldots\}}\right)$$
$$= 1 + \sum_{k=1}^{n} \frac{(-1)^k}{k!} \sum_{x_1,\ldots,x_k=n+t}^{\infty} \det_{i,j=1}^{n}\left(\mathbf{M}_n^{m-n+1}(x_i,x_j)\right). \qquad (5.90)$$

By performing an asymptotic analysis of the Meixner polynomials and Meixner kernel, using an appropriate contour integral representation and the saddle point method, one can use (5.90) to prove the following result analogous to Theorem 5.1.

Theorem 5.31 (Limit law for the multicorner growth process passage times) *Let $0 < p < 1$. For $x, y > 0$ define*

$$\Phi_p(x,y) = \frac{1}{p}\left(x + y + 2\sqrt{(1-p)xy}\right),$$
$$\eta_p(x,y) = \frac{(1-p)^{1/6}}{p}(xy)^{-1/6}\left(\sqrt{x} + \sqrt{(1-p)y}\right)^{2/3}\left(\sqrt{y} + \sqrt{(1-p)x}\right)^{2/3}.$$

Let $(m_k)_{k=1}^{\infty}, (n_k)_{k=1}^{\infty}$ be sequences of positive integers with the properties that

$$m_k, n_k \to \infty \text{ as } k \to \infty,$$
$$0 < \liminf_{k\to\infty} \frac{m_k}{n_k} < \limsup_{k\to\infty} \frac{m_k}{n_k} < \infty.$$

Then as $k \to \infty$, the passage times $G(m_k, n_k)$ associated with the multicorner growth process converge in distribution after rescaling to the Tracy–Widom distribution F_2. More precisely, we have

$$\mathbb{P}\left(\frac{G(m_k,n_k) - \Phi_p(m_k,n_k)}{\eta_p(m_k,n_k)} \leq t\right) \xrightarrow[k\to\infty]{} F_2(t) \qquad (t \in \mathbb{R}).$$

5.14 Complex Wishart matrices

To conclude this chapter, we point out another intriguing connection between the corner growth process and a seemingly unrelated problem in probability theory involving random matrices known as **complex Wishart matrices**. We will not treat this topic in detail; for the proofs of Theorems 5.32 and 5.33 below, see chapter 4 of [4].

Given integers $m \geq n \geq 1$, the complex Wishart matrix $W_{m,n}$ is a random $n \times n$ positive-definite Hermitian matrix constructed as follows. Let

$$A = (X_{j,k} + iY_{j,k})_{1 \leq j \leq n, 1 \leq k \leq m}$$

be a random $n \times m$ rectangular matrix whose real and imaginary parts are a family of independent and identically distributed random variables $X_{j,k}, Y_{j,k}$ all having the normal distribution $N(0, 1/2)$. (The distribution of the complex entries $Z_{j,k} = X_{j,k} + iY_{j,k}$ is referred to as the standard complex normal distribution and denoted $N_{\mathbb{C}}(0, 1)$. Note that they satisfy $\mathbb{E}|Z_{j,k}|^2 = 1$.) Now let

$$W_{m,n} = AA^*,$$

which by construction is automatically Hermitian and positive-semidefinite (and, it is not hard to see, actually positive-definite with probability 1).

The distribution of $W_{m,n}$ is of some importance in statistics, where it is known as (a special case of) the **complex Wishart distribution.**[3]

To understand the Wishart distribution, we first parametrize the space of matrices in which $W_{m,n}$ takes its values. Denote by H_n the set of $n \times n$ Hermitian matrices with complex entries, and denote by $H_n^+ \subset H_n$ the subset of positive-semidefinite matrices in H_n. We write a generic element $M \in H_n$ as

$$M = (x_{j,k} + iy_{j,k})_{1 \leq j,k \leq n}.$$

Since $M^* = M$, one can consider as free parameters the elements lying on or above the main diagonal (with the diagonal elements being purely real numbers), namely $(x_{j,j})_{1 \leq j \leq n} \cup (x_{j,k}, y_{j,k})_{1 \leq j < k \leq n}$. For convenience we identify Hermitian matrices with their sets of free parameters, which allows us to think of H_n as the n^2-dimensional space $\mathbb{R}^{n^2}$. With this identification, the standard Lebesgue measure on $\mathbb{R}^{n^2}$ can be written as

$$dM = \prod_{j=1}^{n} dx_{j,j} \prod_{1 \leq j < k \leq n} dx_{j,k}\, dy_{j,k}. \tag{5.91}$$

Note that H_n^+ is an open subset of H_n.

The matrix $W_{m,n}$ can now be thought of as an n^2-dimensional random vector. The simplest description of its distribution is in terms of its joint density (with respect to the Lebesgue measure (5.91)), given in the following result.

Theorem 5.32 (Density of complex Wishart matrices) *The complex Wishart matrix $W_{m,n}$ has the n^2-dimensional joint probability density function*

$$f(M) = \frac{1}{\gamma_{m,n}} \det(M)^{m-n} e^{-\operatorname{tr} M} \qquad (M \in H_n^+), \tag{5.92}$$

where $\gamma_{m,n}$ is a normalization constant given by

$$\gamma_{m,n} = \pi^n \prod_{j=0}^{n} (m-n+j)!.$$

It is interesting to note that the density (5.92) is a function only of $\det M$ and $\operatorname{tr} M$, both of which are invariants under conjugation by a unitary matrix. This reflects the fact (which is not hard to prove, either from (5.92) or directly from the definition) that the distribution of $W_{m,n}$ is invariant under unitary conjugation, that is, for any $n \times n$ unitary matrix U, the matrix $U^* W_{m,n} U$ is equal in distribution to $W_{m,n}$.

Another way of formulating the preceding invariance property is as the statement that the density (5.92) can be expressed as a function of the eigenvalues of $W_{m,n}$. Denote the eigenvalues (which are real and positive, since $W_{m,n} \in H_n^+$), arranged in increasing order, by $\xi_1 < \ldots < \xi_n$. In terms of the eigenvalues, the density (5.92) becomes

$$f(M) = \frac{1}{\gamma_{m,n}} \prod_{j=1}^{n} \xi_j^{m-n} e^{-\sum_{j=1}^{n} \xi_j} \qquad (M \in H_n^+). \tag{5.93}$$

Using (5.93) one can prove the following result giving a formula for the joint density of the eigevalues $\xi_1, \ldots, \xi_n$.

Theorem 5.33 (Eigenvalue distribution of complex Wishart matrices) *Let $m \geq n \geq 1$, and let $D_{m,n}$ be defined as in* (5.13). *The joint density of the eigenvalues $\xi_1 < \ldots < \xi_n$ of the Wishart random matrix $W_{m,n}$ is*

$$f(x_1, \ldots, x_n) = n! D_{m,n} \prod_{j=1}^{n} (e^{-x_j} x_j^{m-n}) \prod_{1 \leq i < j \leq n} (x_i - x_j)^2 \quad (0 \leq x_1 \leq \ldots \leq x_n). \tag{5.94}$$

This expression looks familiar from our studies of the corner growth process. Indeed, a quick comparison with (5.25) identifies (5.94) as being precisely the density function of the orthogonal polynomial ensemble associated with the weight function $W_{m-n}(x) = e^{-x} x^{m-n}$ and the related Laguerre polynomials, which in turn we saw was related to the distribution of the passage times $G(m, n)$. (Note that (5.94) gives the density for the *ordered* eigenvalues, whereas (5.25) treats the points of the orthogonal polynomial ensemble as an *unordered* set.) The precise connection is summarized in the following result.

Theorem 5.34 *The passage time $G(m, n)$ in the corner growth process is equal in distribution to the maximal eigenvalue ξ_n of the complex Wishart random matrix $W_{m,n}$.*

Proof Let $\lambda_1, \dots, \lambda_n$ be the eigenvalues of $W_{m,n}$ where we have randomized the order of $\xi_1, \dots, \xi_n$, that is, we set

$$\lambda_k = \xi_{\sigma(k)}, \qquad k = 1, \dots, n,$$

where σ is a uniformly random permutation in S_n, chosen independently of $W_{m,n}$. Then since the formula on the right-hand side of (5.94) is symmetric in $x_1, \dots, x_n$, it is easy to see that $\lambda_1, \dots, \lambda_n$ have joint density function

$$g(x_1, \dots, x_n) = D_{m,n} \prod_{j=1}^{n} (e^{-x_j} x_j^{m-n}) \prod_{1 \le i < j \le n} (x_i - x_j)^2 \qquad (x_1, \dots, x_n \in \mathbb{R}).$$

It follows that

$$\begin{aligned}
\mathbb{P}(\mu_n \le t) &= \mathbb{P}\big((\lambda_1, \dots, \lambda_n) \in (-\infty, t]^n\big) \\
&= \int \dots \int_{(-\infty,t]^n} g(x_1, \dots, x_n)\, dx_1 \dots dx_n \\
&= D_{m,n} \int_{-\infty}^{t} \dots \int_{-\infty}^{t} \prod_{j=1}^{n} (e^{-x_j} x_j^{m-n}) \prod_{1 \le i < j \le n} (x_i - x_j)^2 \, dx_1 \dots dx_n,
\end{aligned}$$

which by (5.15) is precisely $\mathbb{P}(G(m, n) \le t)$. □

In light of the above result, Theorem 5.1 takes on a new meaning as a result on the limiting distribution of the maximal eigenvalue of a large complex Wishart matrix $W_{m,n}$ when $m, n \to \infty$ in such a way that the ratio m/n is bounded away from infinity.[45]

Exercises

5.1 (☕☕) Show that in Theorem 5.1, the convergence (5.5) need not hold if the condition (5.4) is omitted.

5.2 (☕☕☕) Prove Lemma 5.5.

5.3 (☕) Find the triple (λ, P, Q) associated to the generalized permutation

$$\sigma = \begin{pmatrix} 1 & 2 & 2 & 2 & 4 & 5 & 5 & 6 & 8 & 8 & 10 \\ 6 & 3 & 3 & 4 & 2 & 1 & 2 & 3 & 2 & 3 & 1 \end{pmatrix}$$

by the RSK algorithm.

5.4 (☕) Find the generalized permutation associated via the (inverse) RSK algorithm to the Young diagram $\lambda = (5, 5, 1, 1)$ and semistandard tableaux

$$P = \begin{array}{|c|c|c|c|c|} \hline 1 & 2 & 3 & 3 & 3 \\ \hline 2 & 3 & 4 & 4 & 5 \\ \hline 5 \\ \cline{1-1} 8 \\ \cline{1-1} \end{array}, \qquad Q = \begin{array}{|c|c|c|c|c|} \hline 1 & 1 & 1 & 2 & 3 \\ \hline 2 & 2 & 3 & 4 & 5 \\ \hline 4 \\ \cline{1-1} 6 \\ \cline{1-1} \end{array}.$$

5.5 (☕) Prove Lemma 5.9.

5.6 (☕☕) Prove that if $f(x_1, \ldots, x_k)$ is an antisymmetric polynomial of total degree $k(k-1)/2$, then there is a constant c such that

$$f(x_1, \ldots, x_n) = c \prod_{1 \le i < j \le n} (x_i - x_j).$$

5.7 (☕) Show that the number of sequences $(a_1, \ldots, a_d)$ with nonnegative integer coordinates that satisfy $\sum_{j=1}^{d} a_j = k$ is equal to $\binom{d-1+k}{k}$. Deduce that the number of generalized permutations of length k and row bounds (m, n) is

$$|\mathcal{P}_{m,n}^k| = |\mathcal{M}_{m,n}^k| = \binom{mn - 1 + k}{k}.$$

5.8 (☕☕) Use Lemma 5.16 to prove the **Cauchy–Binet formula**, which states that if a square matrix $C = (c_{i,j})_{1 \le i,j \le n} = AB$ is the product of two rectangular matrices $A = (a_{i,j})_{1 \le i \le n, 1 \le j \le m}$ and $B = (b_{i,j})_{1 \le i \le m, 1 \le j \le n}$, then

$$\det(C) = \sum_{1 \le k_1 < \ldots < k_n \le m} \det_{i,\ell=1}^{n} \left(a_{i,k_\ell}\right) \det_{\ell,j=1}^{n} \left(b_{k_\ell,j}\right).$$

5.9 (☕☕) Let $\mathbf{K} \colon \mathbb{R} \times \mathbb{R} \to \mathbb{R}$ be a kernel with the property that

$$\int_{-\infty}^{\infty} \mathbf{K}(x, y)\mathbf{K}(y, z)\, dy = K(x, z) \qquad (x, z \in \mathbb{R}).$$

Denote $r = \int_{-\infty}^{\infty} \mathbf{K}(x, x)\, dx$. Prove that for any $m \ge 1$ and $x_1, \ldots, x_m \in \mathbb{R}$ we have

$$\int_{-\infty}^{\infty} \det_{i,j=1}^{m} \left(\mathbf{K}(x_i, x_j)\right) dx_n = (r - m + 1) \det_{i,j=1}^{m} \left(\mathbf{K}(x_i, x_j)\right).$$

5.10 (☕☕) Use the result of Exercise 5.9 above to prove Theorem 5.20.

5.11 (☕) Use the result of Exercise 5.9 above to get a new proof of Corollary 5.18.

5.12 (☕) Use Corollary 5.18 to rederive the expression (5.13) for the normalization constants $D_{m,n}$ in (5.15).

5.13 (☕☕) Show that the Laguerre polynomial $\ell_n^\alpha(x)$ satisfies the differential equation

$$xf''(x) + (\alpha + 1 - x)f'(x) + nf(x) = 0.$$

5.14 (☕☕) Prove the generating function identity

$$\sum_{n=0}^{\infty} L_n^\alpha(x)t^n = (1-t)^{-(1+\alpha)} \exp\left(-\frac{tx}{1-t}\right).$$

5.15 (☕☕) Show that the Laguerre polynomial $\ell_n^\alpha(x)$ has n real and positive roots.

5.16 (☕☕☕) Prove that the joint density (5.94) of the eigenvalues of the Wishart random matrix $W_{m,n}$ takes its unique maximum at the point $(\zeta_1, \dots, \zeta_n)$, where $\zeta_1 < \dots < \zeta_n$ are the roots of the Laguerre polynomial $\ell_n^{m-n-1}(x)$.

5.17 (☕☕☕) The **Hermite polynomials** $(h_n(x))_{n=0}^\infty$ are the sequence of orthogonal polynomials associated with the weight function $w(x) = e^{-x^2/2}$; that is, for each $n \geq 0$, $h_n(x)$ is a polynomial of degree 0 with positive leading coefficient, and the polynomials $h_n(x)$ satisfy the orthonormality relation

$$\int_{-\infty}^{\infty} h_n(x)h_m(x)e^{-x^2/2}\,dx = \delta_{nm} \qquad (n, m \geq 0).$$

Prove that the Hermite polynomials have the following properties:

(a) $h_n(x) = \dfrac{(-1)^n}{\sqrt{\sqrt{2\pi}n!}} e^{x^2/2} \dfrac{d^n}{dx^n}\left(e^{-x^2/2}\right)$.

(b) $h_n(x) = \dfrac{(n!)^{1/2}}{(2\pi)^{1/4}} \displaystyle\sum_{k=0}^{\lfloor n/2 \rfloor} \frac{(-1)^k}{2^k k!(n-2k)!} x^{n-2k}$.

(c) The leading coefficient of $h_n(x)$ is $\kappa_n = \left(\sqrt{2\pi}\,n!\right)^{-1/2}$.

(d) The exponential generating function of the h_n (normalized to be monic polynomials) is

$$\sum_{n=0}^{\infty} \frac{h_n(x)}{\kappa_n n!} t^n = e^{xt - t^2/2}.$$

5.18 (☕☕) Use Corollary 5.18 and the properties of the Hermite polynomials from Exercise 5.17 above to prove for any $n \geq 1$ the n-dimensional integration identity

$$\int \dots \int_{\mathbb{R}^n} \prod_{1 \leq i < j \leq n} (x_i - x_j)^2 e^{-\frac{1}{2}\sum_{j=1}^n x_j^2}\, dx_1 \dots dx_n = (2\pi)^{n/2} \prod_{j=1}^{n-1} j!.$$

Deduce that the formula given on p. 84 for the normalization constant Z_n in the joint density of the eigenvalues of a random Gaussian Unitary Ensemble (GUE) matrix is correct.

5.19 (☕☕) Prove (5.71).

5.20 (a) (☕☕) Use the saddle point method to prove that

$$\binom{2n}{n} = (1 + o(1))\frac{4^n}{\sqrt{\pi n}}.$$

(b) (☕☕) The nth **central trinomial coefficient** c_n is defined as the coefficient of x^n in the expansion of the polynomial $(1+x+x^2)^n$ in powers of x. Use the saddle point method to derive and prove an asymptotic formula for c_n as $n \to \infty$.

5.21 (☕☕☕) (Chowla–Herstein–Moore [25]) Let I_n denote the number of involutions of order n, that is, permutations $\sigma \in S_n$ such that $\sigma^2 = \text{id}$. Use the saddle point method together with the contour integral representation

$$I_n = \frac{1}{2\pi i}\oint_{|z|=r} \frac{e^{z+z^2/2}}{z^{n+1}}\,dz$$

(an immediate corollary of the result of Exercise 1.11(c)) to get a new proof of the asymptotic formula (1.61).

5.22 (Angel–Holroyd–Romik [7]) Given permutations $\pi, \sigma \in S_n$, we say that σ **covers** π, and denote $\pi \nearrow \sigma$, if σ is for some $1 \le j \le n-1$ such that $\pi(j) < \pi(j+1)$, σ is obtained from π by transposing the adjacent elements $\pi(j), \pi(j+1)$ of π and leaving all other elements of π in place (in other words, $\sigma = \pi \circ \tau_j$ where τ_j denotes the jth adjacent transposition; see Exercise 3.8). Fix $n \ge 1$. Denote $N = \binom{n}{2}$. Define a random walk $(\sigma_k^n)_{0\le k\le N}$ on S_n, called the **directed adjacent transposition random walk**, by setting $\sigma_0^n = \text{id}$ and, conditionally on the event that $\sigma_k^n = \pi$, choosing σ_{k+1}^n to be a uniformly random permutation σ chosen from among the permutations such that $\pi \nearrow \sigma$. If we define the **inversion number** of a permutation σ by

$$\text{inv}(\sigma) = \#\{1 \le j < k \le n : \sigma(j) > \sigma(k)\},$$

then it is easy to see that $\text{inv}(\sigma_k^n) = k$ and in particular at time $N = \binom{n}{2}$ the random walk reaches the unique permutation with maximal inversion number $N = \binom{n}{2}$, namely the reverse permutation $\text{rev}_n = (n, \dots, 2, 1)$ defined in Exercise 3.8; once it reaches rev_n the random walk terminates.

It is natural to interpret this process as a particle system, which can be described as follows: n particles numbered $1, \dots, n$ are arranged in a line. Initially particle k is in position k. Now successively swap the positions of an adjacent pair (i, j) that is chosen uniformly at random from among the pairs

$$\begin{array}{rccccc}
\sigma_0^5 = & 1 & 2 & \mathbf{3} & 4 & 5 \\
\sigma_1^5 = & 2 & 1 & \mathbf{3} & 4 & 5 \\
\sigma_2^5 = & 2 & 1 & \mathbf{3} & 5 & 4 \\
\sigma_3^5 = & 2 & 1 & 5 & \mathbf{3} & 4 \\
\sigma_4^5 = & 2 & 1 & 5 & 4 & \mathbf{3} \\
\sigma_5^5 = & 2 & 5 & 1 & 4 & \mathbf{3} \\
\sigma_6^5 = & 2 & 5 & 4 & 1 & \mathbf{3} \\
\sigma_7^5 = & 2 & 5 & 4 & \mathbf{3} & 1 \\
\sigma_8^5 = & 5 & 2 & 4 & \mathbf{3} & 1 \\
\sigma_9^5 = & 5 & 4 & 2 & \mathbf{3} & 1 \\
\sigma_{10}^5 = & 5 & 4 & \mathbf{3} & 2 & 1
\end{array}$$

Figure 5.1 An example of a directed adjacent transposition random walk with $n = 5$. The numbers in bold highlight the trajectory of particle number 3.

for which $i < j$ until for each $1 \le k \le n$ the particle that started in position k is in position $n + 1 - k$. See Fig. 5.1 for an example, which also shows how one can graphically track the trajectory of individual particles.

Next, construct a continuous-time version of the random walk $(\sigma_k^n)_{0 \le k \le N}$ in a manner analogous to the construction of the continuous-time corner growth process from its discrete-time counterpart. More precisely, associate with each permutation $\sigma \in S_n$ a random variable X_σ with the exponential distribution $\text{Exp}(\text{out}(\sigma))$, where $\text{out}(\sigma)$ denotes the number of permutations τ such that $\sigma \nearrow \tau$ (that is, the out-degree of σ in the directed graph with adjacency relation "$\nearrow$"). The random variables $(X_\sigma)_{\sigma \in S_n}$ are taken to be independent of each other and of the random walk $(\sigma_k^n)_k$. Define a sequence of random times $0 = S_0 < S_1 < \ldots < S_N < S_{N+1} = \infty$ by setting

$$S_{k+1} = S_k + X_\sigma \quad \text{on the event } \{\sigma_k^n = \sigma\}$$

for $0 \le k \le N - 1$. The continous-time random walk is the process $(\pi_t^n)_{t \ge 0}$ of S_n-valued random variables defined by setting

$$\pi_t^n = \sigma_k^n \quad \text{on the event } \{S_k \le t < S_{k+1}\}.$$

(a) (☕☕☕) Show that the continuous-time process has the following equivalent description. With each pair $(k, k+1)$ of adjacent positions associate a Poisson process of unit intensity on $[0, \infty)$, which can be written as an increasing sequence $E_k = \{0 = T_0^k < T_1^k < T_2^k < \ldots\}$ of times such that the increments $(T_{m+1}^k - T_m^k)_{m\geq 0}$ are i.i.d. with the exponential distribution Exp(1). The different processes E_k are taken to be independent. The random permutation process $(\pi_t^n)_{t\geq 0}$ is defined by setting $\pi_0^n = \mathrm{id}$ and modifying the permutations precisely at the times T_m^k $(1 \leq k \leq n-1,$ $m > 0)$, according to the rule that at each time T_m^k, the particles at (i, j) at positions $(k, k+1)$ swap their positions if and only if $i < j$.

(b) (☕☕☕☕) For $1 \leq j \leq n$, define the **finishing time of particle** j to be the last time when particle j moved, that is,

$$T_j^n = \sup\left\{t > 0 \,:\, (\pi_t^n)^{-1}(j) \neq n + 1 - j\right\}.$$

and define the **total finishing time** by

$$T_*^n = \max\{T_j^n \,:\, 1 \leq j \leq n\} = \inf\{t > 0 \,:\, \pi_t^n = \mathrm{rev}_n\}.$$

Prove that the joint distribution of the finishing times can be expressed in terms of the passage times $G(a, b; i, j)$ in the corner growth process. More precisely, we have the equality in distribution

$$\left\{T_j^n \,:\, 1 \leq j \leq n\right\} \stackrel{d}{=} \left\{R_{j-1}^n \vee R_j^n \,:\, 1 \leq j \leq n\right\},$$

where we define

$$R_j^n = G(j, 1; n-1, j) \qquad (1 \leq j \leq n-1),$$
$$R_0^n = R_n^n = 0.$$

As a result we also have the representation

$$T_*^n \stackrel{d}{=} \max\left\{G(j, 1; n-1, j) \,:\, 1 \leq j \leq n-1\right\}$$

for the distribution of the total finishing time.

(c) (☕☕) Deduce from part (b) above together with the results we proved in Chapters 4 and 5 the following results about the asymptotic behavior of the finishing times.

Theorem 5.35 (Uniform asymptotics for the finishing times) *Denote $g(x) = 1 + 2\sqrt{x(1-x)}$. As $n \to \infty$, we have the convergence in probability*

$$\max_{1\leq j\leq n} \left|\frac{1}{n} T_j^n - g(j/n)\right| \xrightarrow[n\to\infty]{P} 0,$$

and, as a consequence,

$$\frac{1}{n}T_*^n \xrightarrow[n\to\infty]{P} 2.$$

Theorem 5.36 (Fluctuations of the finishing times) *Let $(j_k)_{k=1}^\infty$ and $(n_k)_{k=1}^\infty$ be sequences of positive integers with the properties*

$$1 \le j_k \le n_k,$$
$$j_k, n_k \to \infty,$$
$$0 < \liminf_{k\to\infty} \frac{j_k}{n_k} < \limsup_{k\to\infty} \frac{j_k}{n_k} < 1.$$

Then as $k \to \infty$ the finishing time $T_j^n = T_{j_k}^{n_k}$ converges in distribution after scaling to the Tracy–Widom distribution F_2. More precisely, we have

$$\frac{T_{j_k}^{n_k} - \Psi(j_k, n_k + 1 - j_k)}{\sigma(j_k, n + 1 - j_k)} \xrightarrow{d} F_2 \quad \text{as } n \to \infty,$$

where $\Psi(\cdot,\cdot)$ and $\sigma(\cdot,\cdot)$ are defined in (5.1) *and* (5.2).

(d) (☕☕☕☕) Prove that asymptotically as $n \to \infty$ we have for any $t \ge 0$ that

$$\binom{n}{2}^{-1} \mathrm{inv}(\pi_{nt}^n) \xrightarrow[n\to\infty]{P} \begin{cases} \frac{2}{3}t - \frac{1}{15}t^2 & \text{if } 0 \le t < 1, \\ 1 - \frac{2}{15}t^{-1/2}(2-t)^{3/2}(1+2t) & \text{if } 1 \le t < 2, \\ 1 & \text{if } t \ge 2. \end{cases}$$

(e) (☕☕☕☕☕) Find sequences $(a_n)_{n=1}^\infty$, $(b_n)_{n=1}^\infty$ and a nontrivial probability distribution F such that we have the convergence in distribution

$$\frac{T_*^n - a_n}{b_n} \xrightarrow[n\to\infty]{d} F.$$

Appendix

Kingman's subadditive ergodic theorem

In Chapters 1 and 4 we make use of **Kingman's subadditive ergodic theorem**, an advanced result from probability theory. This appendix gives a brief discussion of this important result in a version suited to our needs. The starting point is the following well-known elementary lemma in real analysis, known as Fekete's subadditive lemma.

Lemma A.1 (Fekete's subadditive lemma) *If $(a_n)_{n=1}^{\infty}$ is a sequence of nonnegative numbers that satisfies the subadditivity property*

$$a_{m+n} \leq a_n + a_m \qquad (n, m \geq 1),$$

Then the limit $\lim_{n\to\infty} \frac{a_n}{n}$ exists and is equal to $\inf_{n\geq 1} \frac{a_n}{n}$.

The condition of subadditivity appears naturally in many places, making Fekete's lemma an extremely useful fact. John F. C. Kingman [69] discovered that subadditivity can be applied in a probabilistic setting to a family of random variables, with equally powerful consequences. His theorem (in a version incorporating a small but significant improvement added later by Liggett [76]) is stated as follows.

Theorem A.2 (Kingman's subadditive ergodic theorem) *Let $(X_{m,n})_{0\leq m<n}$ be a family of random variables, defined on some probability space, that satisfies:*

1. $X_{0,n} \leq X_{0,m} + X_{m,n}$ *for all* $m < n$.
2. *For any* $k \geq 1$, *the sequence* $(X_{nk,(n+1)k})_{n=1}^{\infty}$ *is a stationary sequence.*
3. *For any* $m \geq 1$, *the two sequences* $(X_{0,k})_{k=1}^{\infty}$ *and* $(X_{m,m+k})_{k=1}^{\infty}$ *have the same joint distribution.*
4. $\mathbb{E}|X_{0,1}| < \infty$, *and there exists a constant* $M > 0$ *such that for any* $n \geq 1$, $\mathbb{E}X_{0,n} \geq -Mn$.

Then

(a) *limit* $\gamma = \lim_{n\to\infty} \frac{\mathbb{E}(X_{0,n})}{n}$ *exists and is equal to* $\inf_{n\ge 1} \frac{\mathbb{E}(X_{0,n})}{n}$.
(b) *The limit* $X = \lim_{n\to\infty} \frac{X_{0,n}}{n}$ *exists almost surely, and satisfies* $\mathbb{E}(X) = \gamma$.
(c) *If the stationary sequences* $(X_{nk,(n+1)k})_{n=1}^{\infty}$ *are ergodic for any* $k \ge 1$, *then* $X = \gamma$ *almost surely.*

Note that conclusion (a) is a straightforward consequence of Fekete's lemma.

For the proof, see [76] or [33], chapter 7. Both the formulation of the theorem and its proof make reference to concepts and results from ergodic theory, which are beyond the scope of this book and may be unfamiliar to some readers. It is comforting, however, to notice that for our applications in Chapters 1 and 4 we can use the following special case of the theorem that avoids reference to ergodicity, relying instead on more elementary concepts.

Theorem A.3 (Kingman's subadditive ergodic theorem; i.i.d. case) *Let* $(X_{m,n})_{0\le m<n}$ *be a family of random variables, defined on some probability space, that satisfies:*

1 $X_{0,n} \le X_{0,m} + X_{m,n}$ *for all* $m < n$.
2 *For any* $k \ge 1$, *the sequence* $(X_{nk,(n+1)k})_{n=1}^{\infty}$ *is a sequence of independent and identically distributed random variables.*
3 *For any* $m \ge 1$, *the two sequences* $(X_{0,k})_{k=1}^{\infty}$ *and* $(X_{m,m+k})_{k=1}^{\infty}$ *have the same joint distribution.*
4 $\mathbb{E}|X_{0,1}| < \infty$, *and there exists a constant* $M > 0$ *such that for any* $n \ge 1$, $\mathbb{E}X_{0,n} \ge -Mn$.

Then the limit $\gamma = \lim_{n\to\infty} \frac{\mathbb{E}(X_{0,n})}{n}$ *exists and is equal to* $\inf_{n\ge 1} \frac{\mathbb{E}(X_{0,n})}{n}$. *Furthermore, we have the convergence* $\frac{1}{n}X_{0,n} \to \gamma$ *almost surely.*

Theorem A.3 can be proved in an identical manner to the proof of Theorem A.2, except that instead of applying Birkhoff's pointwise ergodic theorem one invokes the strong law of large numbers.

Notes

Chapter 1

1 Hammersley's paper [54], based on his Berkeley symposium talk, is written in an unusual informal and entertaining style and is worth reading also for the insights it offers as to the way mathematical research is done.

2 Actually Logan and Shepp only proved that $\liminf_{n\to\infty} \ell_n/\sqrt{n} \ge 2$, but it is this proof that was the real breakthrough. Vershik and Kerov proved the same result but also found the relatively simple argument (Section 1.19) that proves the upper bound $\limsup_{n\to\infty} \ell_n/\sqrt{n} \le 2$.

3 The Erdős–Szekeres theorem was mentioned by Ulam in his original paper [138] as the motivation for studying the maximal monotone subsequence length of a random permutation.

4 The plural form of *tableau* when used in its math context is *tableaux*; when used in the original (nonmathematical) meaning, *tableaux* is still the standard plural form but *tableaus* is also considered acceptable in American English.

5 Part (a) of Theorem 1.10 is due to Schensted [113], and part (b) is due to Schützenberger [114].

6 The effect on the recording tableau Q of reversing the order of the elements of σ is more subtle so we do not discuss it here; see [71, Section 5.1.4] and Chapter 3 (specifically, Theorem 3.9.4) of [112].

7 Hardy and Ramanujan proved (1.16) as a corollary of a much more detailed asymptotic expansion they obtained for $p(n)$. Their result was later improved by Hans Rademacher, who used similar techniques to derive a remarkable convergent infinite series for $p(n)$. See [8] for more details on this celebrated result.

8 Varadhan was awarded the Abel prize in 2007 for his fundamental contributions to probability theory and in particular to the theory of large deviations.

9 The following simple example demonstrates that the typical behavior in a probabilistic experiment need not always coincide with the "least exponentially unlikely" behavior: when making n independent coin tosses of a coin with bias

$p = 2/3$, the typical sequence of results will contain approximately $2n/3$ "heads" and $n/3$ "tails," but the most likely sequence consists of a succession of n heads, an outcome which as a function of the parameter n is exponentially more likely than any single outcome in the "typical" region.

10 The relation (1.34) corrects a sign error from [79] which (in a nice demonstration of the way errors can propagate through the scientific literature) was repeated in [28] and [101].

11 A version of Lemma 1.21, due to Boris Pittel, appeared in [101].

12 The question of identifying the irreducible representation of maximal dimension in the symmetric group may have been mentioned for the first time in a 1954 paper by Bivins, Metropolis, Stein, and Wells [14] (see also [10], [87]).

13 The Plancherel growth process was first named and considered systematically by Kerov, who analyzed in [66] the limiting dynamics of the growth process, which provides an alternative approach to understanding the limit shape of Plancherel-random Young diagrams. The same growth process is also implicit in earlier works such as [51], [143].

Chapter 2

1 See the papers by Frieze [45], Bollobás-Brightwell [16], Talagrand [132], Deuschel-Zeitouni [32], Seppäläinen [118]; and the related papers by Aldous-Diaconis [2], Johansson [61], and Seppäläinen [116] that gave new proofs of Theorem 1.1.

2 The graph in Fig. 2.1 was generated from numerical data computed by Prähofer and Spohn [102].

3 This extremely elegant proof is one of the outstanding research achievements mentioned in Okounkov's 2006 Fields Medal citation.

4 The heading of Theorem 2.3 refers to the fact that the theorem describes the limiting distribution of the edge, or maximal, points, of the point process of modified Frobenius coordinates associated with the Plancherel-random Young diagram. One can also study the so-called **bulk statistics** corresponding to the behavior of the point process away from the edge. Like the edge statistics, the bulk statistics also reveal interesting parallels with the analogous quantities associated with eigenvalue distributions in random matrix theory; see [19].

5 Determinantal point processes were first defined in 1975 by Macchi [83], who referred to them as fermion processes because they arise naturally as descriptions of many-particle systems in statistical quantum mechanics. See [59], [80], [82], [81], [121] and [29, Section 5.4] for more details on the general theory and many examples.

6 Actually this definition corresponds to what is usually called in the literature a **simple point process**; one can consider more generally nonsimple point processes,

where points are allowed to have multiplicity greater than 1, but all the processes we will consider will be simple.

7 Fredholm determinants play an important role in random matrix theory, integrable systems and other areas of analysis. For more background see [74], [119], [120].

8 Considering the crucial importance of Nicholson's approximation to the developments in Chapter 2, it is unfortunate that the details of the proof of this result are rather obscure and inaccessible. The result in a form similar to ours is derived in Section 4 of the Borodin–Okounkov–Olshanski paper [19] from an earlier and more traditional form of the result presented in Section 8.43 of the classical reference [145] on Bessel functions by G. N. Watson. However, the details of Watson's derivation are difficult to decipher, and appear to be incomplete (one finds in the text remarks such as "*... it has been proved, by exceedingly heavy analysis which will not be reproduced here, that...*"). It would be nice to see a more modern and accessible proof of this important asymptotic result. Exercise 2.22 suggests a possible approach for developing such a proof.

9 The representation (2.79) for the Airy kernel is mentioned in [26] (see equation (4.6) in that paper). The proof we give is due to Tracy and Widom [136].

10 This type of difficulty regarding the question of existence and uniqueness of the Airy process is not uncommon. In probability theory one often encounters the "foundational" question of whether a stochastic process with prescribed statistical behavior exists and is unique. To be sure, such questions are important, but nonetheless one should not be too bothered by this. As a more or less general rule, if the prescribed statistics "make sense" as probabilistic descriptions of a random process (i.e., if they are self-consistent in some obvious sense) then, barring pathological examples that most ordinary practicioners of probability will never actually encounter, existence and uniqueness can be safely assumed. Unfortunately, the rigorous verification of these types of claims is usually quite tedious and can involve heavy machinery from measure theory and functional analysis.

Chapter 3

1 Erdős–Szekeres permutations and the formulas (3.1) and (3.2) are mentioned in [71, Exercise 5.1.4.9] and [125, Example 7.23.19(b)]. The results of Sections 3.2 and 3.3 are due to Romik [108].

2 Theorems 3.5 and 3.6 are due to Pittel and Romik [101].

3 The name "arctic circle" is borrowed from an analogous result due to Jockusch, Propp and Shor [60], known as the **arctic circle theorem**, which deals with random domino tilings of a certain planar region; see Exercises 4.12 and 4.13 at the end of Chapter 4. Two other arctic circle theorems are proved in [28] (see also [64]) and [98].

4 This result is proved in [109]. The idea of interpreting the limit shape theorem for random square Young tableaux in terms of an arctic circle in an associated interacting particle system was proposed by Benedek Valkó.

Chapter 4

1 The terminology we use in this chapter is somewhat simplified and differs slightly from that used in the existing literature on the subject. The process we call the corner growth process is often referred to as the **corner growth model with exponential weights** or as **directed last-passage percolation on $\mathbb{Z}^2$ with exponential weights**; in the context of interacting particle systems, an equivalent process is known in the literature as the **totally asymmetric simple exclusion process with step initial condition**. The process discussed here under the name "multicorner growth process" (Section 4.8) is referred to elsewhere as the **corner growth model with geometric weights** (or **last-passage percolation with geometric weights**).

2 In the existing literature on the corner growth process, the variables $\tau_{i,j}$ are usually referred to as **weights**.

3 The **totally asymmetric simple exclusion process with step initial condition**; see Note 1.

4 Thorem 4.24 is due to Jockusch, Propp, and Shor [60] in the case $p \leq 1/2$, and to Cohn, Elkies, and Propp [27] in the case $p > 1/2$.

5 The developments in Sections 4.3, 4.5, 4.6 and 4.8 follow [115]. Note that the idea of comparing the growth process with a slowed down version of it as a means for deriving the limit shape goes back to Rost's original paper [110]. The particular (rather ingenious) way of slowing down the process described here, which appears to offer the easiest route to the limit shape theorems, was first introduced in [13].

Chapter 5

1 Theorem 5.8 is a standard result in enumerative combinatorics, being, for example, an immediate corollary of equation (7.105) on p. 305 of [125]. The proof given here is due to Elkies, Kuperberg, Larsen, and Propp [37].

2 Our approach to the proof of Theorem 5.1 adapts the saddle point techniques used by Johansson [62] in the case of the Meixner polynomials to the case of the Laguerre polynomials, and is also influenced by the presentation of [115]. Another proof based on different asymptotic analysis techniques appeared in [65].

3 The complex Wishart distribution has a real-valued counterpart that was introduced in 1928 by the British mathematician and statistician John Wishart [147]. The complex variant was first defined and studied in 1963 by Goodman [48].

4 A generalization of this result proved by Soshnikov [122] states that under similar assumptions, for each $k \geq 1$ the distribution of the largest k eigenvalues

$(\xi_{n-k+1}, \ldots, \xi_n)$ will converge as $n \to \infty$ to the joint distribution of the k largest points in the Airy ensemble (see Section 2.2).

5 The question of the limiting distribution of the maximal eigenvalue can be asked more generally for Wishart-type matrices (also called **sample covariance matrices**) $W_{m,n} = AA^*$ in which A is an $n \times m$ matrix with i.i.d. entries distributed according to some given distribution μ over the complex numbers having the same first- and second-order moments as the standard complex Gaussian distribution $N_{\mathbb{C}}(0, 1)$. According to a general principle of random matrix theory known as "universality," one expects to observe convergence to the Tracy–Widom limiting law F_2 for a wide class of entry distributions μ. This was proved in various levels of generality in [97], [99], [122], [144].

References

[1] Aitken, A. C. 1943. The monomial expansion of determinantal symmetric functions. *Proc. Royal Soc. Edinburgh (A)*, **61**, 300–310.

[2] Aldous, D., and Diaconis, P. 1995. Hammersley's interacting particle process and longest increasing subsequences. *Probab. Theory Related Fields*, **103**, 199–213.

[3] Aldous, D., and Diaconis, P. 1999. Longest increasing subsequences: from patience sorting to the Baik–Deift–Johansson theorem. *Bull. Amer. Math. Soc.*, **36**, 413–432.

[4] Anderson, G. W., Guionnet, A., and Zeitouni, O. 2010. *An Introduction to Random Matrices*. Cambridge University Press.

[5] Andrews, G. E., Askey, R., and Roy, R. 2001. *Special Functions*. Cambridge University Press.

[6] Angel, O., Holroyd, A. E., Romik, D., and Virág, B. 2007. Random sorting networks. *Adv. Math.*, **215**, 839–868.

[7] Angel, O., Holroyd, A. E., and Romik, D. 2009. The oriented swap process. *Ann. Probab.*, **37**, 1970–1998.

[8] Apostol, T. M. 1990. *Modular Forms and Dirichlet Series in Number Theory*. 2nd edition. Springer.

[9] Arnold, V. I. 1988. *Mathematical Methods of Classical Mechanics*. 2nd edition. Springer.

[10] Baer, R. M., and Brock, P. 1968. Natural sorting over permutation spaces. *Math. Comp.*, **22**, 385–410.

[11] Baik, J., Deift, P., and Johansson, K. 1999a. On the distribution of the length of the longest increasing subsequence of random permutations. *J. Amer. Math. Soc.*, **12**, 1119–1178.

[12] Baik, J., Deift, P., and Johansson, K. 1999b. On the distribution of the length of the second row of a Young diagram under Plancherel measure. *Geom. Funct. Anal*, **10**, 702–731.

[13] Balázs, M., Cator, E., and Seppäläinen, T. 2006. Cube root fluctuations for the corner growth model associated to the exclusion process. *Electron. J. Probab.*, **11**, 1094–1132.

[14] Bivins, R. L., Metropolis, N., Stein, P. R., and Wells, M. B. 1954. Characters of the symmetric groups of degree 15 and 16. *Math. Comp.*, **8**, 212–216.

[15] Blair-Stahn, N. First passage percolation and competition models. Preprint, `arXiv:1005.0649`, 2010.

[16] Bollobás, B., and Brightwell, G. 1992. The height of a random partial order: concentration of measure. *Ann. Appl. Probab.*, **2**, 1009–1018.

[17] Bollobás, B., and Winkler, P. 1988. The longest chain among random points in Euclidean space. *Proc. Amer. Math. Soc.*, **103**, 347–353.

[18] Bornemann, F. 2010. On the numerical evaluation of distributions in random matrix theory: A review. *Markov Process. Related Fields*, **16**, 803–866.

[19] Borodin, A., Okounkov, A., and Olshanski, G. 2000. Asymptotics of Plancherel measures for symmetric groups. *J. Amer. Math. Soc.*, **13**, 491–515.

[20] Boucheron, S., Lugosi, G., and Bousquet, O. 2004. Concentration inequalities. Pages 208–240 of: Bousquet, O., von Luxburg, U., and Rätsch, G. (eds), *Advanced Lectures in Machine Learning*. Springer.

[21] Bressoud, D. 1999. *Proofs and Confirmations: The Story of the Alternating Sign Matrix Conjecture*. Cambridge University Press.

[22] Brown, J. W., and Churchill, R. 2006. *Fourier Series and Boundary Value Problems*. 7th edition. McGraw-Hill.

[23] Bufetov, A. I. 2012. On the Vershik-Kerov conjecture concerning the Shannon-McMillan-Breiman theorem for the Plancherel family of measures on the space of Young diagrams. *Geom. Funct. Anal.*, **22**, 938–975.

[24] Ceccherini-Silberstein, T., Scarabotti, F., and Tolli, F. 2010. *Representation Theory of the Symmetric Groups: The Okounkov-Vershik Approach, Character Formulas, and Partition Algebras*. Cambridge University Press.

[25] Chowla, S., Herstein, I. N., and Moore, W. K. 1951. On recursions connected with symmetric groups I. *Canad. J. Math.*, **3**, 328–334.

[26] Clarkson, P. A., and McLeod, J. B. 1988. A connection formula for the second Painlevé transcendent. *Arch. Rat. Mech. Anal.*, **103**, 97–138.

[27] Cohn, H., Elkies, N., and Propp, J. 1996. Local statistics for random domino tilings of the Aztec diamond. *Duke Math. J.*, **85**, 117–166.

[28] Cohn, H., Larsen, M., and Propp, J. 1998. The shape of a typical boxed plane partition. *New York J. Math.*, **4**, 137–165.

[29] Daley, D. J., and Vere-Jones, D. 2003. *An Introduction to the Theory of Point Processes,* Vol. I: *Elementary Theory and Methods*. 2nd edition. Springer.

[30] Deift, P. 2000. *Orthogonal Polynomials and Random Matrices: A Riemann-Hilbert Approach*. American Mathematical Society.

[31] Dembo, A., and Zeitouni, O. 1998. *Large Deviations Techniques and Applications*. 2nd edition. Springer.

[32] Deuschel, J.-D., and Zeitouni, O. 1999. On increasing subsequences of I.I.D. samples. *Combin. Probab. Comput.*, **8**, 247–263.

[33] Durrett, R. 2010. *Probability: Theory and Examples*. 4th edition. Cambridge University Press.

[34] Edelman, P., and Greene, C. 1987. Balanced tableaux. *Adv. Math.*, **63**, 42–99.

[35] Edwards, H. M. 2001. *Riemann's Zeta Function*. Dover.

[36] Efron, B., and Stein, C. 1981. The jackknife estimate of variance. *Ann. Stat.*, **9**, 586–596.

[37] Elkies, N., Kuperberg, G., Larsen, M., and Propp, J. 1992. Alternating sign matrices and domino tilings. *J. Algebraic Combin.*, **1**, 111–132; 219–234.

[38] Estrada, R., and Kanwal, R. P. 2000. *Singular Integral Equations*. Birkhäuser.

[39] Feit, W. 1953. The degree formula for the skew-representations of the symmetric group. *Proc. Amer. Math. Soc.*, **4**, 740–744.

[40] Feller, W. 1967. A direct proof of Stirling's formula. *Amer. Math. Monthly*, **74**, 1223–1225.

[41] Feller, W. 1968. *An Introduction to Probability Theory and its Applications*, Vol. 1. 3rd edition. Wiley.

[42] Forrester, P. J. 2010. *Log-Gases and Random Matrices*. Princeton University Press.

[43] Frame, J. S., Robinson, G. de B., and Thrall, R. M. 1954. The hook graphs of the symmetric group. *Canad. J. Math.*, **6**, 316–324.

[44] Franzblau, D. S., and Zeilberger, D. 1982. A bijective proof of the hook-length formula. *J. Algorithms*, **3**, 317–343.

[45] Frieze, A. 1991. On the length of the longest monotone subsequence in a random permutation. *Ann. Appl. Probab.*, **1**, 301–305.

[46] Fristedt, B. 1993. The structure of random partitions of large integers. *Trans. Amer. Math. Soc.*, **337**, 703–735.

[47] Gessel, I. M. 1990. Symmetric functions and P-recursiveness. *J. Combin. Theory Ser. A*, **53**, 257–285.

[48] Goodman, N. R. 1963. Statistical analysis based on a certain multivariate complex gaussian distribution (an introduction). *Ann. Math. Statist.*, **34**, 152–177.

[49] Graham, R. L., Knuth, D. E., and Patashnik, O. 1994. *Concrete Mathematics*. Addison-Wesley.

[50] Greene, C., Nijenhuis, A, and Wilf, H. S. 1979. A probabilistic proof of a formula for the number of Young tableaux of a given shape. *Adv. Math.*, **31**, 104–109.

[51] Greene, C., Nijenhuis, A, and Wilf, H. S. 1984. Another probabilistic method in the theory of Young tableaux. *J. Comb. Theory Ser. A*, **37**, 127–135.

[52] Groeneboom, P. 2002. Hydrodynamical methods for analyzing longest increasing subsequences. *J. Comp. Appl. Math.*, **142**, 83–105.

[53] Haiman, M. D. 1989. On mixed insertion, symmetry, and shifted Young tableaux. *J. Comb. Theory Ser. A*, **50**, 196–225.

[54] Hammersley, J.M. 1972. A few seedlings of research. Pages 345–394 of: LeCam, L. M., Neyman, J., and Scott, E. L. (eds), *Proceedings of the Sixth*

Berkeley Symposium on Mathematical Statistics and Probability, Volume 1: Theory of Statistics. University of California Press.

[55] Hamming, R. W. 1980. The unreasonable effectiveness of mathematics. *Amer. Math. Monthly*, **87**, 81–90.

[56] Hardy, G. H., and Ramanujan, S. 1918. Asymptotic formulae in combinatory analysis. *Proc. London Math. Soc.*, **s2-17**, 75–115.

[57] Hastings, S. P., and McLeod, J. B. 1980. A boundary value problem associated with the second Painlevé transcendent and the Korteweg-de Vries equation. *Arch. Rational Mech. Anal.*, **73**, 35–51.

[58] Hiai, F., and Petz, D. 2000. *The Semicircle Law, Free Random Variables and Entropy*. American Mathematical Society.

[59] Hough, J. B., Krishnapur, M., Peres, Y., and Virág, B. 2009. *Zeros of Gaussian Analytic Functions and Determinantal Point Processes*. American Mathematical Society.

[60] Jockusch, W., Propp, J., and Shor, P. Domino tilings and the arctic circle theorem. Preprint, `arXiv:math/9801068`, 1998.

[61] Johansson, K. 1998. The longest increasing subsequence in a random permutation and a unitary random matrix model. *Math. Res. Letters*, **5**, 63–82.

[62] Johansson, K. 2000. Shape fluctuations and random matrices. *Commun. Math. Phys.*, **209**, 437–476.

[63] Johansson, K. 2001. Discrete orthogonal polynomial ensembles and the Plancherel measure. *Ann. Math.*, **153**, 259–296.

[64] Johansson, K. 2002. Non-intersecting paths, random tilings and random matrices. *Probab. Theory Related Fields*, **123**, 225–280.

[65] Johnstone, I. M. 2001. On the distribution of the largest eigenvalue in principal components analysis. *Ann. Stat.*, **29**, 295–327.

[66] Kerov, S. 1998. A differential model of growth of Young diagrams. Pages 111–130 of: Ladyzhenskaya, O. A. (ed), *Proceedings of the St Petersburg Mathematical Society Volume IV*. American Mathematical Society.

[67] Kim, J. H. 1996. On increasing subsequences of random permutations. *J. Comb. Theory Ser. A*, **76**, 148–155.

[68] King, F. W. 2009. *Hilbert Transforms,* Vols. 1–2. Cambridge University Press.

[69] Kingman, J. F. C. 1968. The ergodic theory of subadditive processes. *J. Roy. Stat. Soc. B*, **30**, 499–510.

[70] Knuth, D. E. 1970. Permutations, matrices, and generalized Young tableaux. *Pacific J. Math.*, **34**, 316–380.

[71] Knuth, D. E. 1998. *The Art of Computer Programming,* Vol. 3: *Sorting and Searching*. 2nd edition. Addison-Wesley.

[72] König, W. 2005. Orthogonal polynomial ensembles in probability theory. *Probab. Surv.*, **2**, 385–447.

[73] Korenev, B. G. 2002. *Bessel Functions and Their Applications*. CRC Press.

[74] Lax, P. D. 2002. *Functional Analysis*. Wiley-Interscience.

[75] Lifschitz, V., and Pittel, B. 1981. The number of increasing subsequences of the random permutation. *J. Comb. Theory Ser. A*, **31**, 1–20.

[76] Liggett, T. M. 1985a. An improved subadditive ergodic theorem. *Ann. Probab.*, **13**, 1279–1285.

[77] Liggett, T. M. 1985b. *Interacting Particle Systems*. Springer.

[78] Liggett, T. M. 1999. *Stochastic Interacting Systems: Contact, Voter and Exclusion Processes*. Springer.

[79] Logan, B. F., and Shepp, L. A. 1977. A variational problem for random Young tableaux. *Adv. Math.*, **26**, 206–222.

[80] Lyons, R. 2003. Determinantal probability measures. *Publ. Math. Inst. Hautes Études Sci.*, **98**, 167–212.

[81] Lyons, R. 2014. Determinantal probability: basic properties and conjectures. To appear in *Proc. International Congress of Mathematicians, Seoul, Korea*.

[82] Lyons, R., and Steif, J. E. 2003. Stationary determinantal processes: phase multiplicity, Bernoullicity, entropy, and domination. *Duke Math. J.*, **120**, 515–575.

[83] Macchi, O. 1975. The coincidence approach to stochastic point processes. *Adv. Appl. Prob.*, **7**, 83–122.

[84] Mallows, C. L. 1963. Problem 62-2, patience sorting. *SIAM Rev.*, **5**, 375–376.

[85] Mallows, C. L. 1973. Patience sorting. *Bull. Inst. Math. Appl.*, **9**, 216–224.

[86] Martin, J. 2006. Last-passage percolation with general weight distribution. *Markov Process. Related Fields*, 273–299.

[87] McKay, J. 1976. The largest degrees of irreducible characters of the symmetric group. *Math. Comp.*, **30**, 624–631.

[88] Mehta, M. L. 2004. *Random Matrices*. 3rd edition. Academic Press.

[89] Miller, K. S., and Ross, B. 1993. *An Introduction to the Fractional Calculus and Fractional Differential Equations*. Wiley-Interscience.

[90] Mountford, T., and Guiol, H. 2005. The motion of a second class particle for the tasep starting from a decreasing shock profile. *Ann. Appl. Probab.*, **15**, 1227–1259.

[91] Newman, D. J. 1997. *Analytic number theory*. Springer.

[92] Novelli, J.-C., Pak, I., and Stoyanovskii, A. V. 1997. A direct bijective proof of the hook-length formula. *Discr. Math. Theor. Comp. Sci.*, **1**, 53–67.

[93] Odlyzko, A. M. 1995. Asymptotic enumeration methods. Pages 1063–1229 of: Graham, R. L., Groetschel, M., and Lovász, L. (eds), *Handbook of Combinatorics*, Vol. 2. Elsevier.

[94] Odlyzko, A. M., and Rains, E. M. 2000. On longest increasing subsequences in random permutations. Pages 439–451 of: Grinberg, E. L., Berhanu, S., Knopp, M., Mendoza, G., and Quinto, E. T. (eds), *Analysis, Geometry, Number Theory: The Mathematics of Leon Ehrenpreis*. American Mathematical Society.

[95] Okounkov, A. 2000. Random matrices and random permutations. *Int. Math. Res. Notices*, **2000**, 1043–1095.

[96] Pak, I. 2001. Hook length formula and geometric combinatorics. *Sém. Lothar. Combin.*, **46**, Article B46f.

[97] Peché, S. 2009. Universality results for the largest eigenvalues of some sample covariance matrix ensembles. *Probab. Theory Related Fields*, **143**, 481–516.

[98] Petersen, T. K., and Speyer, D. 2005. An arctic circle theorem for Groves. *J. Combin. Theory Ser. A*, **111**, 137–164.

[99] Pillai, N., and Yin, J. Universality of covariance matrices. Preprint, `arXiv:1110.2501`, 2011.

[100] Pittel, B., and Romik, D. Limit shapes for random square Young tableaux and plane partitions. Preprint, `arXiv:math/0405190v1`, 2004.

[101] Pittel, B., and Romik, D. 2007. Limit shapes for random square Young tableaux. *Adv. Appl. Math*, **38**, 164–209.

[102] Prähofer, M., and Spohn, H. 2004. Exact scaling functions for one-dimensional stationary KPZ growth. *J. Stat. Phys.*, **115**, 255–279. Numerical tables available at `http://www-m5.ma.tum.de/KPZ`.

[103] Reiner, V. 2005. Note on the expected number of Yang-Baxter moves applicable to reduced decompositions. *Eur. J. Combin.*, **26**, 1019–1021.

[104] Robinson, G. de B. 1938. On the representations of the symmetric group. *Amer. J. Math.*, **60**, 745–760.

[105] Rockafellar, R. T. 1996. *Convex Analysis*. Princeton University Press.

[106] Romik, D. 2000. Stirling's approximation for $n!$: the ultimate short proof? *Amer. Math. Monthly*, **107**, 556–557.

[107] Romik, D. 2005. The number of steps in the Robinson-Schensted algorithm. *Funct. Anal. Appl.*, **39**, 152–155.

[108] Romik, D. 2006. Permutations with short monotone subsequences. *Adv. Appl. Math.*, **37**, 501–510.

[109] Romik, D. 2012. Arctic circles, domino tilings and square Young tableaux. *Ann. Probab.*, **40**, 611–647.

[110] Rost, H. 1981. Nonequilibrium behaviour of a many particle process: density profile and local equilibria. *Z. Wahrsch. Verw. Gebiete*, **58**, 41–53.

[111] Rudin, W. 1986. *Real and Complex Analysis*. 3rd edition. McGraw-Hill.

[112] Sagan, B. E. 2001. *The Symmetric Group: Representations, Combinatorial Algorithms, and Symmetric Functions*. Springer.

[113] Schensted, C. 1961. Longest increasing and decreasing subsequences. *Canad. J. Math.*, **13**, 179–191.

[114] Schützenberger, M.-P. 1963. Quelques remarques sur une construction de Schensted. *Math. Scand.*, **12**, 117–128.

[115] Seppäläinen, T. Lecture notes on the corner growth model. Unpublished notes (2009), available at
`http://www.math.wisc.edu/~seppalai/cornergrowth-book/ajo.pdf`.

[116] Seppäläinen, T. 1996. A microscopic model for the Burgers equation and longest increasing subsequences. *Electron. J. Probab.*, **1-5**, 1–51.

[117] Seppäläinen, T. 1998a. Hydrodynamic scaling, convex duality and asymptotic shapes of growth models. *Markov Process. Related Fields*, **4**, 1–26.

[118] Seppäläinen, T. 1998b. Large deviations for increasing sequences on the plane. *Probab. Theory Related Fields*, **112**, 221–244.

[119] Simon, B. 1977. Notes on infinite determinants of Hilbert space operators. *Adv. Math.*, **24**, 244–273.

[120] Simon, B. 2005. *Trace Ideals and Their Applications*. 2nd edition. American Mathematical Society.

[121] Soshnikov, A. 2000. Determinantal random point fields. *Russian Math. Surveys*, **55**, 923–975.

[122] Soshnikov, A. 2002. A note on universality of the distribution of the largest eigenvalues in certain sample covariance matrices. *J. Stat. Phys.*, **108**, 1033–1056.

[123] Spitzer, F. 1970. Interaction of Markov processes. *Adv. Math.*, **5**, 246–290.

[124] Stanley, R. P. 1984. On the number of reduced decompositions of elements of Coxeter groups. *Eur. J. Combin.*, **5**, 359–372.

[125] Stanley, R. P. 1999. *Enumerative Combinatorics,* Vol. 2. Cambridge University Press.

[126] Stanley, R. P. 2007. Increasing and decreasing subsequences and their variants. Pages 545–579 of: Sanz-Solé, M., Soria, J., Varona, J. L., and Verdera, J. (eds), *Proceedings of the International Congress of Mathematicians, Madrid 2006.* American Mathematical Society.

[127] Stanley, R. P. 2011. *Enumerative Combinatorics,* Vol. 1. 2nd edition. Cambridge University Press.

[128] Steele, J. M. 1986. An Efron-Stein inequality for nonsymmetric statistics. *Ann. Stat.*, **14**, 753–758.

[129] Steele, J. M. 1995. Variations on the monotone subsequence theme of Erdős and Szekeres. Pages 111–131 of: Aldous, D., Diaconis, P., Spencer, J., and Steele, J. M. (eds), *Discrete Probability and Algorithms*. Springer.

[130] Szalay, M., and Turán, P. 1977. On some problems of statistical theory of partitions. I. *Acta Math. Acad. Sci. Hungr.*, **29**, 361–379.

[131] Szegő, G. 1975. *Orthogonal Polynomials*. 4th edition. American Mathematical Society.

[132] Talagrand, M. 1995. Concentration of measure and isoperimetric inequalities in product spaces. *Publ. Math. Inst. Hautes Etud. Sci.*, **81**, 73–205.

[133] Temperley, H. 1952. Statistical mechanics and the partition of numbers. The form of the crystal surfaces. *Proc. Camb. Philos. Soc.*, **48**, 683–697.

[134] Thrall, R. M. 1952. A combinatorial problem. *Michigan Math. J.*, **1**, 81–88.

[135] Titschmarsh, E. C. 1948. *Introduction to the Theory of Fourier Integrals*. 2nd edition. Clarendon Press.

[136] Tracy, C. A., and Widom, H. 1994. Level-spacing distributions and the Airy kernel. *Commun. Math. Phys.*, **159**, 151–174.

[137] Tracy, C. A., and Widom, H. 2009. Asymptotics in ASEP with step initial condition. *Commun. Math. Physics*, **290**, 129–154.

[138] Ulam, S. 1961. Monte Carlo calculations in problems of mathematical physics. Pages 261–281 of: Beckenbach, E. F. (ed), *Modern Mathematics For the Engineer, Second Series*. McGraw-Hill.

[139] Varadhan, S. R. S. 2008. Large deviations. *Ann. Probab.*, **36**, 397–419.

[140] Vershik, A., and Pavlov, D. 2009. Numerical experiments in problems of asymptotic representation theory. *Zap. Nauchn. Sem.*, **373**, 77–93. Translated in *J. Math. Sci.*, 168:351–361, 2010.

[141] Vershik, A. M. 1996. Statistical mechanics of combinatorial partitions and their limit shapes. *Funct. Anal. Appl.*, **30**, 90–105.

[142] Vershik, A. M., and Kerov, S. V. 1977. Asymptotics of the Plancherel measure of the symmetric group and the limiting shape of Young tableaux. *Soviet Math. Dokl.*, **18**, 527–531.

[143] Vershik, A. M., and Kerov, S. V. 1985. The asymptotics of maximal and typical dimensions irreducible representations of the symmetric group. *Funct. Anal. Appl.*, **19**, 21–31.

[144] Wang, K. 2012. Random covariance matrices: universality of local statistics of eigenvalues up to the edge. *Random Matrices: Theory Appl.*, **1**, 1150005.

[145] Watson, G. N. 1995. *A Treatise on the Theory of Bessel Functions*. 2nd edition. Cambridge University Press.

[146] Wigner, E. 1960. The unreasonable effectiveness of mathematics in the natural sciences. *Comm. Pure Appl. Math.*, **13**, 1–14.

[147] Wishart, J. 1928. The generalised product moment distribution in samples from a normal multivariate population. *Biometrika*, **20A**, 32–53.

Index

CPSIA information can be obtained
at www.ICGtesting.com
Printed in the USA
LVOW04*2248140316
479171LV00007B/42/P